Hydrogeology: Principles and Applications

Hydrogeology:
Principles and Applications

Edited by Tony McPherson

SYRAWOOD
PUBLISHING HOUSE

New York

Published by Syrawood Publishing House,
750 Third Avenue, 9th Floor,
New York, NY 10017, USA
www.syrawoodpublishinghouse.com

Hydrogeology: Principles and Applications
Edited by Tony McPherson

International Standard Book Number: 978-1-64740-117-7 (Hardback)

Cataloging-in-Publication Data

Hydrogeology : principles and applications / edited by Tony McPherson.
    p. cm.
Includes bibliographical references and index.
ISBN 978-1-64740-117-7
1. Hydrogeology. 2. Geology. 3. Hydrology. 4. Groundwater. I. McPherson, Tony.
GB1003.2 .H93 2022
551.49--dc23

TABLE OF CONTENTS

PREFACE

Hydrogeology, also known as groundwater engineering, is a field of earth science. It deals with the distribution and movement of groundwater in the soil and rocks of the earth's crust. The main focus areas in hydrogeology include groundwater contamination, conservation of supplies and water quality. It deals with the flow of water through aquifers and other shallow porous media. Darcy's law diffusion equation and Laplace equations are a few mathematical relationships that are extensively studied and applied within this field. The aim of this book is to present researches that have transformed this discipline and aided its advancement. The topics included herein on hydrogeology are of utmost significance and bound to provide incredible insights to readers. This book will prove to be immensely beneficial to students and researchers in this field.

This book is a result of research of several months to collate the most relevant data in the field.

When I was approached with the idea of this book and the proposal to edit it, I was overwhelmed. It gave me an opportunity to reach out to all those who share a common interest with me in this field. I had 3 main parameters for editing this text:

1. Accuracy – The data and information provided in this book should be up-to-date and valuable to the readers.

2. Structure – The data must be presented in a structured format for easy understanding and better grasping of the readers.

3. Universal Approach – This book not only targets students but also experts and innovators in the field, thus my aim was to present topics which are of use to all.

Thus, it took me a couple of months to finish the editing of this book.

I would like to make a special mention of my publisher who considered me worthy of this opportunity and also supported me throughout the editing process. I would also like to thank the editing team at the back-end who extended their help whenever required.

Editor

A groundwater recharge perspective on locating tree plantations within low-rainfall catchments to limit water resource losses

J. F. Dean[1,2,*]**, J. A. Webb**[1,2]**, G. E. Jacobsen**[3]**, R. Chisari**[3]**, and P. E. Dresel**[4]

[1]Agricultural Sciences Department, La Trobe University, Bundoora, Victoria, Australia
[2]National Centre for Groundwater Research and Training, Adelaide, Australia
[3]Institute for Environmental Research, ANSTO, Sydney, Australia
[4]Department of Environment and Primary Industries, Bendigo, Victoria, Australia
[*]now at: Biological and Environmental Sciences, University of Stirling, Scotland, UK

Correspondence to: J. F. Dean (joshua.dean@sitr.ac.uk)

Abstract. Despite the many studies that consider the impacts of plantation forestry on groundwater recharge, and others that explore the spatial heterogeneity of recharge in low-rainfall regions, there is little marriage of the two subjects in forestry management guidelines and legislation. Here we carry out an in-depth analysis of the impact of reforestation on groundwater recharge in a low-rainfall (< 700 mm annually), high-evapotranspiration paired catchment characterized by ephemeral streams. Water table fluctuation (WTF) estimates of modern recharge indicate that little groundwater recharge occurs along the topographic highs of the catchments (average $18\,\text{mm}\,\text{yr}^{-1}$); instead the steeper slopes in these areas direct runoff downslope to the lowland areas, where most recharge occurs (average $78\,\text{mm}\,\text{yr}^{-1}$). Recharge estimates using the chloride mass balance (CMB) method were corrected by replacing the rainfall input Cl^- value with that for streamflow, because most recharge occurs from infiltration of runoff through the streambed and adjacent low gradient slopes. The calculated CMB recharge values (average $10\,\text{mm}\,\text{yr}^{-1}$) are lower than the WTF recharge values (average $47\,\text{mm}\,\text{yr}^{-1}$), because they are representative of groundwater that was mostly recharged prior to European land clearance ($> BP\,200$ years). The tree plantation has caused a progressive drawdown in groundwater levels due to tree water use; the decline is less in the upland areas.

The results of this study show that spatial variations in recharge are important considerations for locating tree plantations. To conserve water resources for downstream users in low-rainfall, high-evapotranspiration regions, tree plant-

ing should be avoided in the dominant zone of recharge, i.e. the topographically low areas and along the drainage lines, and should be concentrated on the upper slopes, although this may negatively impact the economic viability of the plantation.

1 Introduction

Tree plantations are known to have the potential to reduce groundwater recharge and surface water flows (e.g. Bell et al., 1990; Benyon, 2002; Bosch and Hewlett, 1982; Jobbagy and Jackson, 2004; Scanlon et al., 2007; van Dijk et al., 2007), particularly in low-rainfall, high-evapotranspiration regions where the high transpiration demands of the trees make them a significant user in the water balance (e.g. Benyon et al., 2006; Fekeima et al., 2010; Jackson et al., 2005; Schofield, 1992). This is often regarded as a negative aspect of tree plantations, but may be a positive outcome if the aim of a particular forestry project is to reduce groundwater levels, e.g. to decrease groundwater salinization (discussed further below). Groundwater recharge in low-rainfall regions is also affected by a variety of other factors that cause substantial spatial variability – in particular topography, soil characteristics and geology (e.g. Delin et al., 2000; Scanlon et al., 2002; Schilling, 2009; Webb et al., 2008; Winter, 2001). However, the important conclusions made in the recharge studies have not been brought together with the results of tree plantation studies and directly applied to water resource management problems accompanying the establishment of tree plantations (Farley et al., 2005).

Since the earliest work on defining groundwater systems, recharge has been shown to be controlled predominantly by topography: the majority of groundwater recharge occurs at topographic highs, and discharge is mostly in topographic lows where the upward hydraulic gradient prevents recharge from occurring (Domenico and Schwartz, 1998; Schilling, 2009). However, in arid and semi-arid regions, recharge following rainfall events often occurs predominantly in local depressions and along ephemeral streams (diverging from early conceptual models), due to the focusing of overland flow in these areas. Water tables under ephemeral streams are generally below the streambed (except during extended rainfall events), and therefore upwards groundwater gradients do not occur most of the time. Infiltration beneath these areas may also be encouraged by the presence of preferential pathways, along which infiltrating water may more readily reach the water table (Delin et al., 2000; Scanlon et al, 2002; Schilling, 2009; Winter, 2001). In southeastern Australia in particular, it has been observed that recharge can vary significantly within catchments due to multiple modes of recharge (Cartwright et al., 2007).

Vegetation can significantly impact groundwater recharge due to transpiration and by intercepting rainfall and overland flow (Scanlon et al., 2002; Winter, 2001); changing land use can therefore affect recharge patterns. For example, land salinization has occurred in large parts of southeastern Australia due to the replacement of native forest by pasture and crops that use less water; this has led to increased recharge which raised water tables, causing saline groundwater to come to the land surface and discharge into surface water features (Allison et al., 1990; Bennetts et al., 2006, 2007). In contrast, afforestation of cleared farmland is likely to decrease recharge, due to the high rate of transpiration by the actively growing, closely planted trees, as well as the interception of overland flow and evaporation from the canopy (Benyon et al., 2006). In particular, the evergreen eucalyptus tree plantations commonly planted in southeastern Australia take up and transpire significantly more water than pasture, their canopy intercepts more rainfall and allows it to evaporate, and their roots reach greater depths than grasses, meaning they can extract water over a larger volume of the soil column (Bosch and Hewlett, 1982; Feikema et al., 2010; Hibbert, 1967). This recharge reduction is the reason why some studies have suggested using targeted tree plantations to reduce recharge in areas where there are high rates of saline groundwater discharge (e.g. Bennetts et al., 2007). Tree plantations also sequester carbon dioxide, prompting ongoing debate over the trade-off between increased water use by trees versus their increased carbon sequestration potential (Farley et al., 2005). As such, efforts over the past few decades in southeastern Australia to reforest land that was cleared in the late 1800s by European settlers (Schofield, 1992) are now causing difficulties for land managers trying to define sus-

tainable action plans for surface water and groundwater (Dalhaus et al., 2008; Jackson et al., 2005; Nicholson et al., 2006).

A whole-catchment approach is key to managing groundwater recharge in the context of land use change (Cartwright et al., 2007). However, despite the evidence that recharge is often concentrated in topographic lows, groundwater management strategies in southeastern Australia typically operate on the assumption that recharge occurs primarily in the upper parts of catchments, particularly along the ridgelines. Current regulations for tree plantations in Australia focus on the percentage of a given catchment that can be forested, rather than what areas should be planted to maintain or intercept groundwater recharge, depending on the management application.

Here we present the findings from a paired catchment study in southwestern Victoria, Australia, where one catchment is planted with a tree plantation, and the adjacent catchment is covered with pasture. This approach largely removes the variables of climate, topography, soil and geology, with the only major difference between the two catchments being vegetation cover. Previous paired catchment studies on the impact of tree plantations tended to focus on surface water responses to afforestation, while groundwater has been somewhat neglected (Brown et al., 2005). In this study conceptual models of groundwater flow (based on ^{14}C and tritium groundwater dating) and groundwater recharge estimates (based on the water table fluctuation and chloride mass balance methods) are used to assess the impact of a *Eucalyptus globulus* plantation on the hydrologic and hydrogeologic regime. This contextualization is then used to discuss the best areas to site tree plantations within low-rainfall catchments.

2 Background

This study is part of a multi-site, paired-catchment investigation into the impacts of land use and climate change on the quality and quantity of groundwater and surface water resources in western Victoria, Australia (Adelana et al., 2014; Camporese et al., 2013, 2014; Dean et al., 2014; Dresel et al., 2012).

2.1 Site description

The study area consists of a pair of small, adjacent catchments at Mirranatwa in southwestern Victoria – one (referred to as the eucalypt catchment) is covered predominantly in a recently planted (July 2008) *E. globulus* (Blue Gum) plantation (0.8 km^2), the other (referred to as the pasture catchment) is mostly pasture for grazing sheep (0.4 km^2; Fig. 1).

2.1.1 Geology

Both catchments are underlain by the same weathered/fractured aquifer, the Devonian Dwyer Granite (390–

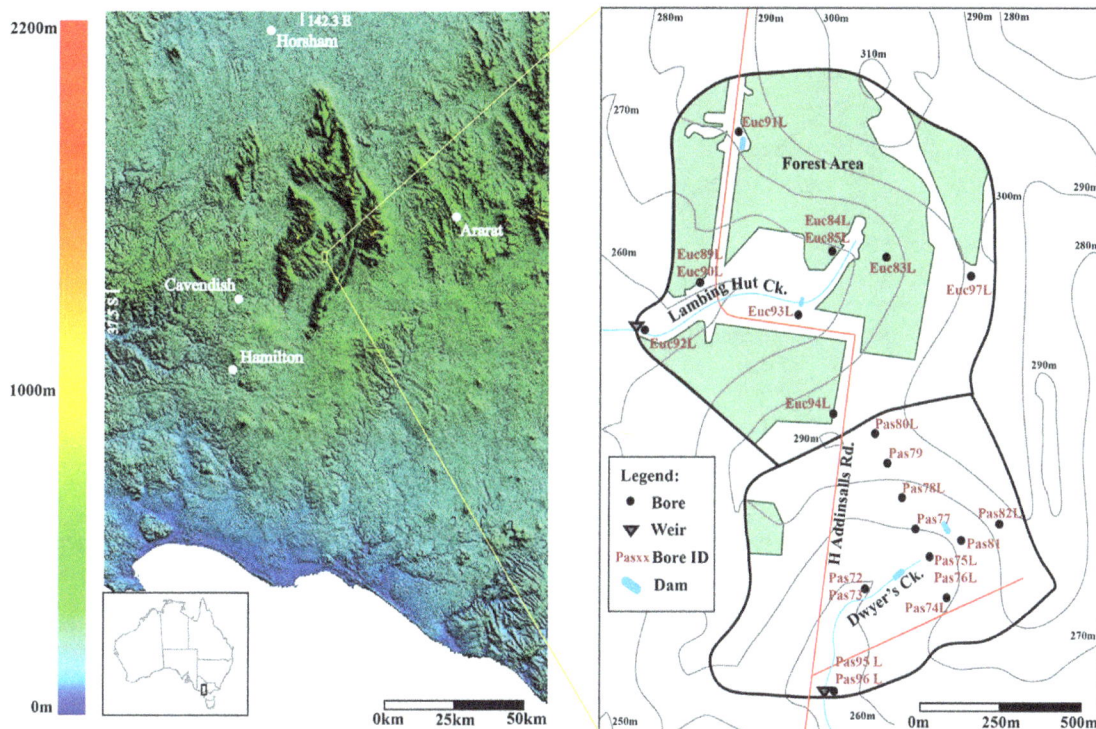

Figure 1. Left: location of the study site in southwestern Victoria, Australia; right: location of the streams, weirs and bores and their reference numbers. "L" denotes the presence of a water level logger in a bore.

395 Ma; Hergt et al., 2007; Van den Berg, 2009). The upper ~ 20 m of the granite is well-weathered, porous and permeable saprolite; below this is relatively fresh, fractured bedrock. The fractured granite aquifer extends no deeper than 150 m, as below this depth the fracture conductivity is negligible due to the high lithostatic pressure (Boutt et al., 2010; Cook, 2003; Dept. Sustainability and Environment, 2012). The granite saprolite is generally thicker beneath the lower parts of the catchment than along the ridges, and is overlain by up to 7 m of alluvial/colluvial material along and adjacent to drainage lines. This alluvium/colluvium is clay-rich and impermeable in places, causing temporally variable artesian behaviour in some of the bores along the drainage lines in both catchments. The topography of the site (hills in the middle of a broad valley, Fig. 1) means that both catchments are local groundwater systems, and there are no regional groundwater inputs. There is 50 m of relief in the eucalypt catchment, and 30 m in the pasture catchment; both catchments comprise reasonably steep hills separated by a marked break in slope from the more or less flat topography along the drainage lines (Fig. 1).

2.1.2 Climate and land use

The climate is Mediterranean, maritime/temperate (Cfb in the Köppen classification); the average annual rainfall since records began in 1901 for the area is 672 mm ($\pm 125\sigma$),

while pan evaporation is around 1350 mm annually, exceeding rainfall for the majority of the year, excepting the winter months of May to September (Dean et al., 2014). Runoff ratios for the pasture and eucalypt catchments are 3.0 and 3.3 % respectively (based on the stream hydrograph records from February 2011 to February 2014), and both streams are ephemeral.

Vegetation of the area prior to European settlement was mostly open eucalypt woodland (Department of Sustainability and Environment, Victoria). Following European settlement there was extensive land clearance, and the catchments were entirely converted to pasture by 1869 (White et al., 2003). 76 % of the northern catchment was subsequently converted to an *E. globulus* plantation in July 2008 (Fig. 1). Prior to the planting of the eucalypts, the eucalypt plantation catchment (Euc – Table 1) was used for grazing, and was virtually identical to the pasture grazing catchment (Pas – Table 1) immediately to the south. During the planting of the trees the eucalypt catchment was ripped to an average depth of 800 mm and mounded to an average height of 300 mm. The tree density is 1010 stems per ha (2.2 m between trees along a row, and 4.5 m between rows), and fertilizer was applied following ripping and mounding at $60\,\mathrm{kg\,ha^{-1}}$ (McEwens Contracting, personal communication, 2011). The tree rows run east–west across the slope in the main northeastern part of the catchment, and north–south ($\sim$ down the slope) to the west of H Addinsalls Road (Fig. 2).

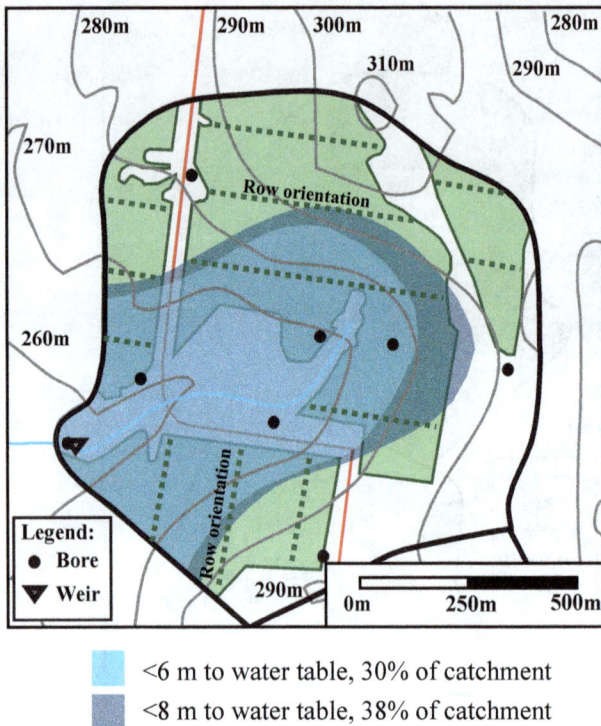

<table>
<tr><td style="background:#aadbe6;width:1em;"></td><td><6 m to water table, 30% of catchment</td></tr>
<tr><td style="background:#8a94b0;width:1em;"></td><td><8 m to water table, 38% of catchment</td></tr>
</table>

Figure 2. Orientation of the tree rows in the eucalypt plantation and the area where tree roots may be able to reach groundwater up to depths of 6 and 8 m below the surface.

2.1.3   Catchment instrumentation

The pasture catchment has 13 bores drilled to different depths, and the eucalypt catchment has 10 bores (the bores may be considered to be piezometers – they are screened towards the bottom of the casing over a discrete 2 m interval; Table 1). Seven bores in the eucalypt catchment and two bores in the pasture catchment were drilled for this project in late 2009; the other bores were installed in the late 1980s in the pasture catchment, and the mid-1990s in the eucalypt catchment. A groundwater logger was installed in every bore in the eucalypt catchment in August 2009, measuring at a minimum 4 h time interval, and eight bores in the pasture catchment have loggers measuring at the same frequency. There is a v-notch weir at the end of each catchment on both streams, with one bore immediately adjacent to the eucalypt catchment weir and two next to the pasture catchment weir (Fig. 1). The bores adjacent to the weirs have Campbell CS450-L pressure transducers (accuracy ±0.01 m) measuring water level and electrical conductivity (EC) at 30 min intervals, while the other bores have Schlumberger Mini Diver loggers (accuracy ±0.025 m) measuring only water level. At the weirs the surface water level was measured using a standard V-notch construction, and electrical conductivity (EC) was recorded using a logger in the weir pool (Dresel et al., 2012). Prior to installation of groundwater log-

gers in the older bores, groundwater levels were generally measured manually bi-monthly.

There are two small dams in each catchment, ranging in size from 10 to 50 m^2; they are not large enough to significantly impact the hydrology of the site (Fig. 1). The roads at the site are single lane and unsealed, and although they are less permeable than the normal ground surface and therefore promote runoff, their very small area means that they have negligible impact on the site hydrology.

3   Methods

Groundwater levels, surface water flow and rainfall data were collected from August 2009 to February 2013 for this study, with some older long-term groundwater level data from manual measurements going back as far as 1986 available from the Victorian Department of Environment and Primary Industries archives. Groundwater and surface chemistry is available from sampling campaigns from August 2010 to August 2011 (Dean et al., 2014).

3.1   Rainfall and streamflow

Daily rainfall measurements were available from a Bureau of Meteorology station (089019) approximately 2 km south of the study site; rainfall was also measured in the study catchments and showed an excellent correlation with the Bureau of Meteorology station. Due to significant gaps in the on-site data, the Bureau of Meteorology station data was used for consistency throughout the study period. To determine rainfall patterns, cumulative deviation from the monthly mean (CDM) values were calculated alongside daily values (Sect. 4.1.1), whereby the difference between a given monthly rainfall total and the average for that month (calculated from the entire station's data record of 1901 to 2012), was cumulatively summed from one month to the next (modified from Craddock, 1979). The CDM values represent the longer-term rainfall patterns, with a sustained negative trend for drought periods and positive values indicating wetter than usual periods, and match well with the longer-term hydrographs (Sect. 4.1).

Streamflow in both catchments is ephemeral, and was measured at 30 min intervals at V-notch weirs at both catchment outlets and summed to annual totals, and a total for the complete study period, 2009–2013. To allow comparison between catchments, volumes were converted to depth equivalents (mm) by dividing by the respective catchment area. Streamflow is derived predominantly from direct runoff, as the proportion of groundwater input into the stream is small (discussed further below).

3.2   Grain size analysis

The grain size of the saprolite was used to estimate the average specific yield value for this aquifer over the whole study

Table 1. Groundwater characteristics and bore construction.

Bore ID	Earliest data from bore	Screen depth (m below surface)	Surface elevation (m AHD)	Radiocarbon age (BP yr)	1σ – error	Activity of ^{3}H (TU)	1σ – error	Logger	Groundwater Cl⁻ (mg L⁻¹)
				Pasture Catchment					
Pas72 – Low	31 Aug 1986	9.4–11.6	259.55	1665	±30	BD		N	3292
Pas73 – Low	31 Aug 1986	4–6.1	259.54	2055	±30			N	3110
Pas75 – Low	31 Aug 1986	12–13.6	263.93	935	±35			Y	2231
Pas76 – Low	31 Aug 1986	2.2–4.2	263.98	575	±30	BD		Y	1595
Pas95 – Low	26 Aug 2009	22.8–24.8	254.13	3540	±30	BD		Y (weir)	2732
Pas96 – Low	26 Aug 2009	5–7.55	254.18	345	±25	1.12	±0.09	Y (weir)	2553
Pas74 – Up	31 Aug 1986	6.2–8.5	268.62	790	±30	0.44	±0.04	Y	306
Pas77 – Up	31 Aug 1986	17.7–19.7[a]	271.11	Modern		2.84	±0.13	N	28
Pas78 – Up	31 Aug 1986	17.3–19.4	277.45	650	±90	BD		Y	1185
Pas79 – Up	31 Aug 1986	23.65–25.65[a]	283.23	Modern		2.55	±0.12	N	38
Pas80 – Up	31 Aug 1986	23.3–24.4	288.23	115	±30	1.24	±0.08	Y	2290
Pas81 – Up	31 Aug 1986	7.1–8.9	272.12	690	±100	0.79	±0.08	N	668
Pas82 – Up	31 Aug 1986	23.2–24.8	283.54	430	±30	0.60	±0.05	Y	329
				Eucalypt catchment					
Euc84 – Low	12 Nov 1996	5.6–7.5	268.67	785	±30			Y	3909
Euc85 – Low	12 Nov 1996	7.9–10	268.66	b		BD		Y	3537
Euc89 – Low	30 Oct 2009	26–28	261.80	7330	±50			Y	2833
Euc90 – Low	30 Oct 2009	13–15	261.93	6980	±45			Y	2788
Euc92 – Low	30 Oct 2009	26.2–29.2	255.43	20770	±90	BD		Y (weir)	1490
Euc93 – Low	2 Mar 2010	11–14	263.31	725	±30	0.73	±0.06	Y	1357
Euc83 – Up	12 Nov 1996	14.8–16.7	274.21	685	±30	BD		Y	2064
Euc91 – Up	30 Oct 2009	33.9–35.9	280.02	415	±30	0.39	±0.04	Y	1114
Euc94 – Up	30 Oct 2009	28–30	286.05	2060	±30	BD		Y	2891
Euc97 – Up	30 Oct 2009	43.1–45.1; 57.6–59.6	291.74	5655	±35	0.30	±0.04	Y	3494

BD: below detectable; [a] assumed screen depths; [b] CO_2 concentration too low for analysis.

site, as the geology of the two catchments is very similar (see Sect. 3.6.1). During drilling of five bores on the eucalypt catchment, samples of the regolith were taken at 1 m intervals to a depth of 10 m, or until bedrock was encountered. Samples were sieved using a 2 mm sieve and the material that passed through was then analysed using a Malvern Mastersizer 2000.

3.3 Groundwater composition

All 23 groundwater bores across the entire site were sampled once each over a period of a year, from August 2010 to August 2011. Seasonal variability in groundwater composition is considered negligible due to the age of the groundwater at the study site (mostly > 200 years; Table 1), and repeat sampling produced virtually identical field parameters (Dean et al., 2014). Subsamples for Cl⁻ were filtered with 0.45 μm filter paper and analysed using Ion Chromatography. Groundwater sampling, Cl⁻ analyses and calculations of volume-weighted, average rainfall Cl⁻ concentrations are described in more detail in Dean et al. (2014).

3.4 ^{14}C analysis and tritium analysis

Dating of the groundwater was carried out to determine the time period over which recharge has occurred. Groundwater samples from all the bores at the study site were ^{14}C dated and no corrections were applied, as there is no indication that

the radiocarbon ages have been compromised by "dead" carbon in the regolith; standard error of groundwater ages is 25–100 years (Dean et al., 2014). In addition, seven bores in the eucalypt catchment and 11 bores in the pasture catchment (including the shallowest and deepest bores and a range in between), were analysed for tritium (standard error in these measurements was 0.04–0.13 tritium units (TU); Dean et al., 2014; Table 1). The methodologies for both are described in more detail in Dean et al. (2014).

3.5 Radon (^{222}Rn)

Radon surveys were carried out on groundwater and surface water samples in both catchments to ascertain whether there is a significant contribution of groundwater to surface water flow. The ^{222}Rn content of surface water and groundwater was measured using the gas-extraction for H_2O accessory of the Durridge RAD-7 radon detector. The RAD-7 is an alpha particle detector that measures the decay of the radon daughters, ^{214}Po and ^{218}Po. Samples from weirs, bores and dams (disconnected surface water bodies; Fig. 1) were collected in 250 mL vials and aerated for 5 min to degas the radon into the air circulation within the instrument, which takes four measurements (5 min each), and then gives the mean ^{222}Rn concentration in Bq L⁻¹; the average standard error for measurements using this instrument is 10 %.

3.6   Groundwater recharge

To ensure robust estimates of groundwater recharge, two different, well-established methods were used, namely the water table fluctuation method and chloride mass balance method. While both methods are in widespread use, they have known deficiencies that are discussed below.

3.6.1   Water table fluctuations

The water table fluctuation (WTF) method for measuring groundwater recharge was first applied in the 1920s (Healy and Cook, 2002; Meinzer, 1923) and has since been refined (e.g. Jie et al., 2011; Scanlon et al., 2005; Sophocleus, 1991). The principle of this method is that rises in the groundwater hydrograph of an unconfined aquifer provide an estimate of recharge to the water table, calculated from

$$R = S_y \frac{\Delta h}{\Delta t}, \tag{1}$$

where recharge (R) is the product of the specific yield of the aquifer (S_y) and the change in hydrograph height (Δh) over a given time interval (Δt). This method assumes that recharge occurs vertically from piston flow and that water discharges continuously from the aquifer, causing a drop in the water table when recharge is not occurring. Therefore the change in hydrograph height from which recharge is calculated is the sum of the rise in the hydrograph, together with the decline in the hydrograph that would have occurred in the absence of recharge over the same time period (Healy and Cook, 2002; Jie et al., 2011). Several techniques have been developed to estimate the hydrograph decline: the graphical approach – in which the exponential decay curve of the hydrograph is manually extended to coincide with the peak of the next recharge event (Delin et al., 2007); the master recession curve approach – in which regression functions are assigned to simulate the potential hydrograph decline for each data time-step (Heppner et al., 2007); and the RISE approach – in which the assumption is made that in the absence of recharge, no decline in the water table occurs (Jie et al., 2011; Rutledge, 1998).

It proved difficult to apply the graphical and master recession curve methods in the present study because these methods focus on the section of the hydrograph recession limb which decays exponentially, whereas the recession limbs in the Mirranatwa hydrographs often had significant sections which were steep and straight (Fig. 3); this can lead to the underestimation of actual groundwater recharge, as has been highlighted elsewhere (Cuthbert, 2014). In addition, because the streams in both study catchments are ephemeral, groundwater discharge as baseflow occurs only occasionally; the majority of groundwater discharge occurs at the bottom of the catchments and downstream of the catchment boundaries. This intermittent baseflow means that the recession curve in the hydrographs following a recharge event may not be

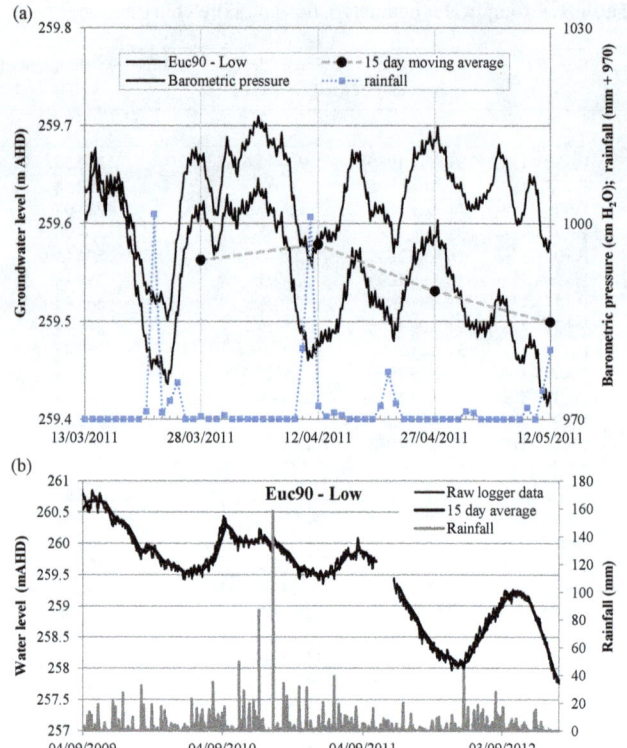

Figure 3. (a) Barometric pressure (in equivalent cm of H$_2$O), groundwater logger data, rainfall and the 15 day moving average used for the water table fluctuation method estimates of groundwater recharge. The black dots represent the average groundwater level for the preceding 15 days. **(b)** Full record for the bore used in **(a)** – Euc90 – showing the complete removal of the large amount of barometric noise, but keeping the overall trend of the 15 day period.

exponential (as observed in the hydrographs). Because the assumption of an exponential recession curve is implicit in the graphical and master recession curve WTF methods, the RISE approach was adopted, i.e. the decay curve of the hydrograph was ignored. Applying the RISE approach means that the values calculated in this study potentially underestimate actual recharge, but when compared with the graphical approach carried out for sections of the hydrographs where exponential recession curves were evident, gave very similar values.

Raw bore hydrograph data collected using data loggers at the site contain small fluctuations due to the impact of barometric pressure on the water column in the bore (Fig. 3a; Rasmussen and Crawford, 1997). The fluctuations in the water level and the barometric pressure are normally inversely correlated (Butler et al., 2011), and can be readily corrected (Rasmussen and Crawford, 1997; Toll and Rasmussen, 2007). At the study site these fluctuations are clearly positively correlated with barometric fluctuations (Fig. 3a), and as a result normal barometric compensation techniques could not be applied. Two types of groundwater level sensors were used: Schlumberger Mini Diver loggers (accuracy

±0.025 m) and Campbell CS450-L pressure transducers (accuracy ±0.01 m); the Campbell sensors are vented and therefore technically do not need compensating for barometric pressure changes, while the Schlumberger sensors require barometric compensation and barometric loggers were installed in the middle of both catchments to collect barometric data for this purpose. The barometric effect shown in Fig. 3a is consistent across all the Schlumberger sensors in both catchments, regardless of landscape position. Figure 3a is based on Fig. 1 from Butler et al. (2011), and the data from this study was prepared in the same manner, so the positive correlation is not an artefact of data processing error. Barometric forcing was evident in the Campbell sensor data also, despite their being vented, so these data were treated in the same way as the Schlumberger data (see below).

A 15-day moving average was used to remove the barometric fluctuations but retain the overall response to rainfall (Fig. 3b). The 15-day time step is a narrow enough time period to incorporate recharge events and reflect the general trend of the hydrograph, but removes the small barometrically forced fluctuations that bear no relationship to rainfall (Fig. 3). Recharge was then calculated using Eq. (1), where Δh was taken as the sum of the increases in groundwater level over the time step, and then summed for the entire length of the record. When there was a drop in groundwater level from one time step to the next, this was taken as zero recharge. The measurement uncertainty of the loggers (±0.025 m) was used as the threshold for recognition of recharge for each 15-day time step. The RISE method was also used to calculate recharge for the longer-term hydrographs (generally bi-monthly measurements taken prior to logger installation).

A specific yield value of 0.095 ± 0.014 was calculated for the unconfined saprolite aquifer from the average grain size (clay to coarse sand; Table 2) of all the bore samples analysed (see Sect. 3.2), using the general relationship between specific yield and grain size in Healy and Cook (2002, Tables 1 and 2). The estimation of specific yield is a potential source of considerable error in recharge calculations as it can vary spatially, although it can be assumed to be independent of time (Healy and Cook, 2002). The specific yield value calculated here is comparable to other values from weathered granites in the region (0.043 – Hekmeijer and Hocking, 2001; 0.075 – Edwards, 2006). When calculating recharge for the study site, this specific yield was applied to bores that are screened within the saprolite, and is assumed to be representative for the whole site because of the relatively uniform nature of the soils (Table 2).

3.6.2 Chloride mass balance

The chloride (Cl^-) mass balance (CMB) method for calculating recharge is based on the relationship between Cl^- in groundwater and in precipitation, assuming that all Cl^- in the groundwater is derived from rainfall and remains in solu-

Table 2. Median grain size compositions for sampled profiles used to estimate a range of values for S_y in Eq. (1).

Bore ID	Clay (%)	Silt (%)	Fine sand (%)	Coarse sand (%)
Euc89 – Low	3	39	38	19
Euc91 – Low	3	39	40	18
Euc93 – Low	3	36	43	18
Euc94 – Up	3	35	44	18
Euc97 – Up	3	34	43	20

tion within the groundwater system, that direct recharge (R, in mm) occurs via piston flow, and that runoff is negligible:

$$R = P \frac{C_p}{C_{gw}}, \qquad (2)$$

where P is the amount of rainfall (mm), C_p is the concentration of Cl^- in P, and C_{gw} is the concentration of Cl^- in groundwater (Allison and Hughes, 1978; Scanlon et al., 2002). R was calculated at all bores using the groundwater Cl^- content (Table 1), and rainfall Cl^- content was the median value from three different sampling periods at nearby sites (Fig. 1): 1954–1955 at Cavendish (Hutton and Leslie, 1958), 2003–2004 at Hamilton (Bormann, 2004), and 2007–2010 at Horsham (Nation, 2009); all Cl^- values were volume weighted based on rainfall during the sampling periods in these studies. These three sampling periods include a wet period (1954–1955) and two dry periods (2003–2004 and 2007–2009). The median rainfall Cl^- from all of these studies is $4.3 \pm 0.9 \, \text{mg L}^{-1}$, and the annual rainfall is $672 \pm 125 \, \text{mm}$ (1σ); the uncertainties associated with each value were used to estimate the overall uncertainty in the recharge values calculated. R is strongly governed by C_p in this equation, so it is important to take into account the variability in C_p.

4 Results and discussion

4.1 Groundwater recharge estimates

Recharge estimates calculated using both the WTF and CMB methods range from 0.8 ± 0.3 to $161 \pm 24 \, \text{mm yr}^{-1}$ (Table 3), a very wide range that matches recharge calculations from similar climatic areas in Australia (5–250 mm yr^{-1}; Allison and Hughes, 1978; Cook et al., 1989), and elsewhere from low-rainfall regions around the world (0.2–35 mm yr^{-1}; Scanlon et al., 2006).

4.1.1 Water table fluctuation method

The groundwater hydrographs vary significantly across the study site (Fig. 4), indicating substantial variation in groundwater recharge. Because hydrographs from the upper parts

Table 3. Recharge (R) values using different methods for all the bores across both catchments.

Bore ID	R (mm yr^{-1})			
	Groundwater Cl$^-$	Groundwater Cl$^-$ with stream input correction	Water table fluctuation method	Long-term hydrograph water table fluctuation method
Pasture catchment – lowland landscape position				
Pas72 – Low*	0.9 ± 0.3	6.8 ± 4.6	L	D
Pas73 – Low*	0.9 ± 0.3	7.2 ± 4.8	L	D
Pas75 – Low	1.3 ± 0.5	3.9 ± 2.6	58 ± 9	38 ± 6
Pas76 – Low	1.8 ± 0.7	5.5 ± 3.7	77 ± 11	D
Pas95 – Low*	1.1 ± 0.4	24 ± 16	C	D
Pas96 – Low	1.1 ± 0.4	26 ± 17	161 ± 24	D
Pasture catchment – upland landscape position				
Pas78 – Up	2.5 ± 0.9	C	36 ± 5	D
Pas80 – Up	1.0 ± 0.4	C	12 ± 2	30 ± 5
Pas82 – Up	8.8 ± 3.3	C	26 ± 4	28 ± 4
Pasture catchment – possible fracture flow				
Pas74 – Up	9.4 ± 3.5	C	65 ± 10	56 ± 8
Pas77 – Up	102 ± 38	C	L	D
Pas79 – Up	76 ± 29	C	L	D
Pas81 – Up	4.3 ± 1.6	C	L	D
Eucalypt catchment – lowland landscape position				
Euc84 – Low*	0.7 ± 0.3	1.7 ± 1.3	C	C
Euc85 – Low*	0.8 ± 0.3	1.9 ± 1.4	C	C
Euc89 – Low	1.0 ± 0.4	5.7 ± 4.3	59 ± 9	D
Euc90 – Low	1.0 ± 0.4	5.8 ± 4.4	74 ± 11	D
Euc93 – Low	2.1 ± 0.8	8.0 ± 6.1	40 ± 6	D
Eucalypt catchment – upland landscape position				
Euc83 – Up	1.4 ± 0.5	C	10 ± 2	19 ± 3
Euc91 – Up	2.6 ± 1.0	C	17 ± 3	D
Euc94 – Up	1.0 ± 0.4	C	1.7 ± 0.2	D
Euc97 – Up	0.8 ± 0.3	C	26 ± 4	D
Eucalypt catchment – possible fracture flow				
Euc92 – Low*	1.9 ± 0.7	C	C	D

* Confined bores; L: no logger present; D: no data; C: this calculation was not done for that bore as it did not meet the required conditions (see Sects. 3.6.1 and 3.6.2).

of the catchment show a limited response to rainfall patterns, both in the detailed groundwater logger data (Fig. 4) and the longer-term monitoring data for the older bores (Fig. 5), recharge values calculated using the WTF method are relatively low for these areas in both catchments (average 18 mm yr^{-1}; 3% of rainfall).

In contrast, bores on or close to drainage lines show a much greater sensitivity to sustained rainfall and stream-flow events (e.g. for bore Pas96, rises in the hydrograph directly correspond to flow in the ephemeral stream channel; Fig. 6). As a result, recharge values calculated from logger data and longer-term hydrographs using the WTF method are relatively high for low-lying areas in both catch-

ments (average 78 mm yr^{-1}; 12% of rainfall; Fig. 4; Table 3). These recharge trends have been consistent over the past 20–30 years (Fig. 5).

The greater recharge in the lower-lying areas is predominantly because the steeper slopes in the upland areas direct runoff downslope to the lowland areas, which are consequently saturated for longer with a greater volume of runoff. In addition, runoff velocities across the lower areas decrease due to the reduction in slope, allowing more infiltration into the soil. Runoff from the upland areas is aided by the low-permeability, silty soils (Table 2), and infiltration in the lower-lying areas, particularly through the streambed, is increased by the greater depth of weathering (9 m depth to

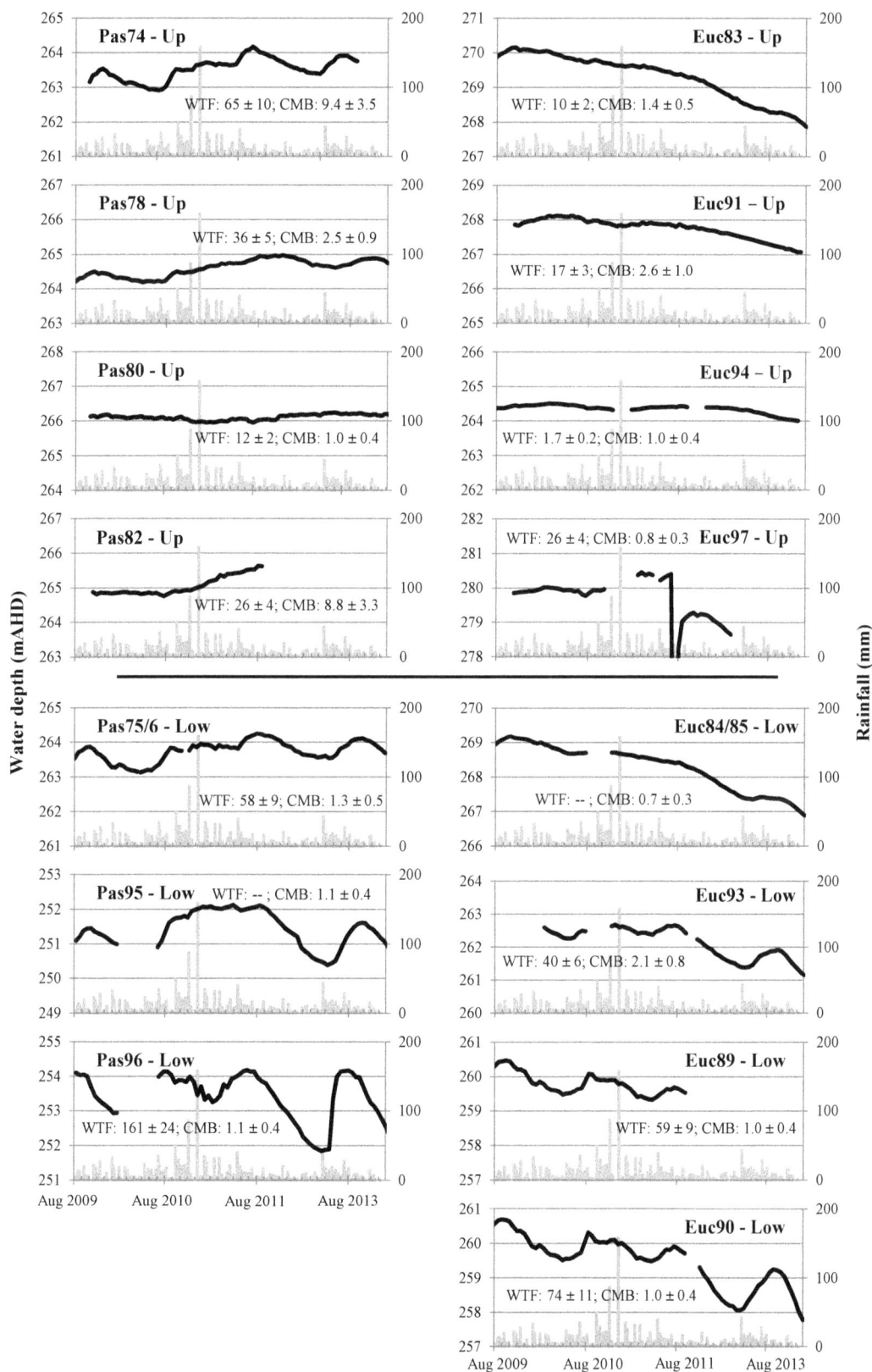

Figure 4. Bore hydrographs, rainfall and recharge estimates (in mm yr^{-1} from Table 3), for the water table fluctuation and chloride mass balance methods. Hydrographs are sorted by landscape position – lowland or upland.

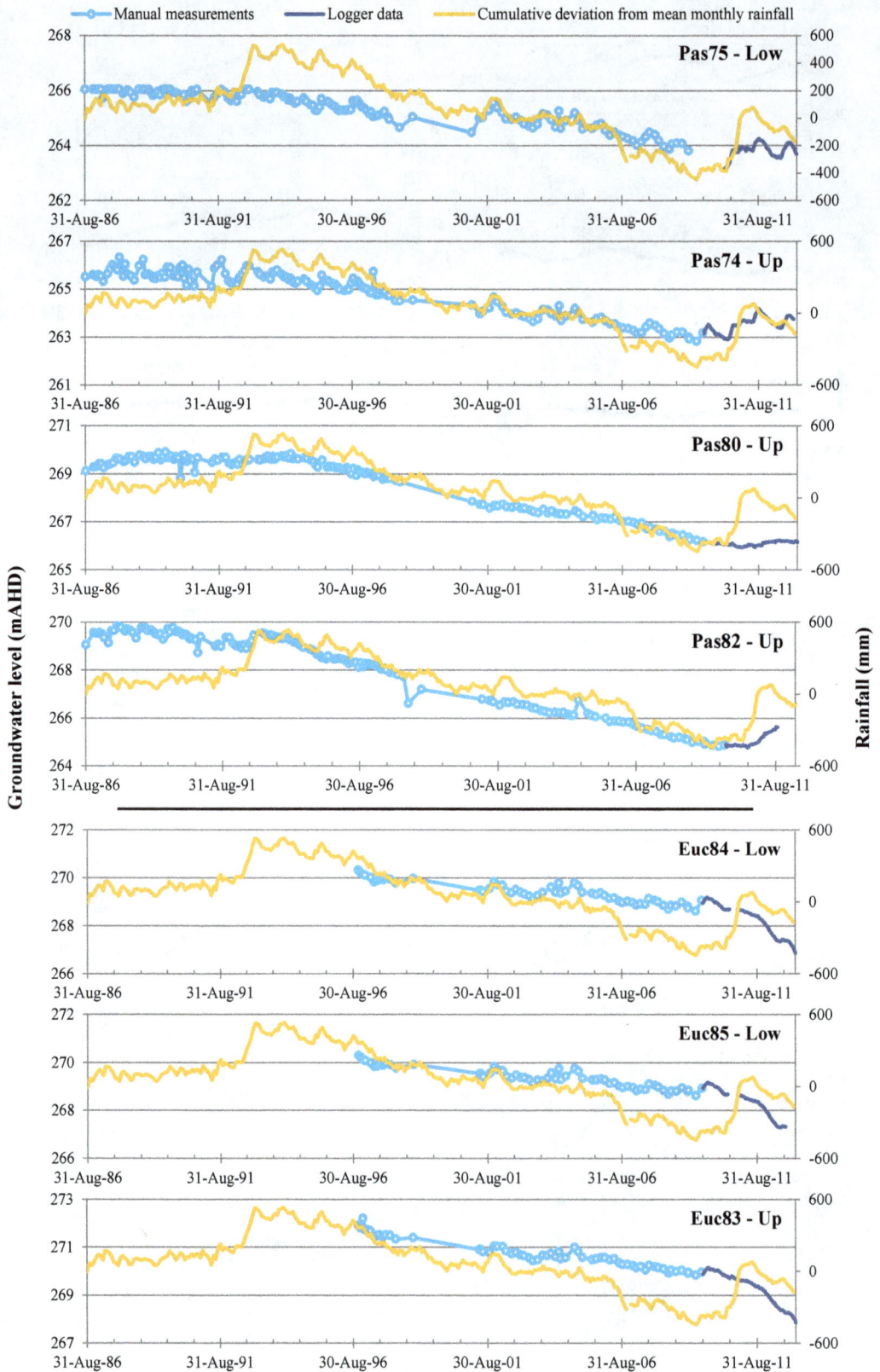

Figure 5. Long-term hydrographs for bores with available data and cumulative deviation from mean monthly rainfall to show the relationship between groundwater levels and long-term rainfall patterns.

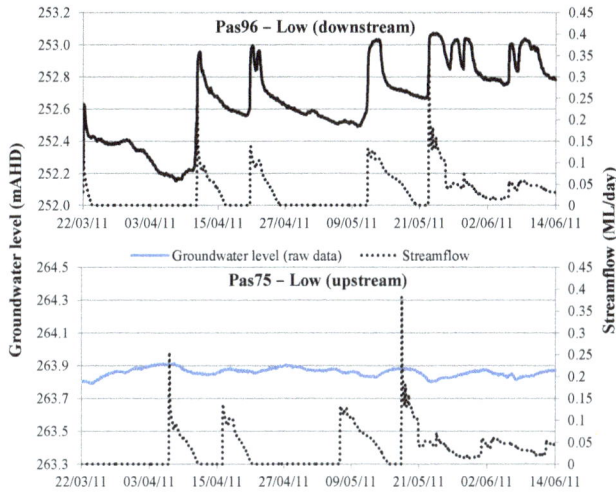

Figure 6. Pasture stream hydrographs (Dwyer's Creek) and bores hydrographs from the bottom of the catchment (Pas96) and midway up the catchment (Pas75).

bedrock in the pasture catchment and 30 m in the eucalypt catchment, except at the very bottom of this catchment).

Two of the lowland bores (Euc84 and Euc85) show very similar recharge patterns to upland bores (e.g. Euc83), i.e. little recharge, due to the presence of a localized confining layer (both bores frequently go artesian; Fig. 4).

Two of the upper-slope bores show high recharge (Pas74 and Pas 78), due to preferential recharge down fractures in the granite (Sect. 4.2; Fig. 5).

4.1.2 Chloride mass balance method

Recharge values calculated from the CMB method (Eq. 2) are much lower than the WTF method values, often by an order of magnitude or more (Table 3): e.g. Pas96 has recharge values of $1.1 \pm 0.4\,\mathrm{mm\,yr^{-1}}$ (CMB) and $161 \pm 24\,\mathrm{mm\,yr^{-1}}$ (WTF), and Pas82 has a CMB value of $8.8 \pm 3.3\,\mathrm{mm\,yr^{-1}}$ and a WTF value of $26 \pm 4\,\mathrm{mm\,yr^{-1}}$. Furthermore, the bore hydrographs used to calculate the WTF recharge values indicate that there is much more recharge occurring in the lowland areas than is indicated by the CMB values.

The most likely explanation for the mismatch between the CMB and WTF results is that the input Cl^- value used in the CMB method should be for runoff/streamflow rather than rainfall, because most recharge occurs from infiltration of surface flow through the streambed and across the low-gradient slopes adjacent to the streams, as previously discussed.

To account for this difference, the CMB values were recalculated using the volume and Cl^- content of streamflow (assumed to be the same as runoff here) in place of rainfall in Eq. (2):

$$R = \mathrm{RO}\frac{C_{\mathrm{ro}}}{C_{\mathrm{gw}}}, \qquad (3)$$

where "RO" (mm) is the estimated amount of runoff that would reach a given bore, and C_{ro} is the estimated Cl^- concentration of the runoff (volume weighted).

The volume of runoff at a particular bore (RO) is calculated using streamflow as a proxy for runoff, by dividing the average streamflow per year by the amount of the catchment that could theoretically provide runoff to the bore location (i.e. a bore in the middle of the catchment is only going to receive approximately half the runoff that could potentially recharge a bore at the bottom of the catchment). The Cl^- concentration of the runoff (C_{ro}) is calculated from the average EC measured at each weir (May 2010 to February 2013), converted to Cl^- using the $EC : Cl^-$ ratio for the study site data set (0.39 and 0.37 for the pasture and eucalypt catchments respectively). Equation (3) was only applied to bores in the lowland parts of the landscape where runoff is likely to recharge the groundwater. Because of the highly variable nature of the streamflow Cl^-, the potential variation in recharge values calculated from Eq. (3) is large, and this is seen in the error values (1σ – Table 3).

The recalculated recharge values generated from Eq. (3) are much closer to the WTF recharge values, but are still generally a factor of 5 to 15 lower. This may reflect the fact that the groundwater across the study site is mostly > 200 years old, indicating that the CMB values are generally representative of recharge rates under native vegetation prior to land clearance during European settlement in the late 1800s, whereas the WTF values represent recent recharge (August 2009 to February 2013). The older, pre-European settlement vegetation caused lower recharge, as these trees transpire much more water from groundwater and the soil zone than modern pasture. This disparity between modern and pre-European recharge rates has been observed elsewhere in southeastern Australia (e.g. Allison et al., 1990; Bennetts et al., 2006, 2007; Cartwright et al., 2007).

The CMB method estimates of recharge do not vary significantly between the two catchments, showing that both catchments behaved in a similar fashion before measurements began, prior to the establishment of the plantation. This corrects for the lack of a calibration period prior to the change in land use, a potential source of considerable error (Brown et al., 2005).

4.2 Topographic controls on recharge

Recharge estimates using the WTF method (Table 3) show that within the local groundwater systems of the study catchments, variations in recharge predominantly reflect differences in topography. Dominant areas of recharge are not along the topographic highs of the catchments, as in the traditional conceptual model of recharge, but are instead analogous with more arid regions, where most recharge occurs in topographic depressions (Scanlon et al., 2002).

Recharge rates increase as surface elevation decreases (Fig. 7). The steeper slopes of the upland areas promote

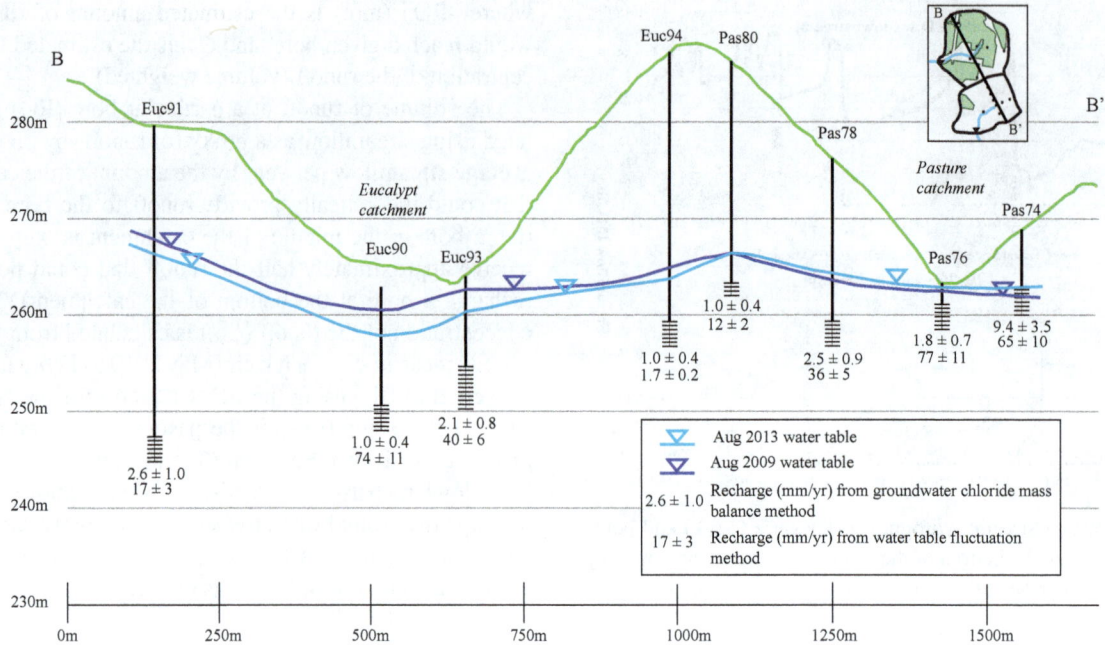

Figure 7. Cross-section from bore Euc91 across both catchments to bore Pas74 showing recharge rates based on both methods used in this study, and the water table change over the course of the study period (see Fig. 1 for bore locations).

runoff rather than infiltration, aided by low-permeability, silty soils (Table 2). Overland flow is focused into topographic lows and along drainage lines. Here the granite is most weathered, as indicated by the greater depth to bedrock here (9 m in the pasture catchment, and 30 m in the eucalypt catchment except at the very bottom of this catchment), encouraging recharge to occur, particularly through the streambed.

4.3   Influence of fractures on groundwater recharge

The ^{14}C data (Table 1) show that most of the groundwater at the study site is older than the tree plantation, but the groundwater in some bores (Pas74, Pas 80, Pas81, Pas 82, Pas 96, Euc91, Euc93 and Euc97) also contains measurable tritium, indicating a component of younger groundwater (< 50 years old). Recharge in fractured rock aquifers like granite is controlled to some extent by the fracture network (Cook, 2003), which forms multiple recharge pathways. In the study area this has allowed mixing of young groundwater (containing tritium) with much older groundwater (as shown by the ^{14}C dates; Table 1). The hydrograph for the upslope bore Pas74 (Fig. 4) shows high recharge following rainfall events (in contrast to most of the other upslope bores), most likely because it is located on a fracture in the granite that allows rapid recharge, as shown by the dilute groundwater with low Cl⁻ concentrations (Dean et al., 2014) and the presence of significant amounts of tritium (Table 1).

This dual porosity (matrix and fracture flow) influence on recharge has been observed elsewhere in southeastern Aus-

tralia where there was disparity between the residence times of groundwater samples (Cartwright et al., 2007). Nevertheless, the dominant recharge control across both catchments is topography rather than fracture heterogeneity, as shown by the relatively flat hydrographs for most of the upland bores, and strongly oscillating hydrographs in the lowland bores (Fig. 4).

4.4   The interplay between ephemeral streamflow and groundwater recharge and discharge

The streams at the study site are ephemeral, flowing on average only 40 % of the time at the catchment outlets. When they are dry, recharge can occur readily along and near the streambeds as upwards groundwater gradients are not present, because the water table is below the base of the stream. As a result, bores in the lower parts of the catchments (e.g. Pas96 near the outlet of the pasture catchment; Fig. 6) show a clear, sometimes instantaneous link between recharge and runoff.

Following extended periods of wet weather, the ephemeral stream at the bottom of the eucalypt catchment is fed by groundwater discharge, as shown by the significant levels of ^{222}Rn measured at the weir (11 Bq L^{-1}; Fig. 8); however, the elevated ^{222}Rn measured in the eucalypt stream could just be due to the close proximity of the granite bedrock to the surface at the bottom of this catchment. This is suggested by the high ^{222}Rn values in Pas95 and Euc92, both screened in granite bedrock, compared to the lower ^{222}Rn values in Euc90 and Pas96, which are screened in the weathered gran-

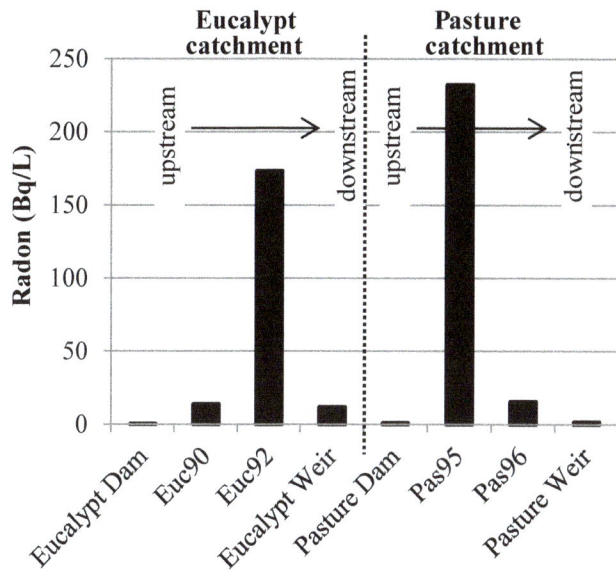

Figure 8. ^{222}Rn concentrations in the streams, measured at the weirs of both sites, and nearby bores. Surface water from further up the catchments is represented by water from dams located upslope in both catchments. The relatively high levels in the groundwater are a result of the decay of uranium present in the allanite and zircon of the granite.

ite saprolite (Fig. 8). Regardless, the shallow granite bedrock at the outlet of the eucalypt catchment (less than 2 m below the surface; Fig. 9) forces groundwater towards the surface here. In contrast, the bedrock at the bottom of the pasture catchment is 9 m deep, so the water table lies more consistently below the base of the stream and there is less groundwater discharge; as a result, the pasture catchment has fewer low flows than the eucalypt catchment (Fig. 9) and lower ^{222}Rn levels (1 Bq L^{-1} at the weir; Fig. 8).

In both catchments, during periods of little or no rainfall, the water table lies below the surface, so recharge can occur along the length of the channel. When it begins to rain and the system wets up, the water table rises at the downstream end of the catchment and groundwater begins to discharge here (this occurs more frequently and to a greater extent in the eucalypt catchment). Continued rain raises the water table so it connects to the stream further upstream, increasing the length of the stream that receives groundwater discharge (Fig. 9; Adelana et al., 2014). When rainfall ceases, the water table drops and progressively disconnects from the stream, starting upstream, until it is completely disconnected throughout the catchment. This means that during smaller rainfall events, when the water table remains below the land surface and does not connect to the stream, recharge occurs along the length of the stream. During larger rainfall events, as the water table comes to the surface along the stream channel, the area of potential recharge decreases.

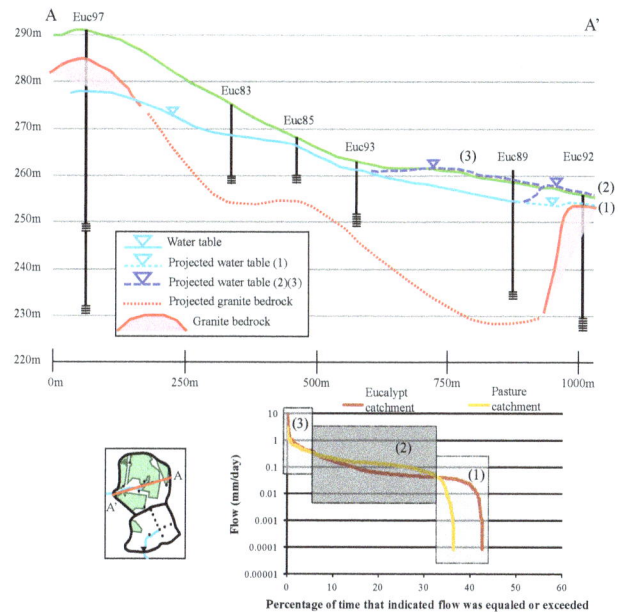

Figure 9. Long section from bores Euc97 to Euc92 showing the effect of the shallow granite on the water table under different flow conditions shown in the flow duration curve below: (1) where low flows in the eucalypt catchment stream are sustained for longer due to some groundwater discharge compared to virtually no groundwater discharge in the pasture catchment, (2) where the water table is at the surface and runoff is transported more quickly out of the eucalypt than in the pasture catchment, and (3) where there are some rare, very high flows, much higher than observed in the pasture catchment.

The groundwater hydrographs indicate that during the study period, recharge occurred readily in the lowland areas of both catchments, particularly when there was enough rainfall to generate consistent flow in the streams, while much less recharge is evident on the upper slopes. There is relatively little groundwater discharge along the streams, as shown by the ^{222}Rn data (Fig. 8), and groundwater within the catchments is lost predominantly through evapotranspiration, particularly when the water table is within 2 m of the ground surface (as commonly occurs in southeastern Australia, e.g. Bennetts et al., 2006, 2007); a small amount flows out at the bottom of the catchment.

4.5   Vegetation controls on recharge

The bore hydrographs in the eucalypt catchment show a clear overall declining trend of up to 3 m during the study period, evident even in artesian bores (Euc84 and Euc85), and regardless of landscape position (Fig. 4). This decline is not evident in hydrographs from the pasture catchment (Fig. 4), where the water table has increased by 0.5–1 m during the whole study period as a result of consecutive wet summers of 2010/2011 and 2011/2012 (Fig. 7). The tree plantation was a little over 1 year old when the main measurements of this study began, and as the age of the plantation increased, a steeper decline in groundwater depth was observed (Fig. 4).

The water level decrease in the eucalypt catchment, with no corresponding drop in the pasture catchment, is attributed to greater water use by the trees, as has been demonstrated elsewhere (e.g. Adelana et al., 2014; Bosch and Hewlett, 1982). The water table decline is less in the upland areas (Fig. 9), probably because recharge rates here are lower, so that the decrease in recharge due to tree water use has had relatively little impact. Furthermore, in the upland areas the water table is too deep for the vegetation to access the groundwater directly; Benyon et al. (2006), in a study in the same region of southeastern Australia, found that deep-rooted eucalypts can only access groundwater up to a depth of 6–8 m. In the lowland areas the trees are able to reach the groundwater (Fig. 2), and this, combined with the interception of potential recharge in the soil zone by the growing plantation, causes the observed decline in groundwater level (Fig. 4). Although tree roots can provide preferential pathways for infiltration of rainfall to the water table (Burgess et al., 2001), any effect of this is masked by the overall impact of eucalypt water use. The increasing rate of decline in groundwater depth over time can be attributed to the greater water usage by the trees as they grow (Fig. 4).

The narrow areas immediately adjacent to the drainage lines in the eucalypt catchment are covered in grass and therefore there is less direct interception of potential recharge, but in fact these areas show the biggest decline in groundwater level (Fig. 7). The highest rates of recharge occur along the drainage lines and the adjacent trees will therefore have a substantial impact there, in particular because they are directly accessing the groundwater.

With groundwater levels in the eucalypt catchment still in decline at the end of the study period, five years since the establishment of the eucalypt plantation, there is no sign that the system is reaching equilibrium under the new land use. Brown et al. (2005) indicate that equilibrium would not be expected until more than 5 years after the land use change occurred.

4.6   Management of tree plantations and recharge

Afforestation of farmland was widespread in southeastern South Australia and southwestern Victoria (known as the Green Triangle) from the 1980s through to the 2000s, with the plantation area expanding by 5–14% to 30 000 ha in Victoria alone (Adelana et al., 2014; Benyon et al., 2006; Ierodiaconou et al., 2005). However, the subsequent development of tree plantations in the region has been hindered by a poor timber market (HVP Plantations, personal communication, 2013) and concerns that plantations use more groundwater and surface water than other land uses like farming. As a result, tree plantations in the state of South Australia must now be licensed as groundwater users (Govt. of South Australia, 2009), while it is hoped that the potential reduction in water

availability resulting from reforestation will be offset by the beneficial gains of the carbon sequestration within the new trees (Schrobback et al., 2011).

A reduction of groundwater recharge by plantations, as documented in this study, lowers the water table and can reduce stream flow. If this is the object of the reforestation, for example to reduce saline groundwater discharge, then this land use change may well serve its purpose (Bennetts et al., 2007). However, the recent drought in southeastern Australia (1997–2010) has exacerbated concerns that trees may be a significant user of local and regional water resources, reducing groundwater recharge, discharge and surface water availability (Jackson et al., 2005).

In order to reduce the impact on water availability, current regulation of tree plantations in southeastern Australia focuses on the percentage of a catchment that may be planted. However, the present study shows that the location of the plantation within the catchment is significant also, with a smaller water table decline seen in the upland areas of the eucalypt catchment. Therefore to reduce the impact of plantations on groundwater recharge, tree planting should be avoided in the dominant zone of recharge, i.e. the topographically low areas and along the drainage lines, and should be concentrated on the upper slopes, where the water tables are deeper and the trees are less likely to access the groundwater and transpire it directly. At present, tree plantations in Victoria cannot be planted within 20 m of drainage lines, to avoid erosion of creek banks when the trees are removed (Dept. of Environment and Primary Industries, Victoria); we suggest that this currently restricted area along the drainage lines be expanded to include as much of the low topography parts of the site as practicable.

The expansion of the drainage line exclusion zone in tree plantations will have an added benefit in many regions of southeastern Australia where the groundwater is saline. This is because the parts of the catchments where the saline groundwater is within a few metres of the land surface (generally the lowland areas) can have a negative effect on tree health; at the study site, the trees closer to the drainage lines are shorter and thinner than those upslope.

However, excluding tree planting from low-elevation areas reduces the number of trees that can be planted within a catchment, and also means that trees are not planted in areas where (good quality) groundwater is shallowest and can be most readily accessed for tree growth. As the primary purpose of many tree plantations is the production of wood and pulp products for economic gain, this restriction will slow economic returns. To overcome this, consideration could be given to planting lower-water-use trees that can better cope with the upslope areas where water supplies for tree growth may be limited.

This management strategy of balancing economic and hydrologic perspectives when locating tree plantations within catchments will be applicable to other low-rainfall, high-evaporation regions, and should be considered for tree plantations in similar climatic areas worldwide.

5 Conclusions

While the importance of topography and ephemeral streams to focused recharge in low-rainfall regions around the world has been known for some time, the implications of this aspect of the groundwater resource literature have not been incorporated into plantation management guidelines and legislation. In this study, it is shown that the majority of modern recharge at the study site, calculated from the water table fluctuation method, occurs in the lower parts of both study catchments (12 % of rainfall versus 3 % in the upland areas). Overland flow is focused into topographic lows and along drainage lines where greater infiltration can occur. Recharge calculations using a corrected chloride mass balance method gave lower values than modern recharge estimates because the groundwater across the study site is mostly > 200 years old, representing recharge under native eucalypt forest prior to European land clearance. Relatively little groundwater discharges into the streams or flows out at the bottom of the catchment; groundwater within the catchments is lost predominantly through evapotranspiration. Overall the tree plantation in this study caused a drawdown in groundwater levels, increasing over time as the trees aged, compared to a slight rise in groundwater levels in the pasture catchment.

The results of this study lead to the conclusion that both the hydrogeological and economic frameworks for commercial forestry need to be considered. If conserving groundwater recharge is a primary objective, tree planting should be avoided in the dominant zone of recharge, and concentrated on the upper slopes, where recharge is low enough that any further reduction will have minimal impact. We suggest expanding present regulations for tree plantations which specify that trees cannot be planted within a certain distance of drainage lines, including as much of the low topography parts of the site as practicable. Consideration should be given to the potential negative impact on the financial viability of a tree plantation. This management strategy is applicable to low-rainfall, high-evaporation regions worldwide.

Acknowledgements. This work would not have been possible without the assistance and support of the land owners, Iven, Iris and Marcia Field, Macquarie Bank Foundation, and the plantation management, McEwen's Contracting Pty Ltd. We would like to acknowledge Phil Cook and Peter Hekmeijer at the Department of Primary Industries Victoria, our collaborators in the National Centre for Groundwater Research and Training (of which this research is part of programme 4 – http://www.groundwater.com.au) for funding and support, the Australian Institute of Nuclear Science and Engineering (AINSE – grant number AINSTU0710) for funding and support, and funding from the State of Victoria through the Department of Environment and Primary Industries, Future Farming Systems Research Division. We would also like to thank Ian Cartwright, Guillaume Bertrand, a third anonymous reviewer and the editor, Przemyslaw Wachniew, for their constructive feedback and helpful comments throughout the review process.

Neither the NCGRT, AINSE nor DEPI Victoria (as funding sources) were involved in the design of this specific study, nor were they involved in the collection or analysis of the resulting data.

Edited by: P. Wachniew

References

Adelana, M., Dresel, E., Hekmeijer, P., Zydor, H., Webb, J., Reynolds, M., and Ryan, M.: A comparison on streamflow and water balances in adjacent farmland and forest catchments in south-western Victoria, Australia, Hydrol. Process., doi:10.1002/hyp.10281, online first, 2014.

Allison, G. B. and Hughes, M. W.: The use of environmental chloride and tritium to estimate total recharge to an unconfined aquifer, Aust. J. Soil Res., 16, 181–195, 1978.

Allison, G. B., Cook, P. G., Barnett, S. R., Walker, G. R., Jolly, I. D., and Hughes, M. W.: Land clearance and river salinisation in the western Murray Basin, Australia, J. Hydrol., 119, 1–20, 1990.

Bell, R. W., Schofield, N. J., Loh, I. C., and Bari, M. A.: Groundwater response to reforestation in the Darling Range of Western Australia, J. Hydrol., 115, 297–317, 1990.

Bennetts, D. A., Webb, J. A., Stone, D. J. M., and Hill, D. M.: Understanding the salinisation process for groundwater in an area of south-eastern Australia, using hydrochemical and isotopic evidence, J. Hydrol., 323, 178–192, 2006.

Bennetts, D. A., Webb, J. A., Stone, D. J. M., and Hill, D. M.: Dryland salinity processes within the discharge zone of a local groundwater system, southeastern Australia, Hydrogeol. J., 15, 1197–1210, 2007.

Benyon, R. G.: Water use by tree plantations in the green triangle, A review of current knowledge, The Glenelg Hopkins Catchment Management Authority, Hamilton, Australia, 2002.

Benyon, R. G., Theiveyanathan, S., and Doody, T. M.: Impacts of tree plantations on groundwater in south-eastern Australia, Aust. J. Bot., 54, 181–192, 2006.

Bormann, M. E.: Temporal and spatial trends in rainwater chemistry across Central and Western Victoria, Honours Thesis, La Trobe University, Melbourne, Australia, 2004.

Bosch, J. M. and Hewlett, J. D.: A review of catchment experiments to determine the effect of vegetation changes on water yield and evapotranspiration, J. Hydrol., 55, 3–23, 1982.

Boutt, D. F., Diggins, P., and Mabee, S.: A field study (Massachusetts, USA) of the factors controlling the depth of groundwater flow systems in crystalline fractured-rock terrain, Hydrogeol. J., 18, 1839–1854, 2010.

Brown, A. E., Zhang, L., McMahon, T. A., Western, A. W., and Vertessy, R. A.: A review of paired catchment studies for determining changes in water yield resulting from alterations in vegetation, J. Hydrol., 310, 28–61, 2005.

Burgess, S. S. O., Adams, M. A., Turner, N. C., White, D. A., and Ong, C. K.: Tree roots: conduits for deep recharge of soil water, Oecologia, 126, 158–165, 2001.

Butler, J. J., Jin, W., Mohammed, G. A., and Reboulet, E. C.: New insights from well responses to fluctuations in barometric pressure, Ground Water, 49, 525–533, 2011.

Camporese, M., Dean, J. F., Dresel, P. E., Webb, J. A., and Daly, E.: Hydrological modelling of paired catchments with competing land uses, Proceedings of the 20th International Congress on Modelling and Simulation, Adelaide, Australia, 1–6 December 2013, 1819–1825, 2013.

Camporese, M., Daly, E., Dresel, P. E., and Webb, J. A.: Simplified modelling of catchment-scale evapotranspiration via boundary condition switching, Adv. Water Resour., 69, 95–105, 2014.

Cartwright, I., Weaver, T. R., Stone, D., and Reid, M.: Constraining modern and historical recharge from bore hydrographs, ^{3}H, ^{14}C, and chloride concentrations: Applications to dual-porosity aquifers in dryland salinity areas, Murray Basin, Australia, J. Hydrol., 332, 69–92, 2007.

Cook, P. G.: A Guide to Regional Groundwater Flow in Fractured Rock Aquifers, CSIRO Land and Water, Adelaide, Australia, 2003.

Cook, P. G., Walker, G. R., and Jolly, I. D.: Spatial variability of groundwater recharge in a semiarid region, J. Hydrol., 111, 195–212, 1989.

Craddock, J. M.: Methods of comparing annual rainfall records for climatic purposes, Weather, 34, 332–346, 1979.

Cuthbert, M. O.: Straight thinking about groundwater recession, Water Resour. Res., 50, 2407–2424, doi:10.1002/2013WR014060, 2014.

Dalhaus, P. G., Cox, J. W., Simmons, C. T., and Smitt, C. M.: Beyond hydrogeologic evidence: challenging the current assumptions about salinity processes in the Corangamite region, Australia, Hydrogol. J., 16, 1283–1298, 2008.

Dean, J. F., Webb, J. A., Jacobsen, G. E., Chisari, R., and Dresel, P. E.: Biomass uptake and fire as controls on groundwater solute evolution on a southeast Australian granite: aboriginal land management hypothesis, Biogeosciences, 11, 4099–4114, doi:10.5194/bg-11-4099-2014, 2014.

Delin, G. G., Healy, R. W., Landon, M. K., and Bhlke, J. K.: Effects of topography and soil properties on recharge at two sites in an agricultural field, J. Am. Water Resour. As., 36, 1401–1416, 200.

Delin, G. N., Healy, R. W., Lorenz, D. L., and Nimmo, J. R.: Comparison of local- to regional-scale ground-water recharge in Minnesota, USA, J. Hydrol., 334, 231–249, 2007.

Dept. Sustainability and Environment: Groundwater SAFE: Secure allocations, future entitlements, Victorian Gov. Dept. of Sustainability and Environment, Melbourne, Australia, 2012.

Domenico, P. A. and Schwartz, F. W.: Physical and Chemical Hydrogeology, John Wiley and Sons Inc., New York, USA, 528 pp., 1998.

Dresel, P. E., Hekmeijer, P., Dean, J. F., Harvey, W., Webb, J. A., and Cook, P.: Use of laser-scan technology to analyse topography and flow in a weir pool, Hydrol. Earth Syst. Sci., 16, 2703–2708, doi:10.5194/hess-16-2703-2012, 2012.

Edwards, M.: A hydrological, hydrogeological and hydrogeochemical study of processes leading to land and water salinisation in the Mount William Creek catchment, southeastern Australia, PhD thesis, La Trobe University, Bundoora, Melbourne, Australia, 2006.

Farley, K. A., Jobbágy, E. G., and Jackson, R. B.: Effects of afforestation on water yield: a global synthesis with implication for policy, Glob. Change Biol., 11, 1565–1576, 2005.

Feikema, P. M., Morris, J. D., and Connell, L. D.: The water balance and water sources of a Eucalyptus plantation over shallow saline groundwater, Plant Soil, 332, 429–449, 2010.

Govt. of South Australia: Managing the water resource impacts of plantation forests, Dept. of Water, Land and Biodiversity Conservation, Adelaide, Australia, 2009.

Healy, R. W. and Cook, P. G.: Using groundwater levels to estimate recharge, Hydrogeol. J., 10, 91–109, 2002.

Hekmeijer, P. and Hocking, M.: Feasibility of groundwater pumping from granite slopes, Centre for Land Protection Research, Bendigo, Victoria, Australia, Technical report no. 71, 2001.

Heppner, C. S., Nimmo, J. R., Folmar, G. J., Gburek, W. J., and Risser, D. W.: Multiple-methods investigation of recharge at a humid-region, fractured rock site, Pennsylvania, USA, Hydrogeol. J., 15, 915–927, 2007.

Hergt, J., Woodhead, J., and Schofield, A.: A-type magmatism in the Western Lachlan Fold Belt? A study of granites and rhyolites from the Grampians region, Western Victoria, Lithos, 97, 122–139, 2007.

Hibbert, A. R.: Forest treatment effects on water yield, in: International Symposium on Forest Hydrology, edited by: Sopper, W. E. and Lull, H. W., Pergamon, Oxford, UK, 30 August–10 September 1965, 527–543, 1967.

Hutton, J. T. and Leslie, T. I.: Accession of non-nitrogenous ions dissolved in rainwater to soils in Victoria, Aust. J. Agr. Res., 9, 59–84, 1958.

Ierodiaconou, D., Laurenson, L., Leblanc, M., Stagnitti, F., Duff, G., Salzman, S., and Versace, V.: Consequences of land use change on nutrient exports: a regional scale assessment in south-west Victoria, Australia, J. Environ. Management, 74, 305–316, 2005.

Jackson, R. B., Jobbagy, E. G., Avissar, R., Baidya Roy, S., Barrett, D. J., Cook, C. W., Farley, K. A., le Maitre, D. C., McCarl, B. A., and Murray, B. C.: Trading water for carbon with biological carbon sequestration, Science, 310, 1944–1947, 2005.

Jie, Z., van Heyden, J., Bendel, D., and Barthel, R.: Combination of soil-water balance models and water-table fluctuation methods for evaluation and improvement of groundwater recharge calculations, Hydrogeol. J., 19, 1487–1502, 2011.

Jobbagy, E. G. and Jackson, R. B.: Groundwater use and salinization with grassland afforestation, Glob. Change Biol., 10, 1299–1312, 2004.

Meinzer, O. E.: The occurrence of groundwater in the United States with a discussion of principles, US Geololgical Survey, Washington, D.C., USA, Water-Supply Paper 489, 1923.

Nation, E.: Groundwater Recharge for Agriculture: Rainfall Chemistry Report, Bureau of Rural Sciences, Canberra, Australia, 62 pp., 2009.

Nicholson, C., Dalhaus, P., Anderson, G., Kelliher, C., and Stephens, M.: Corangamite Salinity Action Plan 2005–2008, Corangamite Catchment Management Authority, Colac, Victoria, Australia, 133 pp., 2006.

Rasmussen, T. C. and Crawford, L. A.: Identifying and Removing Barometric Pressure Effects in Confined and Unconfined Aquifers, Ground Water, 35, 502–511, 1997.

Rutledge, A. T.: Computer programs for describing the recession of ground-water discharge and for estimating mean ground-water recharge and discharge from streamflow records–update, US Geological Survey, Reston, Virginia, USA, Water-Resources Investigations report 98-4148, 43 pp., 1998.

Scanlon, B. R., Healy, R. W., and Cook, P. G.: Choosing appropriate techniques for quantifying groundwater recharge, Hydrogeol. J., 10, 18–39, 2002.

Scanlon, B. R., Reedy, R. C., Stonestrom, D. A., Prudic, D. E., and Dennehy, K. F.: Impact of land use and land cover change on groundwater recharge and quality in the southwestern US, Glob. Change Biol., 11, 1577–1593, 2005.

Scanlon, B. R., Keese, K. E., Flint, A. L., Flint, L. E., Gaye, C. B., Edmunds, M., and Simmers, I.: Global synthesis of groundwater recharge in semiarid and arid regions, Hydrol. Process., 20, 3335–3370, 2006.

Scanlon, B. R., Jolly, I., Sophocleous, M., and Zhang, L.: Global impacts of conversions from natural to agricultural ecosystems on water resources: Quantity versus quality, Water Resour. Res., 43, W03437, doi:10.1029/2006WR005486, 2007.

Schilling, K. E.: Investigating local variation in groundwater recharge along a topographic gradient, Walnut Creek, Iowa, USA, Hydrogeol. J., 17, 397–407, 2009.

Schofield, N. J.: Tree planting for dryland salinity control in Australia, Agroforest. Syst., 20, 1–23, 1992.

Schrobback, P., Adamson, D., and Quiggin, J.: Turning water into carbon: carbon sequestration and water flow in the Murray-Darling Basin, Environ. Resour. Econ, 49, 23–45, 2011.

Sophocleus, M. A.: Combining the soilwater balance and water-level fluctuation methods to estimate natural groundwater recharge: practical aspects, J. Hydrol., 124, 229–241, 1991.

Toll, N. J. and Rasmussen, T. C.: Removal of barometric pressure effects and earth tides from observed water levels, Ground Water, 45, 101–105, 2007.

van Dijk, A. I. J. M., Hairsine, P. B., Arancibia, J. P., and Dowling, T. I.: Reforestation, water availability and stream salinity: A multi-scale analysis in the Murray-Darling Basin, Australia, Forest Ecol. Manag., 251, 94–109, 2007.

VandenBerg, A. H. M.. Rock unit names in western Victoria, Seamless Geology Project, Geological Survey of Victoria Report, Geoscience, Victoria, Australia, 130, 2009.

Webb, J. A., Williams, B. G., Bailue, K., Walker, J., and Anderson, J. W.: Short-term groundwater dynamics at a paddock scale, Proceedings of Water Down Under Adelaide, Australia, 15–17 April 2008, 1493–1500, 2008.

White, M., Oates, A., Barlow, T., Pelikan, M., Brown, J., Rosengren, N., Cheal, D., Sinclair, S., and Stutter, G.: The Vegetation of North-West Victoria, Department of Sustainability and Environment, Melbourne, Australia, 2003.

Winter, T. C.: The concept of hydrologic landscapes, J. Am. Water Resour. As., 37, 335–349, 2001.

2

Evaluation of ORCHIDEE-MICT-simulated soil moisture over China and impacts of different atmospheric forcing data

Zun Yin[1], **Catherine Ottlé**[1], **Philippe Ciais**[1], **Matthieu Guimberteau**[1,2], **Xuhui Wang**[1,3,4], **Dan Zhu**[1], **Fabienne Maignan**[1], **Shushi Peng**[3], **Shilong Piao**[3], **Jan Polcher**[4], **Feng Zhou**[3], **Hyungjun Kim**[5], **and other China-Trend-Stream project members**[*]

[1]Laboratoire des Sciences du Climat et de l'Environnement, CNRS-CEA-UVSQ, Gif-sur-Yvette 91191, France
[2]UMR 7619 METIS, Sorbonne Universités, UPMC, CNRS, EPHE, 4 place Jussieu, Paris 75005, France
[3]Sino-French Institute for Earth System Science, College of Urban and Environmental Sciences, Peking University, Beijing 100871, China
[4]Laboratoire de Météorologie Dynamique, UPMC/CNRS, IPSL, Paris 75005, France
[5]Institute of Industrial Science, University of Tokyo, Tokyo, Japan
[*]A full list of the China-Trend-Stream project members and their affiliations appears at the end of the paper.

Correspondence: Zun Yin (vyin@lsce.ipsl.fr)

Abstract. Soil moisture is a key variable of land surface hydrology, and its correct representation in land surface models is crucial for local to global climate predictions. The errors may come from the model itself (structure and parameterization) but also from the meteorological forcing used. In order to separate the two source of errors, four atmospheric forcing datasets, GSWP3 (Global Soil Wetness Project Phase 3), PGF (Princeton Global meteorological Forcing), CRU-NCEP (Climatic Research Unit-National Center for Environmental Prediction), and WFDEI (WATCH Forcing Data methodology applied to ERA-Interim reanalysis data), were used to drive simulations in China by the land surface model ORCHIDEE-MICT(ORganizing Carbon and Hydrology in Dynamic EcosystEms: aMeliorated Interactions between Carbon and Temperature). Simulated soil moisture was compared with in situ and satellite datasets at different spatial and temporal scales in order to (1) estimate the ability of ORCHIDEE-MICT to represent soil moisture dynamics in China; (2) demonstrate the most suitable forcing dataset for further hydrological studies in Yangtze and Yellow River basins; and (3) understand the discrepancies of simulated soil moisture among simulations. Results showed that ORCHIDEE-MICT can simulate reasonable soil moisture dynamics in China, but the quality varies with forcing data. Simulated soil moisture driven by GSWP3 and WFDEI shows the best performance according to the root mean square error (RMSE) and correlation coefficient, respectively, suggesting that both GSWP3 and WFDEI are good choices for further hydrological studies in the two catchments. The mismatch between simulated and observed soil moisture is mainly explained by the bias of magnitude, suggesting that the parameterization in ORCHIDEE-MICT should be revised for further simulations in China. Underestimated soil moisture in the North China Plain demonstrates possible significant impacts of human activities like irrigation on soil moisture variation, which was not considered in our simulations. Finally, the discrepancies of meteorological variables and simulated soil moisture among the four simulations are analyzed. The result shows that the discrepancy of soil moisture is mainly explained by differences in precipitation frequency and air humidity rather than differences in precipitation amount.

1 Introduction

Climate change strongly influences the hydrological cycle, which in turn affects ecosystems services, food security, and water resources (Bonan, 2008; Piao et al., 2010; Seneviratne et al., 2010; Zhu et al., 2016). More importantly, the main mechanisms governing hydrological processes vary across climate regimes affected by anthropogenic factors (Guimberteau et al., 2012; Wada et al., 2016, 2017). Covering different climate zones and most types of human activities (An et al., 2017; Basheer and Elagib, 2018; Bouwer et al., 2009; Feng et al., 2016; Rogers et al., 2016; Wu et al., 2018), China is a good test bed to investigate the hydrological complexity of climate–water–human interactions. In China, annual precipitation has increased in the south but declined in the north over the last several decades (Ye et al., 2013; Zhai et al., 2005). This dipole of precipitation trends is partly reflected in the discharge trends of Yangtze and Yellow rivers (Piao et al., 2010), but other factors than precipitation changes affect river discharge including changes in rainfall intensity, land surface state or condition, and water management (Ayalew et al., 2014; Grillakis et al., 2016; Williams et al., 2015). A prerequisite to understand how precipitation changes transfer into river discharge changes is to analyze and evaluate the various components of the surface water budget and especially the key variable relationships between precipitation and soil moisture (SM), the result of the partition of precipitation among evapotranspiration, infiltration, and runoff.

SM indeed plays a crucial role in adjusting local climate (Seneviratne et al., 2013; Teuling et al., 2010), regulating productivity and ecosystem dynamics (Schymanski et al., 2008; Yin et al., 2014), and affecting carbon budgets (Calvet et al., 2004). SM controls vegetation photosynthesis through transpiration, which in turn significantly influences surface temperature (Bonan, 2008; Dai et al., 2004). It also impacts the infiltration rate of precipitation in the soil, and its state before rainfall events determines the ratio of surface runoff to precipitation (Grillakis et al., 2016). Therefore SM is not only of importance in understanding land surface processes, but also is a key indicator for predicting and addressing extreme events, such as heatwaves, floods, and droughts (Hirschi et al., 2011; Teuling et al., 2010; Wanders et al., 2014).

In the investigation of spatial and temporal SM variations, in situ measurements (Dorigo et al., 2011; Liu et al., 2001; Piao et al., 2009; Robock et al., 2000) are too sparse and not always representative of larger scales. Although they can provide firsthand records of SM fluctuations, the density of in situ networks cannot meet the requirement for continental-scale studies., and the different measurement techniques make it difficult to combine different datasets. Satellite-based SM products (Dorigo et al., 2015; Njoku et al., 2003; Su et al., 2003; Wagner et al., 2012) provide excellent spatial coverage and temporal sampling, but their accuracy varies between instruments and retrieval algorithms used (Liu et al., 2012). Moreover, these estimations concern only the first few centimeters of soil depth, so the root-zone SM cannot be directly assessed, unless a model simulating the water transfer processes is used. To overcome the uneven coverage of raw data, data assimilation is widely applied to analyze SM from in situ or satellite observations (Draper et al., 2012; Martens

et al., 2016; Reichle et al., 2007). Analyzed products help us to understand SM variation and its relation to climate (Liu et al., 2015b, 2017; Taylor et al., 2012). However, to capture changes of hydrological mechanisms for future projection, such measurements are not enough.

Land surface models (LSMs) are able to simulate the short- and long-term SM dynamics consistently with atmospheric forcing and surface information (Pierdicca et al., 2015; Rebel et al., 2012; Xia et al., 2014) by reproducing physical processes and interactions with other climatic, hydrological, and ecological factors (Seneviratne et al., 2010). The uncertainty of simulated SM depends on the accuracy of atmospheric forcing, in particular precipitation frequency and intensity, as well as radiation. However, LSMs' complexity is a source of structural errors (missing processes) and biased parameters. Thus it is necessary to validate simulated SM by observations in order to diagnose the source of errors and estimate the ability of the chosen LSM to simulate SM dynamics in the area of interest.

In this study, the land surface model ORCHIDEE-MICT (ORganizing Carbon and Hydrology in Dynamic EcosystEms: aMeliorated Interactions between Carbon and Temperature; Guimberteau et al., 2018) is used to simulate SM over China. Besides land surface hydrology, ORCHIDEE-MICT simulates energy budgets and vegetation dynamics (mechanistic phenology, photosynthesis, and ecosystem carbon cycling) which interact with the water cycle and climate (Guimberteau et al., 2012). Moreover, the evaluation of simulated SM controlled by natural processes is useful to identify human effects (e.g., crops, irrigation, dam operation) on the water budget in regions where there is a large misfit between the model and observations.

Four global atmospheric forcing datasets are chosen to drive the simulations in China, including GSWP3 (Global Soil Wetness Project Phase 3), PGF (Princeton Global meteorological Forcing), CRU-NCEP (Climatic Research Unit-National Center for Environmental Prediction), and WFDEI (WATCH Forcing Data methodology applied to ERA-Interim reanalysis data), due to their wide application in numerous hydrological studies (Getirana et al., 2014; Guimberteau et al., 2014, 2017, 2018; Hirschi et al., 2014; Van Den Hurk et al., 2016; Polcher et al., 2016; Müller Schmied et al., 2016; Tangdamrongsub et al., 2018; Yang et al., 2015; Zhao et al., 2017; Zhou et al., 2018). Although they provide gridded surface climate variables at the global scale, their uncertainties of representing regional climate are not clear. Through comparison of simulated SM to various datasets, our study also addresses which forcing has the best performance in SM simulation in China.

Our SM simulations are evaluated with different SM datasets including in situ data, remote sensing measurements, and reanalysis. In situ measurements including ISMN (International Soil Moisture Network; Dorigo et al., 2011) and PKU (in situ SM from Peking University; Piao et al., 2009) are used to evaluate temporal validation of simulated

SM. To evaluate spatiotemporal variations of simulated SM, the satellite-based dataset ESA CCI SM (European Space Agency Climate Change Initiative Soil Moisture; Wagner et al., 2012) is applied in the comparison. Note that both in situ and satellite SM datasets represent the truth to some extent. This implies that real-world SM is influenced by processes that are not modeled such as irrigation and wetlands. Thus mismatches between measured and simulated SM may exist in some regions strongly affected by anthropogenic factors. Moreover, satellite instruments do not measure SM directly; it is derived via a complex modelization of the radiative transfer at the soil–vegetation interface calibrated with in situ data.

Finally, GLEAM SM data (Global Land Evaporation Amsterdam Model; Martens et al., 2017) are compared to the simulated SM. Different from other SM datasets, GLEAM SM results from a land surface model constrained by a number of satellite and in situ observations. This reanalysis product was shown to reproduce reasonable long-term SM dynamics at the global scale (Martens et al., 2017), which is valuable to evaluate ORCHIDEE-MICT simulations for both surface and root-zone SM. Furthermore, GLEAM assimilates CCI SM data, so that evaluation of our model against root-zone SM from GLEAM is consistent with evaluation against surface SM from CCI.

Through the simulations and comparisons, three questions will be addressed:

– Is the model able to provide a reasonable estimation of SM dynamics in China, as a prerequisite for further hydrological studies?

– Which atmospheric forcing gives the best SM simulation according to the comparisons with available observations?

– Which meteorological variable drives the differences of SM among the simulations?

The study area, atmospheric forcing, and SM datasets used in this study are described in Sect. 2. Section 3 presents the model experiments. Evaluation of simulated SM and discussion are given in Sects. 4 and 5, respectively.

2 Study area, forcing, and evaluation datasets

2.1 Study area

China has multiple climate regimes, which creates hydrological situations influenced by different variables in different regions. The land water budget in China is affected by anthropogenic factors, such as irrigation (Puma and Cook, 2010), afforestation (Liu et al., 2015a; Peng et al., 2014), deforestation (Wei et al., 2018), polders (Yan et al., 2016), dams (Deng et al., 2016), and inter-basin water transfer (Li et al., 2015). Two main river basins are of interest: the Yangtze

River basin (YZRB) and the Yellow River basin (YLRB) (red and magenta contours, respectively, in Fig. 1), which cover the main regions of industry and agriculture (gray regions in Fig. 1). The Yangtze River originates in the Qinghai–Tibetan Plateau and flows through two wetted traditional agricultural zones, the Sichuan Basin, and the plain downstream of the Yangtze River (Fig. 1). The Yellow River originates in the Qinghai–Tibetan Plateau as well, but it flows through another two agricultural regions (the Loess Plateau and the North China Plain) under semi-arid and semi-humid zones (Kottek et al., 2006). Our simulations cover the main part of China ($[85–124° E] \times [20–44° N]$) including these two watersheds to assess SM dynamics at the catchment scale. Note that in the analysis, the specific regions of the two river basins are coarser than the exact basin contours shown in Fig. 1 due to the interpolation of routing files at the resolution of our simulations.

2.2 Atmospheric forcing

Four gridded atmospheric forcing datasets are used to force the model over China: GSWP3, PGF, CRU-NCEP, and WFDEI. All input variables needed are the air temperature at 2 m (T_a), rainfall and snowfall rates, atmospheric specific humidity at 2 m (Q_a), surface pressure, downward short-/longwave radiation (R_s and R_l), and wind speed (W). The four forcing datasets are combinations of reanalysis and observation data. These datasets, although built by different methods, are not independent from each other since they share some common inputs. Detailed descriptions are listed below and general information is summarized in Table 1. Preprocessing of the datasets for ORCHIDEE-MICT is described in Sect. 3.2.

2.2.1 GSWP3

The GSWP3 v0 (http://hydro.iis.u-tokyo.ac.jp/GSWP3, last access: 16 October 2018); Kim, 2017) provides 3-hourly climate data at 0.5° resolution from 1901 to 2010. It is based on the 20th Century Reanalysis (20CR; Compo et al., 2011), which is downscaled from 2 to 0.5° by a spectral nudging technique in a Global Spectral Model (Yoshimura and Kanamitsu, 2008), in order to maintain both low- and high-frequency signals at a high spatiotemporal scale. Single ensemble correction and vertically weighted damping are applied to remove known artifacts in high latitude regions (Hong and Chang, 2012; Yoshimura and Kanamitsu, 2013). Moreover, observation data are used for bias correction, such as GPCC v6 (Global Precipitation Climatology Centre; Becker et al., 2013) for precipitation, SRB (Surface Radiation Budget; Stephens et al., 2012) for radiation, and CRU TS v3.21 (Climate Research Unit; Harris et al., 2014) for temperature.

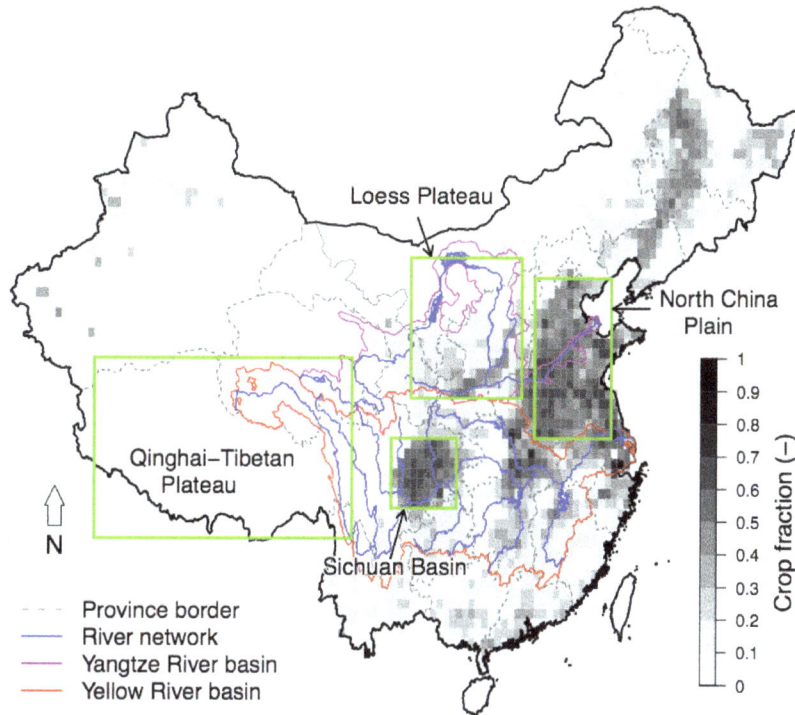

Figure 1. Four important regions mentioned in this paper (green rectangles). The grey background is the cropland fraction.

Table 1. General information of the climate forcing datasets. "Reanalysis" and "Observations" are corresponding datasets used in producing the atmospheric forcing. Detailed description can be found in Sect. 2.2.

Dataset	Resolution		Duration	Reanalysis	Observations
	Spatial	Temporal			
GSWP3	0.5°	3-hourly	1901–2010	20CR	GPCC, CRU TS, SRB
PGF	1°	3-hourly	1901–2012	NCEP-NCAR	CRU TS, GPCP, TRMM, SRB
CRU-NCEP	0.5°	6-hourly	1901–2015	NCEP	CRU TS
WFDEI	0.5°	3-hourly	1979–2009	ERA-Interim	CRU TS, GPCC

2.2.2 PGF

The PGF (http://hydrology.princeton.edu/data.pgf.php, latest version released: 13 July 2014, last access: 16 October 2018) provides 3-hourly data at 1° resolution from 1901 to 2012 (Sheffield et al., 2006). It is constructed by combining the NCEP-NCAR (National Centers for Environmental Prediction-National Center for Atmospheric Research) reanalysis of Kalnay et al. (1996) with several observation datasets. Precipitation is corrected by downscaled CRU TS v3.1, GPCP (Global Precipitation Climatology Project; Adler et al., 2003), and TRMM (Tropical Rainfall Measuring Mission; Huffman et al., 2007) data. SRB and CRU TS data are used in the assimilation of radiation and air temperature, respectively. Other variables (e.g., specific humidity, surface air pressure, wind speed) are just spatially downscaled from NCEP-NCAR according to the local elevation.

2.2.3 CRU-NCEP

The CRU-NCEP v6.1 (ftp://nacp.ornl.gov/synthesis/2009/frescati/model_driver/cru_ncep/analysis/readme.htm, last access: 16 October 2018) provides 6-hourly 0.5° data. It combines the coarse temporal resolution (monthly) CRU TS dataset with the NCEP reanalysis, which has a higher time interval (6-hourly) but is only available at 2.5°. Monthly climate (except for precipitation) is identical to CRU TS, and NCEP is used only to reconstruct the 6-hourly variability within each month after bi-linearly interpolated to 0.5°. For precipitation the original NCEP values are used for temporal linear interpolation in those CRU grid cells (0.5°) covered by the specific NCEP grid cell (2.5°) in each month. CRU-NCEP dataset is available from 1901 to 2015 at the global scale and it is updated every year.

2.2.4 WFDEI

The WFDEI forcing (version 31 July 2012) is generated by applying the WATCH Forcing Data methodology (http://www.eu-watch.org, last access: 16 October 2018; Weedon et al., 2014) to the ERA-Interim reanalysis (Dee et al., 2011) providing 3-hourly data at 0.5° from 1979 to 2009. The ERA-Interim blends Global Climate Model (GCM) modeled variables and a suite of observations using a 4D-Var (four-dimensional variable analysis) data assimilation system (Weedon et al., 2014). All variables are bias-corrected using CRU TS. For precipitation, we use a version that has been bias-corrected by GPCC v5 and v6.

2.3 Soil moisture datasets

2.3.1 International Soil Moisture Network (ISMN)

The ISMN is an international cooperative project providing a global gauged SM database (Dorigo et al., 2011). It is based on in situ measurements from multiple monitoring regional subprojects. Here only data from the CHINA subproject are used (Robock et al., 2000) with in situ volumetric water content (depth of water column over depth of soil in $m^3\,m^{-3}$) from 40 stations between 1981 and 1999. SM profiles on 11 vertical layers were collected three times per month (on 8th, 18th, and 28th of each month). The 11 sampled soil layers are 0–5, 5–10, and then every 10 cm layers until 1 m. Most stations are located in cropland or grasslands, but information about land use types and soil texture of each site is not provided. Moreover, there is no information about management practices affecting SM, such as irrigation or tillage.

In spite of the long length of this dataset, the data availability and monitoring period among stations vary widely. Some stations only recorded SM during the growing season, while others have a full year record. Furthermore, the measurements including the five deep layers (below 50 cm) are fewer in number than those including the top six layers. Only stations with more than 15 years of data were selected, which at least cover the same period (1984–1999). To make sure that at least half of the data are available in the 15-year time series, stations with fewer than 270 measurement points in the top six layers are removed. This leads to selecting a subset of 20 stations, and given the sparseness of data below 50 cm, only SM in the top six layers is used for model evaluation.

2.3.2 In situ SM from Peking University

The SM was measured over 778 stations of agro-meteorological stations over China by the Chinese Meteorological Administration and collected and harmonized by the research team in Peking University (PKU; Piao et al., 2009). The dataset provides 10-day SM variation during the growing season (mainly between May and September) from 1991 to 2007. It provides SM profiles in seven soil layers (0–10, 10–20, 0–30, 30–40, 40–50, 50–70, and 70–100 cm), but the bottom four layers often have missing records. This dataset concerns exclusively croplands but there is no explicit information of soil texture and irrigation. Similar to the ISMN, the monitoring durations among gauging stations are different. A total of 203 stations that cover the period of 1992–2006 are chosen.

2.3.3 ESA CCI SM

The ESA CCI SM is a multi-satellite-based product (Liu et al., 2011, 2012; Wagner et al., 2012) and has been validated both at the global scale (Dorigo et al., 2015) and in China (Peng et al., 2015; An et al., 2016). The daily SM is retrieved from a suite of microwave sensors spreading the period of 1979–2010 with 0.5° resolution. The representative soil layer depth is approximately 0.5–2 cm. Multiple sensors ensure long-term records of SM dynamics; however the uncertainty of the data varies with the change of available sensors and corresponding algorithms. Moreover, the remote sensing technique limits its ability to detect SM in frozen soils or under snow cover. Therefore SM data are not available during winter in high-latitude regions (e.g., northern China). The data availability also varies through the period according to the number of available instruments and the increase of their temporal and spatial resolutions. In China, the fraction of days with available records (Fig. 4 of Dorigo et al., 2015) is lower than 20 % from 1979 to 2006. More importantly, large spatial variation of gaps exists as well before 2006 (Fig. S1). The period after the launch of MetOp-A AS-CAT (Advanced Scatterometer) at the end of 2006 appears to be more stable. To provide a reliable validation, we only use the CCI SM data between 2007 and 2009.

2.3.4 GLEAM v3.0A SM

GLEAM v3.0 is a multiple-algorithm, observation-based model reconstructing the components of the land evaporation process, including daily SM, evapotranspiration, and interception at 0.25° resolution (Martens et al., 2017). It has three subversions. Due to the short duration of version B (2003–2015) and C (2011–2015), only version A, which covers the period 1980–2014, is used here. Radiation and air temperature used in GLEAM 3.0A are from ERA-Interim, and precipitation is from MSWEP (Multi-Source Weighted-Ensemble Precipitation; Beck et al., 2017).

Both surface and root-zone SM from GLEAM, which has been validated by Martens et al. (2017), are used for comparison. The surface SM in the top 0–10 cm is a combination of simulated SM from the GLEAM soil module, SMOS (the Soil Moisture Ocean Salinity satellite mission; Kerr et al., 2001), and ESA CCI SM (ESA Climate Change Initiative Soil Moisture; Liu et al., 2011, 2012; Wagner et al., 2012) through the data assimilation system developed by Martens et al. (2016). The Community Noah land surface model SM fields in GLDAS (Global Land Data Assimilation System; Rodell et al., 2004) were used to estimate the errors of these SM products. Root-zone SM is derived from the GLEAM soil module based on mass balance. GLEAM provides SM in separate land cover tiles of bare soil (0–10 cm), low vegetation (0–100 cm), and tall vegetation (0–250 cm). These tiles are based on MODIS Vegetation Continuous Fields (MOD44B; Hansen et al., 2003).

3 Land surface model, simulation protocol, and model–data comparison metrics

3.1 Land surface model

ORCHIDEE (Organizing Carbon and Hydrology In Dynamic EcosystEms; Krinner et al., 2005) is a physical-based land surface process model. It is mainly composed of two modules. The SECHIBA (surface–vegetation–atmosphere transfer scheme) module calculates the exchange of water and energy between land and the atmosphere with a high-frequency time interval (half an hour), while the STOMATE (Saclay Toulouse Orsay Model for the Analysis of Terrestrial Ecosystems) module estimates the carbon cycle at the daily timescale. The ORCHIDEE-MICT (aMeliorated Interactions between Carbon and Temperature, SVN version 3952; Guimberteau et al., 2018; Zhu et al., 2015) is a recent version of ORCHIDEE including new processes such as the interactions among frozen soil, snow, plants, and soil carbon pools. It accounts for soil freezing, soil carbon discretization, snow processes, and lateral water flows to improve the simulation of the main biogeochemical cycles in permafrost regions. It has been chosen in this study because China has a large permafrost area, especially in the Tibetan Plateau, where both Yellow and Yangtze rivers originate. To simulate the SM dynamics, ORCHIDEE-MICT uses a scheme with 11 soil layers, whose depth increases exponentially down to 2 m. The respective depths (in meters) of the calculation nodes are the following: 0.0005, 0.002, 0.006, 0.014, 0.03, 0.06, 0.12, 0.25, 0.5, 1.0, and 1.75. Each grid cell can include up to three soil tiles: bare soil, trees, and grass/crops, which are filled by the corresponding plant functional types (PFTs) of the 13-PFT scheme of ORCHIDEE-MICT to allow better representation of their specific hydrology. The hydrological budget is calculated separately in each soil tile. The amplitude of SM depends on soil texture, which is a part of boundary conditions. Explicit description of the ORCHIDEE-MICT model can be found in Guimberteau et al. (2018). ORCHIDEE will be referred to ORCHIDEE-MICT for brevity in the following text.

There are two main outputs of SM in ORCHIDEE. The total SM (θ_t) indicates the total amount of soil water volume in the top 2 m soil layer in a grid cell. The SM profile (θ_p) records the vertical distribution of soil water content in the 11 soil layers. Note that the θ_p in each soil layer is an average value among the three soil tiles. The initial unit of ORCHIDEE SM is $m^3 m^{-3}$.

3.2 Simulation protocol

Four simulations were performed driven by different forcing datasets described in Sect. 2.2. In the simulations, CO_2 rise and land use change are taken into account but without human processes like irrigation. The 13-PFT map is from LUH2 (http://luh.umd.edu) and the soil texture map is from Zobler (1986). For the 3-soil texture scheme of Zobler (1986), the minimum residual and maximum saturated SM are 0.065 and 0.43 $m^3 m^{-3}$, respectively. The model domain covers the main part of China ($[85–124° E] \times [20–44° N]$). The spatial resolution is the same as the atmospheric forcing (Table 1). The simulation period covers 39 years, from 1971 to 2009, except for the one driven by WFDEI, which is from 1979 to 2009. To make sure that carbon (LAI and biomass) and water cycle variables can reach equilibrium, a 100-year spin-up was performed by repeating the forcing of the period 1971–1980 10 times (for WFDEI, 50 times the period 1979–1980). Starting from the end of the spin-up, simulations were run from 1981 to 2009. The output driven by PGF forcing was re-gridded at $0.5° \times 0.5°$ to match the resolution of other simulation outputs.

The temporal resolution of forcing datasets is either 3-hourly or 6-hourly (CRU-NCEP), which is larger than the simulation time step of SECHIBA (30 min). To have a reasonable precipitation intensity, and thus a good infiltration of water in the soil, the default precipitation splitting algorithm of ORCHIDEE is applied in our simulations. At the beginning of each forcing time step, if precipitation occurred, the precipitation amount (precipitation rate multiplied by the time interval of specific forcing) will be uniformly distributed to the first half of the forcing time step.

3.3 Model–data comparison methodology and metrics

3.3.1 Comparison protocol

As the soil depths, periods, and spatiotemporal resolutions are different in the four SM datasets (Sect. 2.3), we have to choose corresponding ORCHIDEE outputs for each comparison. For the comparison with the in situ data of ISMN and PKU, we first extracted the modeled daily SM profile (θ_p) from the nearest grid cell for each station. Then the SM above a certain soil depth was chosen (50 cm for ISMN and 20 cm for PKU). PKU SM is provided in degrees of saturation, defined as the volume ratio of actual water content to its maximum value when the soil is saturated. As the soil porosity is unknown, the PKU SM dataset cannot be directly compared with simulated SM from ORCHIDEE, which is defined from modeled porosity. To overcome this problem, normalization was applied in both datasets before comparison. The normalized data at each station and in the corresponding grid cell of

Table 2. Summary of the SM datasets for validation. "M+RS+RA" indicates that the dataset is a model output driven by both remote sensing and reanalysis data. More details can be found in Sect. 2.3.

Dataset	Type	Unit	Resolution	Duration		Contents		Corresponding ORCHIDEE soil layer
				Analysis period		Analysis depth		
ISMN	in situ	$m^3 m^{-3}$	station, 10-day	1981–1999		11 layers; 0–100 cm		1–9 layers (0–75 cm)
				1984–1999		0–50 cm		
PKU	in situ	% of porosity	station, 10-day	1991–2007		7 layers; 0–100 cm		1–8 layers (0–37 cm)
				1992–2006		0–30 cm		
ESA CCI	RS	$m^3 m^{-3}$	0.25°, daily	1979–2010		top layer, depth $\approx$ 0.5–2 cm		1–4 layers (0–2 cm)
				2007–2009				
GLEAM surface	M + RS + RA	$m^3 m^{-3}$	0.25°, daily	1980–2014		0–10 cm		1–6 layers (0–9 cm)
				1981–2009				
GLEAM root zone	M + RS + RA	$m^3 m^{-3}$	0.25°, daily	1980–2014		mixture of bare soil (1–10 cm),		all layers (0–200 cm)
				1981–2009		low vegetation (0–100 cm), and high vegetation (0–250 cm)		

the model are the ratio of the difference between the original value and its mean (during the observation period) to its standard deviation.

According to the sampled depth of the ESA CCI SM, the daily top four-layer (2.2 cm) averaged SM from ORCHIDEE is used. Regarding the definition of GLEAM SM (Sect. 2.3), we used the daily top six-layer (approximately 9.2 cm depth) averaged SM and the total SM of ORCHIDEE for the comparison with GLEAM surface and root-zone SM, respectively. The period length and soil depth of each comparison are shown in Table 2. In addition, the timing of all SM datasets is made uniform according to coordinated universal time (UTC).

3.3.2 Metrics

Pearson's correlation coefficient (r) is calculated to estimate the correlation between simulated and observed SM. Daily SM corresponding to the measurement date reported in ISMN was collected to calculate r. As there is no date information from the 10-day PKU dataset, we used the 10-day averaged SM from ORCHIDEE for comparison.

The root mean square error (RMSE) is applied in order to estimate the temporal differences between simulation and observation. The same data pairs are used for RMSE calculation as the correlation coefficient except for PKU due to the normalization. Note that RMSE is related to the magnitude of SM, which varies significantly in China. To make it comparable in space, the relative RMSE is calculated by dividing the mean of the simulated and observed SM.

According to Kobayashi and Salam (2000), the mean squared deviation (MSD), which is $RMSE^2$, can be decom-

posed into the squared bias (SB), the squared difference between standard deviation (SDSD), and the lack of correlation weighted by the standard deviation (LCS), as

$$MSD = RMSE^2 = SB + SDSD + LCS. \tag{1}$$

SB is the bias between simulations and observations. It is independent from other two components:

$$SB = (\bar{s} - \bar{m})^2, \tag{2}$$

where $\bar{s}$ and $\bar{m}$ are the mean of simulated and measured values, respectively. The SDSD indicates the mismatch of variation magnitude between simulated and observed variables, defined as

$$SDSD = (SD_s - SD_m)^2, \tag{3}$$

where SD_s and SD_m are the standard deviation of simulations and measurements, respectively. High SDSD implies a failure of the model in simulating the degree of fluctuation across the n measurements. Note that SDSD correlates with LCS, which accounts for SD_s and SD_m as well:

$$LCS = 2SD_s SD_m (1 - r), \tag{4}$$

where r is the Pearson's correlation coefficient. The LCS is an indicator of the performance of the model to simulate the pattern of fluctuation of the measurements. The lower the LCS is, the better the model performs.

Finally, to evaluate the characteristic timescale of modeled SM response to hydrological processes, the lag-k autocorrelation coefficient (R_k) is computed. The R_k is the correlation coefficient of a time series with itself but with a k time step lag, as

$$R_k = \frac{\sum_{i=1}^{n-k}(x_i - \overline{x})(x_{i+k} - \overline{x})}{\sum_{i=1}^{n}(x_i - \overline{x})^2}, \qquad (5)$$

where n $(n > k)$ is the length of the specific time series; x is the mean value. For SM time series in a specific grid cell, R_k was computed for different k values. The value of R_k decreases with increasing k and the k-lag time series are considered not auto-correlated if R_k is less than a threshold $1/e$ (Maurer et al., 2001; Rebel et al., 2012). The day number when R_k first drops below a threshold of $1/e$ is called the number of lag days (NLD). The NLD difference is used to compare the overall characteristic timescales between datasets. The difference of R_k profiles gives additional information on the autocorrelations for lag. The R_k comparison was implemented between GLEAM and ORCHIDEE because other datasets do not have complete daily records.

The linear trend of SM change in the 29 years is of interest as well. The Mann–Kendall test (Kendall, 1975; Mann, 1945) is applied to test if simulations capture observed trends of SM, with a p value < 0.05 indicating a significant trend.

3.4 Correlation of uncertainties between SM and meteorological factors

In our simulations, the difference in atmospheric forcing is the only source of difference in simulated SM. We look at different climate variables to explain SM differences among simulations. These variables include monthly precipitation amount (P) and the number of precipitation days in one month (N_p) excluding days with $P < 0.01\,\mathrm{mm\,d^{-1}}$. Precipitation days are categorized into five classes of 0.01–1, 1–5, 5–10, 10–15, and $> 15\,\mathrm{mm\,d^{-1}}$. The number of days with precipitation amount in each class was calculated, denoted by N_p^i with $1 \le i \le 5$. Other meteorological variables are incoming short-/long-wave radiation (R_s/R_l), air temperature (T_a), air humidity (Q_a), and wind speed (W). Regarding SM, both total SM (θ_t) and SM in each soil layer $(\theta_p^i$, i is the index of soil layer) were correlated with these variables. To estimate the difference of a variable x among the four simulations, the averaged MSD (D_x) is computed as

$$D_x = \frac{\frac{1}{n}\sum_{i \ne j}^{N}\sum_{t=1}^{n}(x_{t,i} - x_{t,j})^2}{\binom{n}{2}}, \qquad (6)$$

where $N = 4$ is the number of simulations; i and j $(1 \le i, j \le N)$ are indexes of the four simulations; $\binom{n}{2}$ is the binomial coefficient; n is the length of the time series; t is the time step. Note that we use the absolute value of D_x not the relative D_x $(D_x$ over averaged value of x in the specific grid cell) for the analysis because the relative D_x cannot reflect the linkage of uncertainty between inputs and outputs. Detailed explanation is shown in Sect. S1 in the Supplement.

4 Results

4.1 SM evaluation against multiple datasets

4.1.1 Comparison with ISMN and PKU in situ data

In most cases, the correlations between modeled and measured SM at ISMN stations (see Sect. 2.3) are significantly positive (Fig. 2). High correlations $(r > 0.6)$ are found over the Loess Plateau in the semi-arid zone, which is a region of rainfed agriculture where SM is less affected by anthropogenic processes. In the North China Plain water is limited as well, whereas irrigation is widely applied for agriculture, leading to low r (below 0.5). To further compare the simulated and measured SM, three ISMN stations (marked by squares in Fig. 2a) are chosen to represent different wet conditions, and model–data comparisons are shown in Fig. 3. Xifeng is located in the semi-arid zone (MAP $=$ 556 mm yr^{-1}), where θ_t is low (0.2 m^3 m^{-3} on average) with a large inter-annual variation. The variability of simulated θ_t is consistent with observations $(0.73 < r < 0.87$; when CRU-NCEP is excluded) due to lower human impacts on rainfed agriculture in this region (Li et al., 2014). Xinxian is located in the North China Plain with similar MAP (580 mm yr^{-1}) to Xifeng, but in a traditional irrigation region (Wang et al., 2016). θ_t at Xinxian is underestimated, possibly because irrigation is not included in our simulations. Thus the model cannot capture the seasonal variations of θ_t, given r values ranging between 0.11 and 0.21. Xuzhou is in the North China Plain as well, but with a higher MAP (847 mm yr^{-1}). The fluctuation of simulated and observed θ_t is coherent, leading to r from 0.55 to 0.64. However the magnitude of θ_t is systematically underestimated as well (Fig. 3).

The correlation coefficients of θ_t between simulations and PKU dataset are shown in Fig. 2 as circles. Modeled θ_t has a better performance in the Loess Plateau and the North China Plain than other regions, suggesting that ORCHIDEE is able to capture the variations of SM in semi-arid and temperate zones. In comparison to ISMN, r between ORCHIDEE and PKU θ_t is lower. This may be caused by the shallower depth of the PKU data (20 cm), with a stronger influence from fast infiltration and transpiration processes than in the ISMN records (1 m). Moreover, the PKU dataset only records θ_t during the growing season, leading to lower r in absence of full seasonal variations. Negative correlations are found in several sites located along river networks. The negative r $(-0.4 < r < -0.2)$ coincides with the coupling of wetness anomaly and irrigation: when droughts occur (reflected by low simulated SM), more water will be withdrawn from the river and irrigated on the croplands (reflected by high observed SM). Thus the negative r found in Fig. 2 reveals that SM dynamics cannot be well understood without considering anthropogenic activities.

According to the r shown in Fig. 2, we find that GSWP3 and WFDEI provide better simulated SM than the other two.

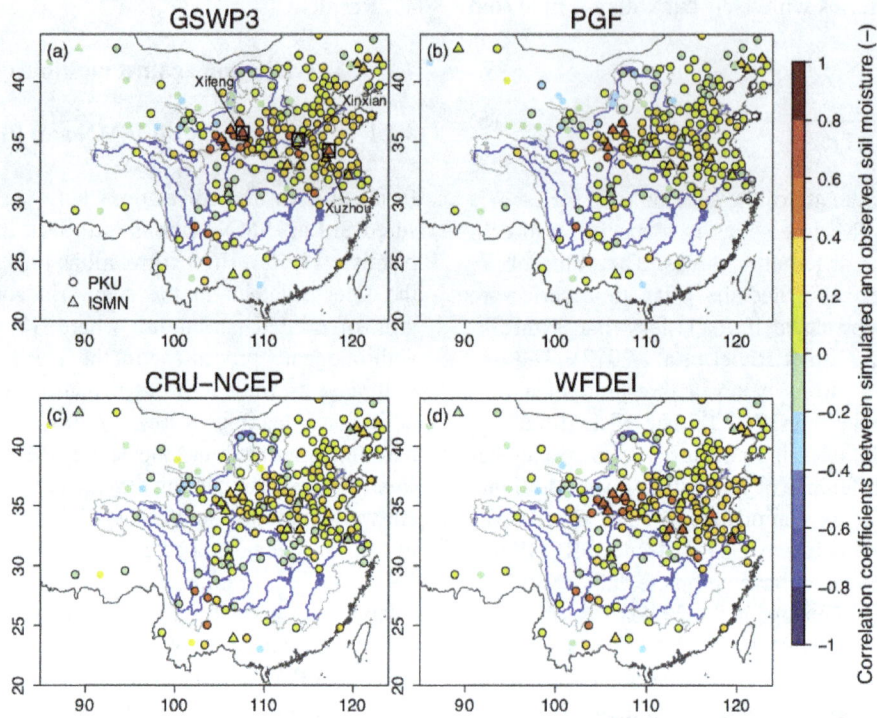

Figure 2. Pearson's correlation coefficients of modeled and measured SM at each gauging station from ISMN (triangles) and PKU (circles). Symbols with a dark border indicate significant correlations ($p < 0.05$). The locations of three ISMN stations shown in Fig. 3 are marked by black squares in panel **(a)**.

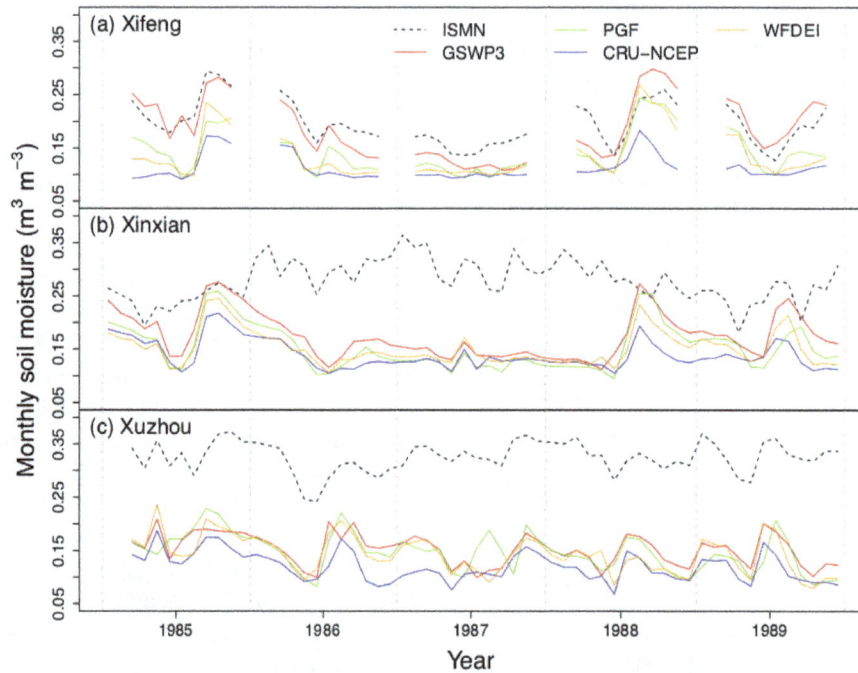

Figure 3. Time series of 10-day SM from **ORCHIDEE** and ISMN at three stations. The station locations are shown in Fig. 2a. The mean annual precipitation values at Xifeng, Xinxian, and Xuzhou (according to GSWP3) are 556, 580, and $847\,\mathrm{mm\,yr^{-1}}$, respectively. Dark dashed lines indicate ISMN SM. Red, green, blue, and orange lines indicate simulated SM based on GSWP3, PGF, CRU-NCEP, and WFDEI, respectively.

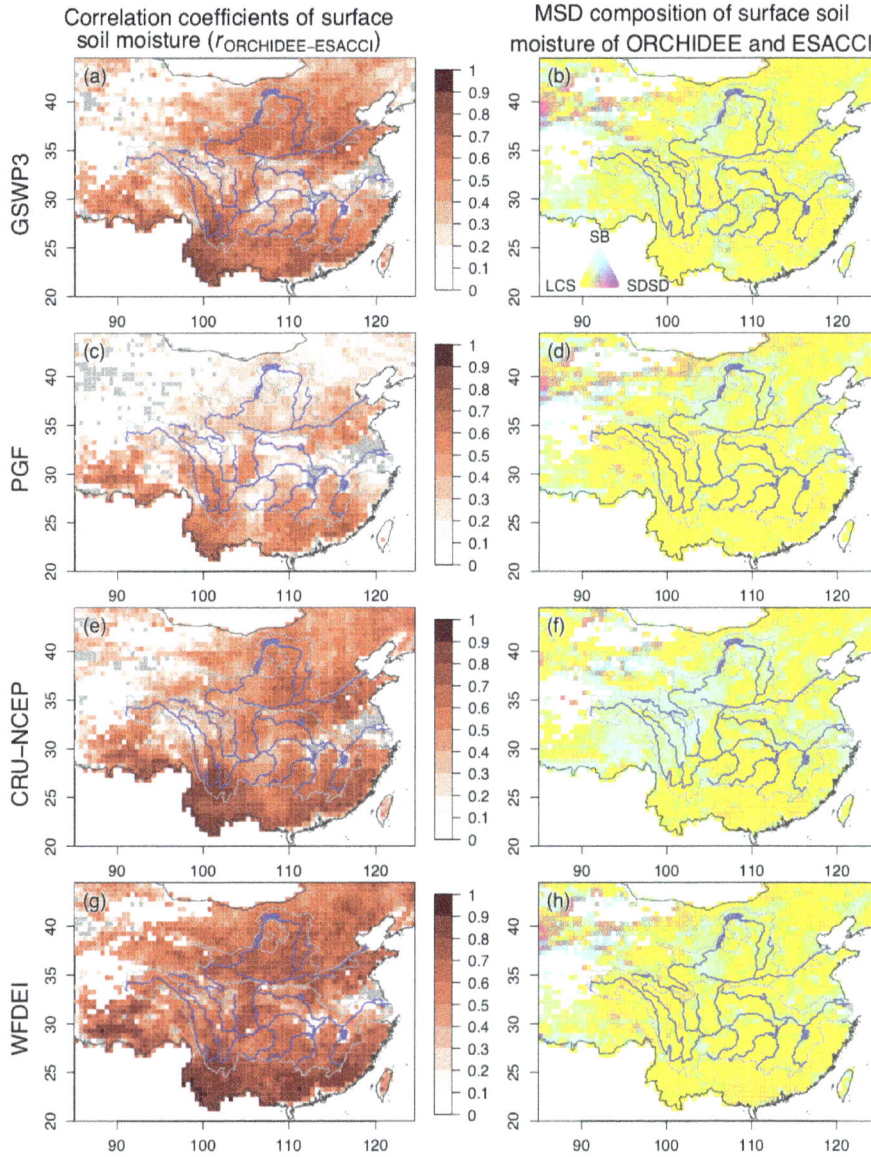

Figure 4. (a, c, e, g) Correlation coefficients of the ESA CCI SM and the corresponding ORCHIDEE SM. Gray pixels indicate no correlation and negative correlation. (b, d, f, h) Decomposition of the MSD between the daily ESA CCI SM and the corresponding ORCHIDEE SM (Eq. 1). Cyan, magenta, and yellow indicate the fractions of SB, SDSD, and LCS, respectively.

The main difference is found in the North China Plain, where the r values of GSWP3 and WFDEI are higher. It indicates that simulation can be improved by selecting suitable atmospheric forcing. Nevertheless, the r is still limited by the lack of measurement information (soil texture, irrigation flag, land cover, etc.) and anthropogenic processes in OR-CHIDEE. The disagreements between simulated and measured SM are caused by the spatial scale as well. The spatial resolutions of forcings ($0.5° ≈ 55$ km) are too coarse to represent the specific climatic conditions of gauging stations. On the other hand, the comparison cannot provide a comprehensive validation in the YZRB, where there are few measurements. Thus remote sensing and hybrid SM datasets are required to evaluate the simulations.

4.1.2 Comparison with ESA CCI SM data

Figure 4 (left panel) shows r between CCI and ORCHIDEE θ_s from 2007 to 2009. High r is found in both the North China Plain and southern China. In southern China, SM is less disturbed by anthropogenic factors due to its wet conditions. Thus SM has little variation driven by climate, which can be well simulated by the model. On the other hand, human activities strongly affect SM in the North China Plain, whereas their impacts on r are neutralized by the large annual variation due to seasonality. Weak correlations only exist in the transition zone from the south to the north along the Yangtze River network, where there are both human disturbances and little annual variation.

Figures S2 and 4 (right panel) show the relative RMSE and the MSD decomposition of θ_s between CCI and ORCHIDEE. Low relative RMSE (< 0.3) is found in the YZRB, but in the YLRB the value is higher (> 0.4). The mean source of MSD is the LCS (phase mismatch). It implies that the magnitude of the simulated θ_s is reasonable, but the timing of the fluctuations differs between ORCHIDEE and CCI. The coincidence of magnitude is reflected in the relative difference (Fig. S3), the absolute value of which is less than 0.1 in 76% grid cells, excluding the CRU-NCEP case. A large LCS might be due to human activities and the discretization of the CCI SM time series. Irrigation in northern China may significantly affect the fluctuation of θ_s, which leads to underestimation of simulated SM and contributes to the LCS. Simultaneously, due to the incomplete records of CCI SM (Fig. S1), the seasonal variation of SM cannot be fully taken into account in the comparison. The r is consequently declined and the LCS increases (Eq. 4).

The availability and uncertainty of CCI SM vary with space and time (Sect. 2.3 and Fig. S1). To provide a reliable estimation, we performed the analysis exclusively in the period 2007–2009. In fact, there is little difference when the comparison covered the whole period of 1981–2009. The patterns of r and MSD decompositions (Fig. S4) are similar to those of the comparison of 2007–2009. The r of 1981–2009 is lower with no doubt because a longer period contains more errors due to the fragmentary records of CCI SM data.

4.1.3   Comparison with GLEAM v3.0A data

The left panel of Fig. 5 shows correlation coefficients between GLEAM surface SM (θ_s) and corresponding modeled SM in the surface layer (0–10 cm). Simulated θ_s is significantly correlated with GLEAM (median $r = 0.54$). In the Sichuan Basin, r is lower than its surroundings. According to the spatially averaged r of θ_s, GSWP3 (0.55) and WFDEI (0.66) lead to better performances with ORCHIDEE than PGF (0.43) and CRU-NCEP (0.51). Note that both WFDEI and GLEAM v3.0A used ERA-Interim reanalysis to reconstruct the time series of precipitation, which can explain the higher r when ORCHIDEE is forced by WFDEI. The correlation coefficients of simulated and GLEAM root-zone SM (θ_r, Fig. S5) have similar patterns to the θ_s but higher values (median $r = 0.57$) due to the lower variability of θ_r, which smoothes out misfits related to differences in individual rainfall events between ORCHIDEE and GLEAM for θ_s. Compared to CCI θ_s, the r between ORCHIDEE and GLEAM θ_s is much higher. It is probably due to the shallower depth of the CCI θ_s, which is more sensitive to surface processes and forcing data errors. Moreover, CCI θ_s is a purely satellite product while GLEAM θ_s (v3.0A) is a combination of modeled, in situ, and satellite SM. The latter is not totally independent of the forcing datasets and therefore more comparable to our simulations.

Figure S6 and the right panel of Fig. 5 show the relative RMSE and the MSD decomposition of θ_s between GLEAM and ORCHIDEE. A low relative RMSE (< 0.3) covers most regions except for the North China Plain (> 0.5), where the MSD is dominated by the squared bias values (SB, Fig. 5b). This is clearly shown in the relative difference (Fig. S7) between GLEAM and ORCHIDEE, where simulated θ_s is approximately 30% lower than in GLEAM. Southern China has a lower relative RMSE (< 0.2), and MSD is dominated by SB as well. Different from the North China Plain, SB in southern China may be due to the mismatch of land cover and soil parameterization between ORCHIDEE and GLEAM. For instance, the saturated SM in southern China is 0.36 m^3 m^{-3} while the maximum SM in GLEAM is 0.45 m^3 m^{-3}. A high contribution of LCS to MSD is found in the Qinghai–Tibetan Plateau, the upper part of the YZRB and the YLRB, suggesting a mismatch of the phase of SM variability. The MSD is dominated by SDSD in northwestern China ($P < 200$ mm yr^{-1}), suggesting different magnitudes of SM fluctuations. Nevertheless, the relative RMSE in the Qinghai–Tibetan Plateau and northwestern China is as low as in southern China ($< 20\%$). Overall, ORCHIDEE is able to give a reasonable estimation of θ_s in regions where irrigation is not widespread.

Figure 6a–e show NLD of ORCHIDEE and GLEAM θ_s computed based on the k-lag autocorrelation coefficient R_k. High NLD implies that θ_s has a longer memory in response to rainfall inputs. However, the spatial distribution of NLD depends not only on rainfall frequency and intensity but also on evapotranspiration and runoff losses after SM recharge by rainfall. The NLD patterns of GLEAM and ORCHIDEE θ_s are similar, which is encouraging in terms of how ORCHIDEE simulates the processes controlling the decrease of SM after each rainfall. Both southern and southeastern China have higher NLD, like in GLEAM. A lower NLD (≈ 20 days) prevails around $30°$ N in eastern China, whilst the North China Plain has NLD values of 40 days. The main difference of NLD between GLEAM and ORCHIDEE is in Inner Mongolia and over the Loess Plateau, where the ORCHIDEE NLD has values of 20 days, against 40 days in GLEAM. R_k of spatially averaged θ_s in three regions is shown in Fig. 6f–h. Overall, R_k of ORCHIDEE θ_s is consistent with that of GLEAM. The GLEAM R_k is close to the ORCHIDEE R_k in the YZRB, with a difference of less than 6 days. In the YLRB, GLEAM R_k is larger than ORCHIDEE R_k, suggesting that modeled θ_s has a faster response to rainfall input. Such bias can be explained by higher simulated evapotranspiration in the YLRB compared to GLEAM (Fig. S8), suggesting that the decline of ORCHIDEE θ_s is faster after rainfall events than in GLEAM and leads to a lower R_k.

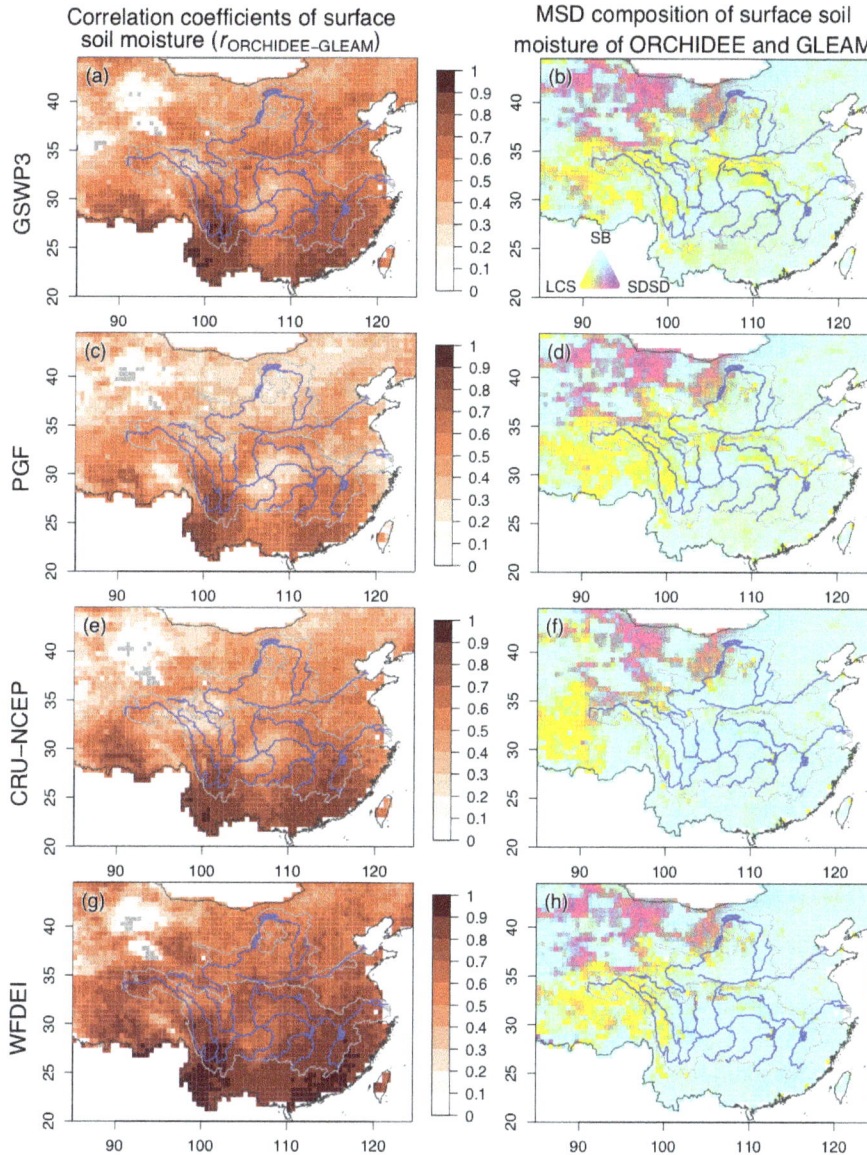

Figure 5. (a, c, e, g) Pearson's correlation coefficients of the GLEAM surface SM and the corresponding ORCHIDEE SM. Gray indicates no correlation and negative correlation. **(b, d, f, h)** Decomposition of the MSD between the daily GLEAM surface SM and the corresponding ORCHIDEE SM (Eq. 1). Cyan, magenta, and yellow indicate the fractions of SB, SDSD, and LCS, respectively.

The trend of ORCHIDEE θ_s (Fig. S9) is less significant than that of GLEAM θ_s (Fig. S9). In northwestern China, increasing θ_s is found in simulations ($< 0.2 \times 10^{-3} \, \text{m}^3 \, \text{m}^{-3} \, \text{yr}^{-1}$) and GLEAM ($0.2$–$0.4 \times 10^{-3} \, \text{m}^3 \, \text{m}^{-3} \, \text{yr}^{-1}$). The trend may be due to increasing P (Fig. S10). GLEAM θ_s decreased dramatically in eastern China ([103–122° E] × [20–35° N]), while the trends of ORCHIDEE θ_s are not homogeneous in this region. In addition, all forcing datasets show an increasing P in the North China Plain, which leads to a slight increase of simulated θ_s. However, GLEAM shows decreasing θ_s in most areas of the North China Plain. The mismatch of θ_s and

P trends suggests that the change of precipitation amount is not the only driver of the trend of SM.

4.2 Comparison of the four forcing datasets

To find the most realistic forcing dataset for SM performance given the ORCHIDEE model, several metrics were calculated and are shown in Fig. 7. Radar charts show the correlation coefficients (r) and RMSE of simulated SM in comparison to different datasets. Histograms show MSD and its three components. The median of specific metrics is listed in Table 3. GSWP3 has the best performance in estimating the magnitude of SM (lowest MSD), while WFDEI shows

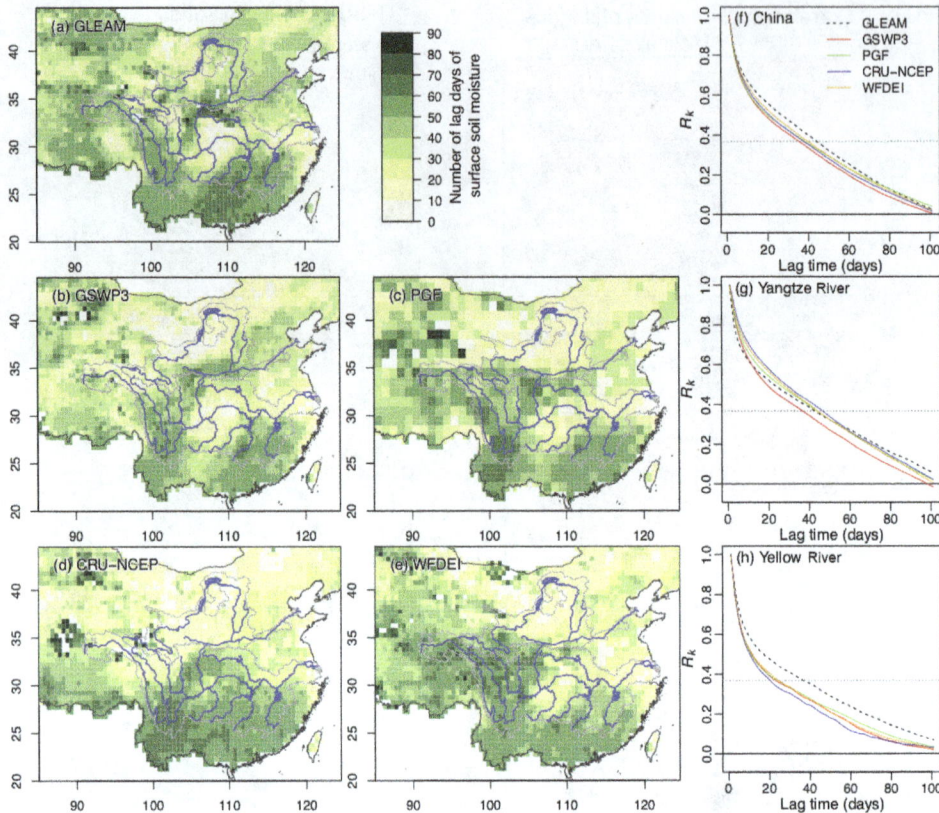

Figure 6. (a) Number of lag days (NLD) of GLEAM surface SM. **(b–e)** Difference of NLD between GLEAM and ORCHIDEE surface SM. **(f–h)** Autocorrelation coefficient R_k of spatial averaged surface SM as a function of NLD. The dashed line ($y = 1/e$) is the threshold of significant correlation.

the best score in simulating SM variation (highest r). PGF provides as good an estimation as GSWP3 in the YZRB, but performs more poorly in capturing SM variation in the YLRB, which is also reflected in the components of MSD. The largest MSD is found in CRU-NCEP in most of comparisons, which is mainly contributed by SB. The SDSD and LCS of CRU-NCEP are also larger than others but the differences are not as significant as SB. In addition, we performed the comparison over the full period (1981–2009). Corresponding metrics are shown in Table S1. The values vary slightly, but they do not change our conclusions.

Thus we conclude that both GSWP3 and WFDEI are suitable to simulate SM dynamics in China with ORCHIDEE. The best choice can be made based on the main focus of specific research. For estimating magnitude of SM, GSWP3 is preferable; for investigating SM variation, WFDEI is the best choice. Note that this study only provides the evaluation of SM, but other hydrological components should be compared with observations to confirm the superiority of GSWP3 and WFDEI.

4.3 Source of SM difference among simulations

By investigating the D of meteorological variables and simulated SM among the four simulations (D_x for variable x; Eq. 6), two questions are addressed. (1) How is D of simulated SM and forcing variables spatially distributed? (2) Can spatial patterns of D of SM be explained by that of meteorological variables? Note that the relative value of D, D over the magnitude of specific variable in each grid cell, is not suitable for the analysis (detailed explanation is in Sect. S1).

Figure S11 shows maps of D of θ_t and meteorological variables. As the unit of D depends on specific variables, it can only be used to compare spatial distributions, not values. High D_{θ_t} is found in the southwest of China ([92–104° E] × [28–35° N]). However, similar patterns do not exist in the D_P (Fig. S11b), suggesting that the difference of simulated θ_t is not caused by the difference of precipitation amount of forcing data. Similarly, in southwestern China, no high D is found in meteorological variables except for the number of precipitation days (N_p) and air humidity (Q_a), although the patterns of D_{N_p} and D_{Q_a} overlap with D_{θ_t} but extend to zones with low D_{θ_t} as well (Fig. S11c and g).

Table 3. Median of metrics in specific comparisons. The subscripts of correlation coefficients indicate the quantile of stations (samples) with significant correlation (p value < 0.05).

Dataset	Simulations	Correlation			RMSE (m^3 m^{-3})		
ISMN	GSWP3	$0.52_{0.85}$			0.07		
	PGF	$0.46_{0.90}$			0.07		
	CRU-NCEP	$0.36_{0.95}$			0.10		
	WFDEI	$0.55_{0.95}$			0.08		
PKU	GSWP3	$0.38_{0.91}$			NA		
	PGF	$0.31_{0.85}$			NA		
	CRU-NCEP	$0.31_{0.86}$			NA		
	WFDEI	$0.45_{0.93}$			NA		
		China	Yangtze	Yellow	China	Yangtze	Yellow
ESA CCI	GSWP3	$0.47_{0.93}$	$0.42_{0.94}$	$0.58_{0.99}$	0.06	0.06	0.06
	PGF	$0.26_{0.83}$	$0.32_{0.91}$	$0.28_{0.96}$	0.06	0.07	0.07
	CRU-NCEP	$0.51_{0.94}$	$0.50_{0.94}$	$0.54_{0.99}$	0.06	0.07	0.07
	WFDEI	$0.61_{0.97}$	$0.60_{0.96}$	0.68_{1}	0.05	0.05	0.06
GLEAM surface SM	GSWP3	0.54_{1}	0.60_{1}	0.52_{1}	0.07	0.07	0.10
	PGF	0.42_{1}	0.51_{1}	0.35_{1}	0.08	0.08	0.10
	CRU-NCEP	$0.49_{0.99}$	0.61_{1}	0.49_{1}	0.10	0.10	0.12
	WFDEI	$0.68_{0.99}$	0.77_{1}	0.63_{1}	0.08	0.09	0.10
GLEAM root-zone SM	GSWP3	$0.60_{0.98}$	$0.67_{0.99}$	$0.60_{0.99}$	0.05	0.04	0.08
	PGF	$0.57_{0.98}$	0.69_{1}	$0.57_{0.99}$	0.06	0.04	0.09
	CRU-NCEP	$0.40_{0.96}$	$0.48_{0.97}$	$0.37_{0.97}$	0.08	0.08	0.11
	WFDEI	$0.63_{0.98}$	0.74_{1}	0.59_{1}	0.06	0.04	0.10

NA: not available.

To look for clearer links between input and SM D, we decompose N_p and θ_t by scales of P and soil layers, respectively (Sect. 3.4). The r values of D between simulated SM and meteorological variables are shown in Fig. 8. D_{Q_a}, D_{N_p}, and $D_{N_p^2}$ are highly correlated with D_{θ_t}, implying that the difference of simulated θ_t can be explained by the differences of Q_a and N_p among the four forcing datasets. The r between D_{θ_t} and D_P is less than 0.3. All in all, the results suggest that the uncertainty of precipitation frequency and intensity is more important than that of precipitation amount in influencing SM differences among the simulations.

5 Discussion

5.1 Performance of the model to simulate SM

Due to the spatiotemporal complexity of SM and its vertical profile, four datasets were selected to drive the simulations, and modeled SM at different depths was validated against multiple datasets. The results showed that ORCHIDEE SM coincides well with CCI (median $r = 0.48$; median RMSE $= 0.06$) and GLEAM SM (median $r = 0.55$; median RMSE $= 0.07$) in comparison to other model studies (Lai et al., 2016).

Higher r values were systematically found in southern China, the Loess Plateau, and the North China Plain; lower r values were found in northwestern China, the western Tibetan Plateau, the eastern Sichuan Basin, and downstream of the YZRB. SM is underestimated significantly in the Loess Plateau and the North China Plain, with modeled values being 20% and 30% less than in CCI and GLEAM, respectively (Figs. S7 and S3). It is not only due to model parameterization but also due to irrigation activities in those agricultural regions (Fig. 1), which are not considered in the simulations.

Because the in situ SM measurements were only collected for croplands and grasslands (Piao et al., 2009; Robock et al., 2000), implying potential disturbances from human activities, r was low in the comparison to ISMN and PKU datasets (median $r = 0.37$, Fig. 2). For instance, drought occurred in northern China during 1987–1988 (Yang et al., 2012), which is reflected in the variation of measured SM at Xifeng and Xinxian (Fig. 3a–b). ORCHIDEE successfully reproduced the drought-induced SM decline at the two stations. But SM measured at Xinxian was maintained at a high level. A possible explanation is that the soil at Xinxian was irrigated. Consequently SM at Xinxian did not vary with precipitation, leading to a low r (< 0.23). Another possible reason leading to the mismatch between simulations and in situ measurements is scale effects. Local measurements can only be an ideal choice for model validation if the atmospheric forcing was provided at the same scale due to the spatial variability

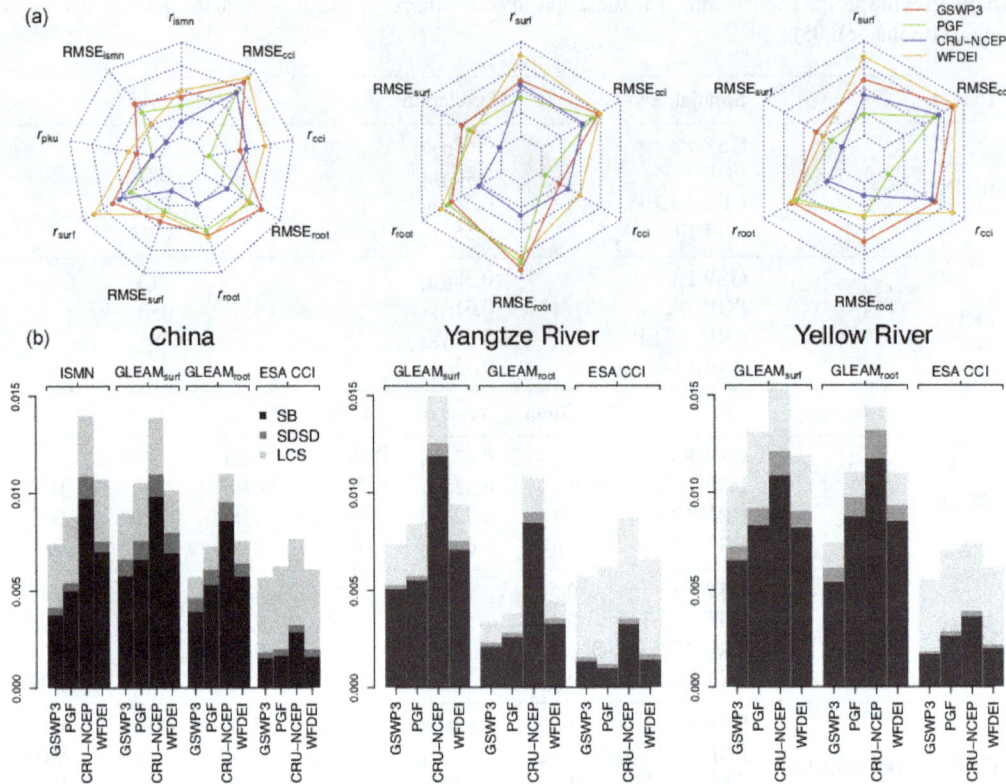

Figure 7. Evaluation of the forcing datasets for simulating SM dynamics in China, YZRB, and YLRB. **(a)** Radar charts of criteria of the four forcing datasets. The center implies bad criteria. Red, green, blue, and orange lines indicate GSWP3, PGF, CRU-NCEP, and WFDEI, respectively; "surf" and "root" indicate surface and root-zone SM of GLEAM 3.0A. **(b)** Composition of MSD from each comparison. The x axis indicates the drivers of specific simulations; top labels indicate the dataset used in the specific comparison.

of precipitation and of landscape. Otherwise, remote sensing products derived from multiple observations averaged or aggregated at a daily time step are probably more comparable to model simulations obtained using meteorological reanalysis than local in situ measurements.

In the comparison to CCI and GLEAM SM, low r did not occur in northern China, such as the Loess Plateau and the North China Plain, but was found in the climatic transition zone between southern and northern China (Figs. 4, S5, and 5). Irrigation may strongly influence SM dynamics in northern China, and in turn reduce r. However, such an effect on r is not significant because of the large seasonality of SM in this region. Instead of r, the impacts of irrigation are mainly reflected in the RMSE and relative difference (Figs. S2, S3, S6, and S7). Thus for a region with both irrigation and strong seasonality, bias and RMSE are recommended to trace the footprint of irrigation rather than correlation coefficients. In the climatic transition zone (e.g., Sichuan Basin, mid- and downstream of the YZRB), climatic seasonality is not as large as in northern China. Meanwhile irrigation is still needed for agriculture, which consequently results in low r between simulated and observed SM.

From the results we conclude that ORCHIDEE provides a satisfactory simulation of SM dynamics in China, except in areas subject to irrigation. This calls for inclusion of irrigation and realistic crop phenology (Wang et al., 2017) as a priority for future application of this model for SM and river discharge dynamics.

5.2 Linkage of discrepancies between meteorological factors and SM through ORCHIDEE

In Sect. 4.3, we showed that the spatial differences of simulated SM among the four forcing datasets were highly correlated with forcing differences in N_p and Q_a. This suggests that the uncertainty of precipitation frequency is more critical than that of precipitation amount in determining variation of SM patterns, as pointed out by other studies, especially in arid and semi-arid regions (Baudena and Provenzale, 2008; Cissé et al., 2016; Piao et al., 2009). To precise the result, we studied the correlation coefficients between the spatial averaged D of SM in different soil layers and of N_p categorized by classes of precipitation intensity (Fig. 8). The result showed that differences in small rainfall events N_p^i with $1 < P < 5$ mm d^{-1} are more important than other precipita-

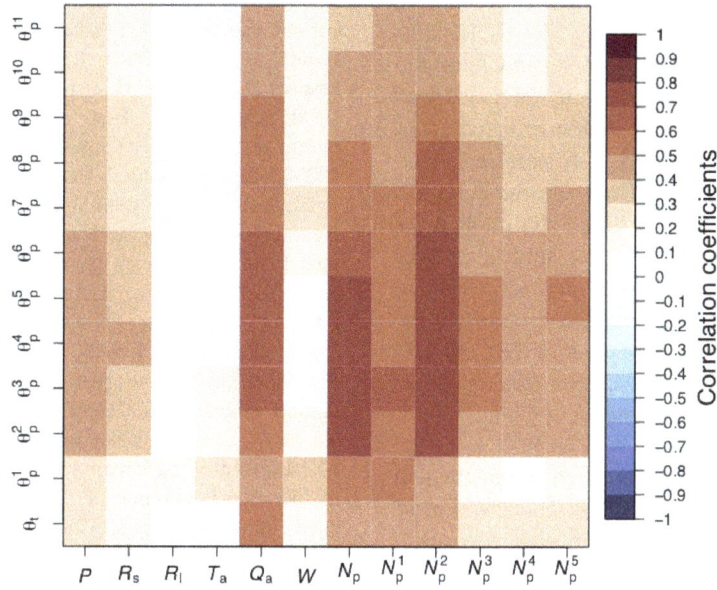

Figure 8. Matrix of correlation coefficients between the D of meteorological variables and the D of simulated SM. D is the averaged MSD defined by Eq. (6). θ_t indicates total SM. θ_p^i indicates SM in the ith layer. P indicates annual precipitation. R_s and R_l indicate short- and long-wave incoming radiation, respectively. T_a indicates air temperature. Q_a indicates air humidity. W indicates wind speed. N_p indicates the number of days with precipitation no less than $0.01\ \mathrm{mm\ d^{-1}}$. N_{p^i} indicates the number of days with a specific precipitation range (Sect. 3.4).

tion classes in explaining SM differences due to atmospheric forcing datasets.

Differences in Q_a were also shown to explain a large fraction of the simulated SM differences across different forcings. Q_a determines vapor pressure deficit, which in turn controls transpiration (Farquhar and Sharkey, 1982) and evaporation (Monteith, 1965), suggesting a strong control by atmospheric dryness of the differences in SM found among the four forcing datasets. Both Q_a and N_p have positive impacts on SM, which enhances the correlation in Fig. 8.

Estimating impacts of meteorological factors on SM dynamics is difficult. First of all, the importance of a meteorological variable on SM may vary with climate regimes. For instance, the importance of precipitation and radiation on SM changes from water- to energy-limited regions. Secondly, impacts of meteorological variables can be nonlinear through interactions with local ecosystems (Seneviratne et al., 2010), suggesting that even with the same meteorological variables, the simulated SM can be totally different (e.g., with different soil texture or land cover types). Moreover, SM can be strongly coupled with the atmosphere (Koster, 2004; Taylor et al., 2012), implying that meteorological factors can be influenced by SM as well (such as cloudiness, precipitation, and air humidity), which is not included in this study. However, the logic of our importance analysis is simple. If the model inputs (forcing data) are the same, the outputs (SM) should be the same. In other words, the differences of outputs can only be caused by the difference of inputs in our simulation results. It does not matter whether the quality of atmospheric forcing is good. On the contrary, the more differences that exist among these forcing datasets, the better our analysis is. To keep the analysis simple, we did not investigate temporal correlations in each pixel, but focused on spatial patterns of D at the continental scale. Therefore, our results provided a general estimation of the importance of meteorological variable uncertainties to SM simulation through ORCHIDEE.

Indeed this approach is not able to demonstrate explicit links between meteorological variables and SM. We underlined the impacts of N_p and Q_a uncertainties, but it does not mean that other factors are unimportant. For instance, assuming that a variable can strongly influence simulated SM, if there was not much difference of the variable among forcing datasets, its importance cannot be detected in this work. Moreover, only one model was used in this study. Although ORCHIDEE performed very well in SM simulation, the lack of unknown mechanisms may weaken the linkage between SM and specific atmospheric variables. In a word, our analysis only focused on the inputs and outputs of the model and tried to diagnose the relationship between their differences.

6 Conclusions

Simulations in China were performed in ORCHIDEE-MICT driven by different forcing datasets, GSWP3, PGF, CRU-

NCEP, and WFDEI. Simulated soil moisture was compared to several datasets to evaluate the ability of ORCHIDEE-MICT in reproducing soil moisture dynamics in China. Results showed that ORCHIDEE soil moisture coincided well with other datasets in wet areas and in non-irrigated areas. It was suggested that ORCHIDEE-MICT is suitable for further hydrological studies in China. However, the abnormal variation of observed SM in North China Plain implied potential impacts of irrigation, which was recommended to be considered in further simulations. Moreover, results showed that bias was mainly from model parameterization and atmospheric forcing. Thus parameterizations in ORCHIDEE-MICT should be calibrated, and atmospheric forcing should be carefully selected to reflect the situation of China.

Several criteria were chosen and compared among the four simulations in China, YZRB, and YLRB. Results showed that GSWP3 and WFDEI, which had the best performances in correlation coefficients and RMSE, respectively, were ideal choices for hydrological study in China. However, higher MSD in the Yellow River basin reflected the complicated climate conditions in northern China, which might be significantly influenced by human activities as well. Finally, we used the differences of simulated soil moisture and meteorological variables to simply investigate the linkage between them. Results showed that the differences of simulated soil moisture were mainly explained by the differences of air humidity and precipitation frequency among the four atmospheric forcing datasets. However, this coarse analysis cannot give explicit explanations about related mechanisms. Further study is needed to discover the interactions between soil water and climate through tracing the surface hydrological cycles and energy balances.

Code and data availability. The SVN version of ORCHIDEE-MICT used in this study is 3952, which is available at https://forge.ipsl.jussieu.fr/orchidee/wiki/DevelopmentActivities/ORCHIDEE-MICT-IMBALANCE-P (last access: 19 October 2018). The ORCHIDEE code and scripts of analysis are available by contacting the corresponding author. The GSWP3, PGF, WFDEI, ISMN, GLEAM, and ESA CCI datasets are freely available online; for the CRU-NCEP and PKU datasets, please contact the corresponding author and Shilong Piao (slpiao@pku.edu.cn) respectively.

Team list. Philippe Ciais (Laboratoire des Sciences du Climat et de l'Environnement, CNRS-CEA-UVSQ, Gif-sur-Yvette 91191, France), Patrice Dumas (Centre de Coopération Internationale en Recherche Agronomique pour le Développement, Avenue Agropolis, 34398 Montpellier CEDEX 5, France), Xiaoming Feng (State Key Laboratory of Urban and Regional Ecology, Research Center for Eco-Environmental Sciences, Chinese Academy of Sciences, Beijing 100085, China), Matthieu Guimberteau (Laboratoire des Sciences du Climat et de l'Environnement, CNRS-CEA-UVSQ, Gif-sur-Yvette 91191, France; UMR 7619 METIS, Sorbonne Universités, UPMC, CNRS, EPHE, 4 place Jussieu, Paris 75005, France), Laurent Li (Laboratoire de Météorologie Dynamique, UPMC/CNRS, IPSL, Paris 75005, France), Catherine Ottlé (aboratoire des Sciences du Climat et de l'Environnement, CNRS-CEA-UVSQ, Gif-sur-Yvette 91191, France), Shushi Peng (Sino-French Institute for Earth System Science, College of Urban and Environmental Sciences, Peking University, Beijing 100871, China), Shilong Piao (Sino-French Institute for Earth System Science, College of Urban and Environmental Sciences, Peking University, Beijing 100871, China), Jan Polcher (Laboratoire de Météorologie Dynamique, UPMC/CNRS, IPSL, Paris 75005, France), Pengfei Shi (State Key Laboratory of Hydrology-Water Resources and Hydraulic Engineering, Center for Global Change and Water Cycle, Hohai University, Nanjing 210098, China), Shuai Wang (State Key Laboratory of Urban and Regional Ecology, Research Center for Eco-Environmental Sciences, Chinese Academy of Sciences, Beijing 100085, China), Xuhui Wang (Laboratoire des Sciences du Climat et de l'Environnement, CNRS-CEA-UVSQ, Gif-sur-Yvette 91191, France; Laboratoire de Météorologie Dynamique, UPMC/CNRS, IPSL, Paris 75005, France; Laboratoire de Météorologie Dynamique, UPMC/CNRS, IPSL, Paris 75005, France; Sino-French Institute for Earth System Science, College of Urban and Environmental Sciences, Peking University, Beijing 100871, China), Yi Xi (Sino-French Institute for Earth System Science, College of Urban and Environmental Sciences, Peking University, Beijing 100871, China), Hui Yang (Sino-French Institute for Earth System Science, College of Urban and Environmental Sciences, Peking University, Beijing 100871, China), Tao Yang (State Key Laboratory of Hydrology-Water Resources and Hydraulic Engineering, Center for Global Change and Water Cycle, Hohai University, Nanjing 210098, China), Zun Yin (Laboratoire des Sciences du Climat et de l'Environnement, CNRS-CEA-UVSQ, Gif-sur-Yvette 91191, France), Xuanze Zhang (Sino-French Institute for Earth System Science, College of Urban and Environmental Sciences, Peking University, Beijing 100871, China), Feng Zhou (Sino-French Institute for Earth System Science, College of Urban and Environmental Sciences, Peking University, Beijing 100871, China), and Xudong Zhou (State Key Laboratory of Hydrology-Water Resources and Hydraulic Engineering, Center for Global Change and Water Cycle, Hohai University, Nanjing 210098, China).

Author contributions. PC, CO, and ZY designed the research. ZY performed the research, analyzed the data, and wrote the draft; all authors contributed to interpreting results, discussing findings, and improving the manuscript.

Competing interests. The authors declare that they have no conflict of interest.

Acknowledgements. This study was supported by the National Natural Science Foundation of China (grant number 41561134016) and by the CHINA-TREND-STREAM French national project (ANR grant no. ANR-15-CE01-00L1-0L). Matthieu Guimberteau and

Philippe Ciais acknowledge support from the European Research Council Synergy grant ERC-2013-SyG-610028 IMBALANCE-P. Hyungjun Kim was supported by Japan Society for the Promotion of Science KAKENHI (16H06291). We thank Brecht Martens and Suxia Liu for helpful discussion about GLEAM and ISMN datasets. We gratefully acknowledge two anonymous referees and the editor for their helpful comments and efforts.

Edited by: Miriam Coenders-Gerrits

References

Adler, R. F., Huffman, G. J., Chang, A., Ferraro, R., Xie, P. P., Janowiak, J., Rudolf, B., Schneider, U., Curtis, S., Bolvin, D., Gruber, A., Susskind, J., Arkin, P., and Nelkin, E.: The Version-2 Global Precipitation Climatology Project (GPCP) Monthly Precipitation Analysis (1979–Present), J. Hydrometeorol., 4, 1147–1167, https://doi.org/10.1175/1525-7541(2003)004<1147:TVGPCP>2.0.CO;2, 2003.

An, R., Zhang, L., Wang, Z., Quaye-Ballard, J. A., You, J. J., Shen, X. J., Gao, W., Huang, L. J., Zhao, Y. H., and Ke, Z. Y.: Validation of the ESA CCI soil moisture product in China, Int. J. Appl. Earth Obs., 48, 28–36, https://doi.org/10.1016/j.jag.2015.09.009, 2016.

An, W. M., Li, Z. S., Wang, S., Wu, X., Lu, Y. H., Liu, G. H., and Fu, B. J.: Exploring the effects of the "Grain for Green" program on the differences in soil water in the semi-arid Loess Plateau of China, Ecol. Eng., 107, 144–151, https://doi.org/10.1016/j.ecoleng.2017.07.017, 2017.

Ayalew, T. B., Krajewski, W. F., Mantilla, R., and Small, S. J.: Exploring the effects of hillslope-channel link dynamics and excess rainfall properties on the scaling structure of peak-discharge, Adv. Water Resour., 64, 9–20, https://doi.org/10.1016/j.advwatres.2013.11.010, 2014.

Basheer, M. and Elagib, N. A.: Sensitivity of Water-Energy Nexus to dam operation: A Water-Energy Productivity concept, Sci. Total Environ., 616–617, 918–926, https://doi.org/10.1016/j.scitotenv.2017.10.228, 2018.

Baudena, M. and Provenzale, A.: Rainfall intermittency and vegetation feedbacks in drylands, Hydrol. Earth Syst. Sci., 12, 679–689, https://doi.org/10.5194/hess-12-679-2008, 2008.

Beck, H. E., van Dijk, A. I. J. M., Levizzani, V., Schellekens, J., Miralles, D. G., Martens, B., and de Roo, A.: MSWEP: 3-hourly 0.25° global gridded precipitation (1979–2015) by merging gauge, satellite, and reanalysis data, Hydrol. Earth Syst. Sci., 21, 589–615, https://doi.org/10.5194/hess-21-589-2017, 2017.

Becker, A., Finger, P., Meyer-Christoffer, A., Rudolf, B., Schamm, K., Schneider, U., and Ziese, M.: A description of the global land-surface precipitation data products of the Global Precipitation Climatology Centre with sample applications including centennial (trend) analysis from 1901–present, Earth Syst. Sci. Data, 5, 71–99, https://doi.org/10.5194/essd-5-71-2013, 2013.

Bonan, G. B.: Forests and Climate Change: Forcings, Feedbacks, and the Climate Benefits of Forests, Science, 320, 1444–1449, https://doi.org/10.1126/science.1155121, 2008.

Bouwer, L. M., Bubeck, P., Wagtendonk, A. J., and Aerts, J. C. J. H.: Inundation scenarios for flood damage evaluation in polder areas, Nat. Hazards Earth Syst. Sci., 9, 1995–2007, https://doi.org/10.5194/nhess-9-1995-2009, 2009.

Calvet, J. C., Rivalland, V., Picon-Cochard, C., and Guehl, J. M.: Modelling forest transpiration and CO_2 fluxes–response to soil moisture stress, Agr. Forest Meteorol., 124, 143–156, https://doi.org/10.1016/j.agrformet.2004.01.007, 2004.

Cissé, S., Eymard, L., Ottlé, C., Ndione, J., Gaye, A., and Pinsard, F.: Rainfall Intra-Seasonal Variability and Vegetation Growth in the Ferlo Basin (Senegal), Remote Sensing, 8, 66, https://doi.org/10.3390/rs8010066, 2016.

Compo, G. P., Whitaker, J. S., Sardeshmukh, P. D., Matsui, N., Allan, R. J., Yin, X., Gleason, B. E., Vose, R. S., Rutledge, G., Bessemoulin, P., Brönnimann, S., Brunet, M., Crouthamel, R. I., Grant, A. N., Groisman, P. Y., Jones, P. D., Kruk, M. C., Kruger, A. C., Marshall, G. J., Maugeri, M., Mok, H. Y., Nordli, Ø., Ross, T. F., Trigo, R. M., Wang, X. L., Woodruff, S. D., and Worley, S. J.: The Twentieth Century Reanalysis Project, Q. J. Roy. Meteor. Soc., 137, 1–28, https://doi.org/10.1002/qj.776, 2011.

Dai, A. G., Trenberth, K. E., and Qian, T. T.: A Global Dataset of Palmer Drought Severity Index for 1870–2002: Relationship with Soil Moisture and Effects of Surface Warming, J. Hydrometeorol., 5, 1117–1130, https://doi.org/10.1175/JHM-386.1, 2004.

Dee, D. P., Uppala, S. M., Simmons, A. J., Berrisford, P., Poli, P., Kobayashi, S., Andrae, U., Balmaseda, M. A., Balsamo, G., Bauer, P., Bechtold, P., Beljaars, A. C. M., van de Berg, L., Bidlot, J., Bormann, N., Delsol, C., Dragani, R., Fuentes, M., Geer, A. J., Haimberger, L., Healy, S. B., Hersbach, H., Hólm, E. V., Isaksen, L., Kållberg, P., Köhler, M., Matricardi, M., McNally, A. P., Monge-Sanz, B. M., Morcrette, J.-J., Park, B.-K., Peubey, C., de Rosnay, P., Tavolato, C., Thépaut, J.-N., and Vitart, F.: The ERA-Interim reanalysis: configuration and performance of the data assimilation system, Q. J. Roy. Meteor. Soc., 137, 553–597, https://doi.org/10.1002/qj.828, 2011.

Deng, K., Yang, S. Y., Lian, E. G., Li, C., Yang, C. F., and Wei, H. L.: Three Gorges Dam alters the Changjiang (Yangtze) river water cycle in the dry seasons: Evidence from H-O isotopes, Sci. Total Environ., 562, 89–97, https://doi.org/10.1016/j.scitotenv.2016.03.213, 2016.

Dorigo, W. A., Wagner, W., Hohensinn, R., Hahn, S., Paulik, C., Xaver, A., Gruber, A., Drusch, M., Mecklenburg, S., van Oevelen, P., Robock, A., and Jackson, T.: The International Soil Moisture Network: a data hosting facility for global in situ soil moisture measurements, Hydrol. Earth Syst. Sci., 15, 1675–1698, https://doi.org/10.5194/hess-15-1675-2011, 2011.

Dorigo, W. A., Gruber, A., De Jeu, R. A. M., Wagner, W., Stacke, T., Loew, A., Albergel, C., Brocca, L., Chung, D., Parinussa, R. M., and Kidd, R.: Evaluation of the ESA CCI soil moisture product using ground-based observations, Remote Sens. Environ., 162, 380–395, https://doi.org/10.1016/j.rse.2014.07.023, 2015.

Draper, C. S., Reichle, R. H., De Lannoy, G. J. M., and Liu, Q.: Assimilation of passive and active microwave soil moisture retrievals, Geophys. Res. Lett., 39, L04401, https://doi.org/10.1029/2011GL050655, 2012.

Farquhar, G. D. and Sharkey, T. D.: Stomatal Conductance and Photosynthesis, Ann. Rev. Plant Physio., 33, 317–345, https://doi.org/10.1146/annurev.pp.33.060182.001533, 1982.

Feng, X. M., Fu, B. J., Piao, S. L., Wang, S., Ciais, P., Zeng, Z. Z., Lü, Y. H., Zeng, Y., Li, Y., Jiang, X. H., and Wu, B. F.: Revegetation in China's Loess Plateau is approaching sustainable water resource limits, Nat. Clim. Change, 6, 1019–1022, https://doi.org/10.1038/nclimate3092, 2016.

Getirana, A. C. V., Dutra, E., Guimberteau, M., Kam, J., Li, H. Y., Decharme, B., Zhang, Z. Q., Ducharne, A., Boone, A., Balsamo, G., Rodell, M., Toure, A. M., Xue, Y. K., Peters-Lidard, C. D., Kumar, S. V., Arsenault, K., Drapeau, G., Ruby Leung, L., Ronchail, J., and Sheffield, J.: Water Balance in the Amazon Basin from a Land Surface Model Ensemble, J. Hydrometeorol., 15, 2586–2614, https://doi.org/10.1175/JHM-D-14-0068.1, 2014.

Grillakis, M. G., Koutroulis, A. G., Komma, J., Tsanis, I. K., Wagner, W., and Blöschl, G.: Initial soil moisture effects on flash flood generation? A comparison between basins of contrasting hydro-climatic conditions, J. Hydrol., 541, 206–217, https://doi.org/10.1016/j.jhydrol.2016.03.007, 2016.

Guimberteau, M., Laval, K., Perrier, A., and Polcher, J.: Global effect of irrigation and its impact on the onset of the Indian summer monsoon, Clim. Dynam., 39, 1329–1348, https://doi.org/10.1007/s00382-011-1252-5, 2012.

Guimberteau, M., Ducharne, A., Ciais, P., Boisier, J. P., Peng, S., De Weirdt, M., and Verbeeck, H.: Testing conceptual and physically based soil hydrology schemes against observations for the Amazon Basin, Geosci. Model Dev., 7, 1115–1136, https://doi.org/10.5194/gmd-7-1115-2014, 2014.

Guimberteau, M., Ciais, P., Ducharne, A., Boisier, J. P., Dutra Aguiar, A. P., Biemans, H., De Deurwaerder, H., Galbraith, D., Kruijt, B., Langerwisch, F., Poveda, G., Rammig, A., Rodriguez, D. A., Tejada, G., Thonicke, K., Von Randow, C., Von Randow, R. C. S., Zhang, K., and Verbeeck, H.: Impacts of future deforestation and climate change on the hydrology of the Amazon Basin: a multi-model analysis with a new set of land-cover change scenarios, Hydrol. Earth Syst. Sci., 21, 1455–1475, https://doi.org/10.5194/hess-21-1455-2017, 2017.

Guimberteau, M., Zhu, D., Maignan, F., Huang, Y., Yue, C., Dantec-Nédélec, S., Ottlé, C., Jornet-Puig, A., Bastos, A., Laurent, P., Goll, D., Bowring, S., Chang, J., Guenet, B., Tifafi, M., Peng, S., Krinner, G., Ducharne, A., Wang, F., Wang, T., Wang, X., Wang, Y., Yin, Z., Lauerwald, R., Joetzjer, E., Qiu, C., Kim, H., and Ciais, P.: ORCHIDEE-MICT (v8.4.1), a land surface model for the high latitudes: model description and validation, Geosci. Model Dev., 11, 121–163, https://doi.org/10.5194/gmd-11-121-2018, 2018.

Hansen, M. C., DeFries, R. S., Townshend, J. R. G., Carroll, M., Dimiceli, C., and Sohlberg, R. A.: Global Percent Tree Cover at a Spatial Resolution of 500 Meters: First Results of the MODIS Vegetation Continuous Fields Algorithm, Earth Interact., 7, 1–15, https://doi.org/10.1175/1087-3562(2003)007<0001:GPTCAA>2.0.CO;2, 2003.

Harris, I., Jones, P. D., Osborn, T. J., and Lister, D. H.: Updated high-resolution grids of monthly climatic observations – the CRU TS3.10 Dataset, Int. J. Climatol., 34, 623–642, https://doi.org/10.1002/joc.3711, 2014.

Hirschi, M., Seneviratne, S. I., Alexandrov, V., Boberg, F., Boroneant, C., Christensen, O. B., Formayer, H., Orlowsky, B., and Stepanek, P.: Observational evidence for soil-moisture impact on hot extremes in southeastern Europe, Nat. Geosci., 4, 17–21, https://doi.org/10.1038/ngeo1032, 2011.

Hirschi, M., Mueller, B., Dorigo, W., and Seneviratne, S. I.: Using remotely sensed soil moisture for land-atmosphere coupling diagnostics: The role of surface vs. root-zone soil moisture variability, Remote Sens. Environ., 154, 246–252, https://doi.org/10.1016/j.rse.2014.08.030, 2014.

Hong, S. Y. and Chang, E. C.: Spectral nudging sensitivity experiments in a regional climate model, Asia-Pac. J. Atmos. Sci., 48, 345–355, https://doi.org/10.1007/s13143-012-0033-3, 2012.

Huffman, G. J., Bolvin, D. T., Nelkin, E. J., Wolff, D. B., Adler, R. F., Gu, G. J., Hong, Y., Bowman, K. P., and Stocker, E. F.: The TRMM Multisatellite Precipitation Analysis (TMPA): Quasi-Global, Multiyear, Combined-Sensor Precipitation Estimates at Fine Scales, J. Hydrometeorol., 8, 38–55, https://doi.org/10.1175/JHM560.1, 2007.

Kalnay, E., Kanamitsu, M., Kistler, R., Collins, W., Deaven, D., Gandin, L., Iredell, M., Saha, S., White, G., Woollen, J., Zhu, Y., Leetmaa, A., Reynolds, R., Chelliah, M., Ebisuzaki, W., Higgins, W., Janowiak, J., Mo, K. C., Ropelewski, C., Wang, J., Jenne, R., and Joseph, D.: The NCEP/NCAR 40-Year Reanalysis Project, B. Am. Meteor. Soc., 77, 437–471, https://doi.org/10.1175/1520-0477(1996)077<0437:TNYRP>2.0.CO;2, 1996.

Kendall, M. G.: Rank correlation methods, 4th edn., Charles Griffin, London, 1975.

Kerr, Y. H., Waldteufel, P., Wigneron, J. P., Martinuzzi, J., Font, J., and Berger, M.: Soil moisture retrieval from space: the Soil Moisture and Ocean Salinity (SMOS) mission, IEEE T. Geosci. Remote, 39, 1729–1735, https://doi.org/10.1109/36.942551, 2001.

Kim, H.: Global Soil Wetness Project Phase 3 Atmospheric Boundary Conditions (Experiment 1) [Data set], Data Integration and Analysis System (DIAS), https://doi.org/10.20783/DIAS.501, 2017.

Kobayashi, K. and Salam, M. U.: Comparing Simulated and Measured Values Using Mean Squared Deviation and its Components, Agron. J., 92, 345–352, https://doi.org/10.2134/agronj2000.922345x, 2000.

Koster, R. D.: Regions of Strong Coupling Between Soil Moisture and Precipitation, Science, 305, 1138–1140, https://doi.org/10.1126/science.1100217, 2004.

Kottek, M., Grieser, J., Beck, C., Rudolf, B., and Rubel, F.: World Map of the Köppen–Geiger climate classification updated, Meteorol. Z., 15, 259–263, https://doi.org/10.1127/0941-2948/2006/0130, 2006.

Krinner, G., Viovy, N., de Noblet-Ducoudre, N., Ogée, J., Polcher, J., Friedlingstein, P., Ciais, P., Sitch, S., and Prentice, I. C.: A dynamic global vegetation model for studies of the coupled atmosphere-biosphere system, Global Biogeochem. Cy., 19, 1–33, https://doi.org/10.1029/2003GB002199, 2005.

Lai, X., Wen, J., Cen, S. X., Huang, X., Tian, H., and Shi, X. K.: Spatial and Temporal Soil Moisture Variations over China from Simulations and Observations, Adv. Meteorol., 2016, 1–14, https://doi.org/10.1155/2016/4587687, 2016.

Li, L. C., Zhang, L. P., Xia, J., Gippel, C. J., Wang, R. C., and Zeng, S. D.: Implications of Modelled Climate and Land Cover Changes on Runoff in the Middle Route of the South to North Water Transfer Project in China, Water Resour. Manag., 29, 2563–2579, https://doi.org/10.1007/s11269-015-0957-3, 2015.

Li, L. L., Zhang, R. Z., Luo, Z. Z., Liang, W. L., Xie, J. H., Cai, L. Q., and Bellotti, B.: Evolution of soil and water conservation in rain-fed areas of China, International Soil and Water Conservation Research, 2, 78–90, https://doi.org/10.1016/S2095-6339(15)30015-0, 2014.

Liu, B., Xu, M., and Henderson, M.: Where have all the showers gone? Regional declines in light precipitation events in China, 1960-2000, Int. J. Climatol., 31, 1177–1191, https://doi.org/10.1002/joc.2144, 2011.

Liu, S. X., Mo, X. G., Li, H. B., Peng, G. B., and Robock, A.: Spatial Variation of Soil Moisture in China: Geostatistical Characterization, J. Meteorol. Soc. Jpn., 79, 555–574, https://doi.org/10.2151/jmsj.79.555, 2001.

Liu, W. F., Wei, X. H., Fan, H. B., Guo, X. M., Liu, Y. Q., Zhang, M. F., and Li, Q.: Response of flow regimes to deforestation and reforestation in a rain-dominated large watershed of subtropical China, Hydrol. Process., 29, 5003–5015, https://doi.org/10.1002/hyp.10459, 2015a.

Liu, Y. L., Pan, Z. H., Zhuang, Q. L., Miralles, D. G., Teuling, A. J., Zhang, T. L., An, P. L., Dong, Z. Q., Zhang, J. T., He, D., Wang, L. W., Pan, X. B., Bai, W., and Niyogi, D.: Agriculture intensifies soil moisture decline in Northern China, Scientific Reports, 5, 11261, https://doi.org/10.1038/srep11261, 2015b.

Liu, Y. W., Wang, W., and Hu, Y. M.: Investigating the impact of surface soil moisture assimilation on state and parameter estimation in SWAT model based on the ensemble Kalman filter in upper Huai River basin, J. Hydrol. Hydromech., 65, 123–133, https://doi.org/10.1515/johh-2017-0011, 2017.

Liu, Y. Y., Dorigo, W. A., Parinussa, R. M., de Jeu, R. A. M., Wagner, W., McCabe, M. F., Evans, J. P., and van Dijk, A. I. J. M.: Trend-preserving blending of passive and active microwave soil moisture retrievals, Remote Sens. Environ., 123, 280–297, https://doi.org/10.1016/j.rse.2012.03.014, 2012.

Mann, H. B.: Nonparametric Tests Against Trend, Econometrica, 13, 245–259, https://doi.org/10.2307/1907187, 1945.

Martens, B., Miralles, D., Lievens, H., Fernández-Prieto, D., and Verhoest, N. E. C.: Improving terrestrial evaporation estimates over continental Australia through assimilation of SMOS soil moisture, Int. J. Appl. Earth Obs., 48, 146–162, https://doi.org/10.1016/j.jag.2015.09.012, 2016.

Martens, B., Miralles, D. G., Lievens, H., van der Schalie, R., de Jeu, R. A. M., Fernández-Prieto, D., Beck, H. E., Dorigo, W. A., and Verhoest, N. E. C.: GLEAM v3: satellite-based land evaporation and root-zone soil moisture, Geosci. Model Dev., 10, 1903–1925, https://doi.org/10.5194/gmd-10-1903-2017, 2017.

Maurer, E. P., O'Donnell, G. M., Lettenmaier, D. P., and Roads, J. O.: Evaluation of NCEP/NCAR reanalysis water and energy budgets using macroscale hydrologic model simulations, in: Land Surface Hydrology, Meteorology, and Climate: Observations and Modeling, edited by: Lakshmi, V., Albertson, J., and Schaake, J., Wiley Online Library, 137–158, https://doi.org/10.1029/WS003p0137, 2001.

Monteith, J. L.: Evaporation and environment, Sym. Soc. Exp. Biol., 19, 205–234, https://doi.org/10.1613/jair.301, 1965.

Njoku, E. G., Jackson, T. J., Lakshmi, V., Chan, T. K., and Nghiem, S. V.: Soil moisture retrieval from AMSR-E, IEEE T. Geoscience Remote, 41, 215–229, https://doi.org/10.1109/TGRS.2002.808243, 2003.

Peng, J., Niesel, J., Loew, A., Zhang, S. Q., and Wang, J.: Evaluation of Satellite and Reanalysis Soil Moisture Products over Southwest China Using Ground-Based Measurements, Remote Sensing, 7, 15729–15747, https://doi.org/10.3390/rs71115729, 2015.

Peng, S. S., Piao, S., Zeng, Z., Ciais, P., Zhou, L., Li, L. Z. X., Myneni, R. B., Yin, Y., and Zeng, H.: Afforestation in China cools local land surface temperature, P. Natl. Acad. Sci. USA, 111, 2915–2919, https://doi.org/10.1073/pnas.1315126111, 2014.

Piao, S. L., Yin, L., Wang, X. H., Ciais, P., Peng, S. S., Shen, Z. H., and Seneviratne, S. I.: Summer soil moisture regulated by precipitation frequency in China, Environ. Res. Lett., 4, 044012, https://doi.org/10.1088/1748-9326/4/4/044012, 2009.

Piao, S. L., Ciais, P., Huang, Y., Shen, Z. H., Peng, S. S., Li, J. S., Zhou, L. P., Liu, H. Y., Ma, Y. C., Ding, Y. H., Friedlingstein, P., Liu, C. Z., Tan, K., Yu, Y. Q., Zhang, T. Y., and Fang, J. Y.: The impacts of climate change on water resources and agriculture in China, Nature, 467, 43–51, https://doi.org/10.1038/nature09364, 2010.

Pierdicca, N., Fascetti, F., Pulvirenti, L., Crapolicchio, R., and Muñoz-Sabater, J.: Analysis of ASCAT, SMOS, in-situ and land model soil moisture as a regionalized variable over Europe and North Africa, Remote Sens. Environ., 170, 280–289, https://doi.org/10.1016/j.rse.2015.09.005, 2015.

Polcher, J., Piles, M., Gelati, E., Barella-Ortiz, A., and Tello, M.: Comparing surface-soil moisture from the SMOS mission and the ORCHIDEE land-surface model over the Iberian Peninsula, Remote Sens. Environ., 174, 69–81, https://doi.org/10.1016/j.rse.2015.12.004, 2016.

Puma, M. J. and Cook, B. I.: Effects of irrigation on global climate during the 20th century, J. Geophys. Res., 115, D16120, https://doi.org/10.1029/2010JD014122, 2010.

Rebel, K. T., de Jeu, R. A. M., Ciais, P., Viovy, N., Piao, S. L., Kiely, G., and Dolman, A. J.: A global analysis of soil moisture derived from satellite observations and a land surface model, Hydrol. Earth Syst. Sci., 16, 833–847, https://doi.org/10.5194/hess-16-833-2012, 2012.

Reichle, R. H., Koster, R. D., Liu, P., Mahanama, S. P. P., Njoku, E. G., and Owe, M.: Comparison and assimilation of global soil moisture retrievals from the Advanced Microwave Scanning Radiometer for the Earth Observing System (AMSR-E) and the Scanning Multichannel Microwave Radiometer (SMMR), J. Geophys. Res., 112, D09108, https://doi.org/10.1029/2006JD008033, 2007.

Robock, A., Vinnikov, K. Y., Srinivasan, G., Entin, J. K., Hollinger, S. E., Speranskaya, N. A., Liu, S. X., and Namkhai, A.: The Global Soil Moisture Data Bank, B. Am. Meteor. Soc., 81, 1281–1299, https://doi.org/10.1175/1520-0477(2000)081<1281:TGSMDB>2.3.CO;2, 2000.

Rodell, M., Houser, P. R., Jambor, U., Gottschalck, J., Mitchell, K., Meng, C. J., Arsenault, K., Cosgrove, B., Radakovich, J., Bosilovich, M., Entin*, J. K., Walker, J. P., Lohmann, D., and Toll, D.: The Global Land Data Assimilation System, B. Am. Meteor. Soc., 85, 381–394, https://doi.org/10.1175/BAMS-85-3-381, 2004.

Rogers, S., Barnett, J., Webber, M., Finlayson, B., and Wang, M.: Governmentality and the conduct of water: China's South–North Water Transfer Project, T. I. Brit. Geogr., 41, 429–441, https://doi.org/10.1111/tran.12141, 2016.

Müller Schmied, H., Adam, L., Eisner, S., Fink, G., Flörke, M., Kim, H., Oki, T., Portmann, F. T., Reinecke, R., Riedel, C., Song, Q., Zhang, J., and Döll, P.: Impact of climate forcing uncertainty and human water use on global and continental water balance components, Proc. IAHS, 374, 53–62, https://doi.org/10.5194/piahs-374-53-2016, 2016.

Schymanski, S. J., Roderick, M. L., Sivapalan, M., Hutley, L. B., and Beringer, J.: A canopy-scale test of the optimal water-use hypothesis, Plant Cell Environ., 31, 97–111, https://doi.org/10.1111/j.1365-3040.2007.01740.x, 2008.

Seneviratne, S. I., Corti, T., Davin, E. L., Hirschi, M., Jaeger, E. B., Lehner, I., Orlowsky, B., and Teuling, A. J.: Investigating soil moisture–climate interactions in a changing climate: A review, Earth-Sci. Rev., 99, 125–161, https://doi.org/10.1016/j.earscirev.2010.02.004, 2010.

Seneviratne, S. I., Wilhelm, M., Stanelle, T., van den Hurk, B. J. J. M., Hagemann, S., Berg, A., Cheruy, F., Higgins, M. E., Meier, A., Brovkin, V., Claussen, M., Ducharne, A., Dufresne, J. L., Findell, K. L., Ghattas, J., Lawrence, D. M., Malyshev, S., Rummukainen, M., and Smith, B.: Impact of soil moisture-climate feedbacks on CMIP5 projections: First results from the GLACE-CMIP5 experiment, Geophys. Res. Lett., 40, 5212–5217, https://doi.org/10.1002/grl.50956, 2013.

Sheffield, J., Goteti, G., and Wood, E. F.: Development of a 50-year high-resolution global dataset of meteorological forcings for land surface modeling, J. Climate, 19, 3088–3111, https://doi.org/10.1175/JCLI3790.1, 2006.

Stephens, G. L., Wild, M., Stackhouse, P. W., L'Ecuyer, T., Kato, S., and Henderson, D. S.: The Global Character of the Flux of Downward Longwave Radiation, J. Climate, 25, 2329–2340, https://doi.org/10.1175/JCLI-D-11-00262.1, 2012.

Su, Z. B., Yacob, A., Wen, J., Roerink, G., He, Y. B., Gao, B. H., Boogaard, H., and van Diepen, C.: Assessing relative soil moisture with remote sensing data: theory, experimental validation, and application to drought monitoring over the North China Plain, Phys. Chem. Earth Pt A/B/C, 28, 89–101, https://doi.org/10.1016/S1474-7065(03)00010-X, 2003.

Tangdamrongsub, N., Han, S.-C., Decker, M., Yeo, I.-Y., and Kim, H.: On the use of the GRACE normal equation of inter-satellite tracking data for estimation of soil moisture and groundwater in Australia, Hydrol. Earth Syst. Sci., 22, 1811–1829, https://doi.org/10.5194/hess-22-1811-2018, 2018.

Taylor, K. E., Stouffer, R. J., and Meehl, G. A.: An Overview of CMIP5 and the Experiment Design, B. Am. Meteor. Soc., 93, 485–498, https://doi.org/10.1175/BAMS-D-11-00094.1, 2012.

Teuling, A. J., Seneviratne, S. I., Stöckli, R., Reichstein, M., Moors, E., Ciais, P., Luyssaert, S., van den Hurk, B. J. J. M., Ammann, C., Bernhofer, C., Dellwik, E., Gianelle, D., Gielen, B., Grünwald, T., Klumpp, K., Montagnani, L., Moureaux, C., Sottocornola, M., and Wohlfahrt, G.: Contrasting response of European forest and grassland energy exchange to heatwaves, Nat. Geosci., 3, 722–727, https://doi.org/10.1038/ngeo950, 2010.

van den Hurk, B., Kim, H., Krinner, G., Seneviratne, S. I., Derksen, C., Oki, T., Douville, H., Colin, J., Ducharne, A., Cheruy, F., Viovy, N., Puma, M. J., Wada, Y., Li, W., Jia, B., Alessandri, A., Lawrence, D. M., Weedon, G. P., Ellis, R., Hagemann, S., Mao, J., Flanner, M. G., Zampieri, M., Materia, S., Law, R. M., and Sheffield, J.: LS3MIP (v1.0) contribution to CMIP6: the Land Surface, Snow and Soil moisture Model Intercomparison Project – aims, setup and expected outcome, Geosci. Model Dev., 9, 2809–2832, https://doi.org/10.5194/gmd-9-2809-2016, 2016.

Wada, Y., de Graaf, I. E. M., and van Beek, L. P. H.: High-resolution modeling of human and climate impacts on global water resources, Journal of Advances in Modeling Earth Systems, 8, 735–763, https://doi.org/10.1002/2015MS000618, 2016.

Wada, Y., Bierkens, M. F. P., de Roo, A., Dirmeyer, P. A., Famiglietti, J. S., Hanasaki, N., Konar, M., Liu, J., Müller Schmied, H., Oki, T., Pokhrel, Y., Sivapalan, M., Troy, T. J., van Dijk, A. I. J. M., van Emmerik, T., Van Huijgevoort, M. H. J., Van Lanen, H. A. J., Vörösmarty, C. J., Wanders, N., and Wheater, H.: Human–water interface in hydrological modelling: current status and future directions, Hydrol. Earth Syst. Sci., 21, 4169–4193, https://doi.org/10.5194/hess-21-4169-2017, 2017.

Wagner, W., Dorigo, W., de Jeu, R., Fernandez, D., Benveniste, J., Haas, E., and Ertl, M.: Fusion of Active and Passive Microwave Observations To Create an Essential Climate Variable Data Record on Soil Moisture, XXII ISPRS Congress, 25 August–1 September 2012, Melbourne, Australia, ISPRS Annals of Photogrammetry, Remote Sensing and Spatial Information Sciences, I-7, 315–321, https://doi.org/10.5194/isprsannals-I-7-315-2012, 2012.

Wanders, N., Karssenberg, D., de Roo, A., de Jong, S. M., and Bierkens, M. F. P.: The suitability of remotely sensed soil moisture for improving operational flood forecasting, Hydrol. Earth Syst. Sci., 18, 2343–2357, https://doi.org/10.5194/hess-18-2343-2014, 2014.

Wang, S. S., Mo, X. G., Liu, S. X., Lin, Z. H., and Hu, S.: Validation and trend analysis of ECV soil moisture data on cropland in North China Plain during 1981–2010, Int. J. Appl. Earth Obs., 48, 110–121, https://doi.org/10.1016/j.jag.2015.10.010, 2016.

Wang, X. H., Ciais, P., Li, L., Ruget, F., Vuichard, N., Viovy, N., Zhou, F., Chang, J. F., Wu, X. C., Zhao, H. F., and Piao, S. L.: Management outweighs climate change on affecting length of rice growing period for early rice and single rice in China during 1991–2012, Agr. Forest Meteorol., 233, 1–11, https://doi.org/10.1016/j.agrformet.2016.10.016, 2017.

Weedon, G. P., Balsamo, G., Bellouin, N., Gomes, S., Best, M. J., and Viterbo, P.: The WFDEI meteorological forcing data set: WATCH Forcing Data methodology applied to ERA-Interim reanalysis data, Water Resour. Res., 50, 7505–7514, https://doi.org/10.1002/2014WR015638, 2014.

Wei, X. H., Li, Q., Zhang, M. F., Giles-Hansen, K., Liu, W. F., Fan, H. B., Wang, Y., Zhou, G. Y., Piao, S. L., and Liu, S. R.: Vegetation cover – another dominant factor in determining global water resources in forested regions, Glob. Change Biol., 24, 786–795, https://doi.org/10.1111/gcb.13983, 2018.

Williams, M. R., King, K. W., and Fausey, N. R.: Drainage water management effects on tile discharge and water quality, Agr. Water Manage., 148, 43–51, https://doi.org/10.1016/j.agwat.2014.09.017, 2015.

Wu, Y. L., Peng, S., Ciais, P., Guimberteau, M., Piao, S. L., Polcher, J., and Zhou, F.: Estimating water withdrawals and its impacts on water budget of an eutrophic Lake Dianchi, China, J. Hydrol., 565, 39–48, 2018.

Xia, Y., Sheffield, J., Ek, M. B., Dong, J., Chaney, N., Wei, H., Meng, J., and Wood, E. F.: Evaluation of multi-model simulated soil moisture in NLDAS-2, J. Hydrol., 512, 107–125, https://doi.org/10.1016/j.jhydrol.2014.02.027, 2014.

Yan, R., Gao, J., and Li, L.: Modeling the hydrological effects of climate and land use/cover changes in Chinese lowland polder using an improved WALRUS model, Hydrol. Res., 47, 84–101, https://doi.org/10.2166/nh.2016.204, 2016.

Yang, H., Piao, S. L., Zeng, Z. Z., Ciais, P., Yin, Y., Friedlingstein, P., Sitch, S., Ahlström, A., Guimberteau, M., Huntingford, C., Levis, S., Levy, P. E., Huang, M. T., Li, Y., Li, X. R., Lomas, M. R., Peylin, P., Poulter, B., Viovy, N., Zaehle, S., Zeng, N., Zhao, F., and Wang, L.: Multicriteria evaluation of discharge simulation in Dynamic Global Vegetation Models, J. Geophys. Res.-Atmos., 120, 7488–7505, https://doi.org/10.1002/2015JD023129, 2015.

Yang, J., Gong, D. Y., Wang, W. S., Hu, M., and Mao, R.: Extreme drought event of 2009/2010 over southwestern China, Meteorol. Atmos. Phys., 115, 173–184, https://doi.org/10.1007/s00703-011-0172-6, 2012.

Ye, J. S., Li, W. H., Li, L. F., and Zhang, F.: "North drying and south wetting" summer precipitation trend over China and its potential linkage with aerosol loading, Atmos. Res., 125–126, 12–19, https://doi.org/10.1016/j.atmosres.2013.01.007, 2013.

Yin, Z., Dekker, S. C., van den Hurk, B. J. J. M., and Dijkstra, H. A.: Bimodality of woody cover and biomass across the precipitation gradient in West Africa, Earth Syst. Dynam., 5, 257–270, https://doi.org/10.5194/esd-5-257-2014, 2014.

Yoshimura, K. and Kanamitsu, M.: Dynamical Global Downscaling of Global Reanalysis, Mon. Weather Rev., 136, 2983–2998, https://doi.org/10.1175/2008MWR2281.1, 2008.

Yoshimura, K. and Kanamitsu, M.: Incremental Correction for the Dynamical Downscaling of Ensemble Mean Atmospheric Fields, Mon. Weather Rev., 141, 3087–3101, https://doi.org/10.1175/MWR-D-12-00271.1, 2013.

Zhai, P. M., Zhang, X. B., Wan, H., and Pan, X. H.: Trends in Total Precipitation and Frequency of Daily Precipitation Extremes over China, J. Climate, 18, 1096–1108, https://doi.org/10.1175/JCLI-3318.1, 2005.

Zhao, F., Veldkamp, T. I. E., Frieler, K., Schewe, J., Ostberg, S., Willner, S., Schauberger, B., Gosling, S. N., Schmied, H. M., Portmann, F. T., Leng, G. Y., Huang, M. Y., Liu, X. C., Tang, Q. H., Hanasaki, N., Biemans, H., Gerten, D., Satoh, Y., Pokhrel, Y., Stacke, T., Ciais, P., Chang, J. F., Ducharne, A., Guimberteau, M., Wada, Y., Kim, H., and Yamazaki, D.: The critical role of the routing scheme in simulating peak river discharge in global hydrological models, Environ. Res. Lett., 12, 075003, https://doi.org/10.1088/1748-9326/aa7250, 2017.

Zhou, X., Polcher, J., Yang, T., Hirabayashi, Y., and Nguyen-Quang, T.: Understanding the water cycle over the upper Tarim basin: retrospect the estimated discharge bias to atmospheric variables and model structure, Hydrol. Earth Syst. Sci. Discuss., https://doi.org/10.5194/hess-2018-88, in review, 2018.

Zhu, D., Peng, S. S., Ciais, P., Viovy, N., Druel, A., Kageyama, M., Krinner, G., Peylin, P., Ottlé, C., Piao, S. L., Poulter, B., Schepaschenko, D., and Shvidenko, A.: Improving the dynamics of Northern Hemisphere high-latitude vegetation in the ORCHIDEE ecosystem model, Geosci. Model Dev., 8, 2263–2283, https://doi.org/10.5194/gmd-8-2263-2015, 2015.

Zhu, Z. C., Piao, S. L., Myneni, R. B., Huang, M. T., Zeng, Z. Z., Canadell, J. G., Ciais, P., Sitch, S., Friedlingstein, P., Arneth, A., Cao, C. X., Cheng, L., Kato, E., Koven, C., Li, Y., Lian, X., Liu, Y. W., Liu, R., Mao, J. F., Pan, Y. Z., Peng, S. S., Peñuelas, J., Poulter, B., Pugh, T. A. M., Stocker, B. D., Viovy, N., Wang, X. H., Wang, Y. P., Xiao, Z. Q., Yang, H., Zaehle, S., and Zeng, N.: Greening of the Earth and its drivers, Nat. Clim. Change, 6, 791–795, https://doi.org/10.1038/nclimate3004, 2016.

Zobler, L.: A world soil file for global climate modeling, NASA TM-87802, National Aeronautics and Space Administration, Washington, D.C., 1986.

3

Groundwater withdrawal in randomly heterogeneous coastal aquifers

Martina Siena and Monica Riva

Dipartimento di Ingegneria Civile e Ambientale, Politecnico di Milano, Piazza L. Da Vinci 32, 20133 Milan, Italy

Correspondence: Martina Siena (martina.siena@polimi.it)

Abstract. We analyze the combined effects of aquifer heterogeneity and pumping operations on seawater intrusion (SWI), a phenomenon which is threatening coastal aquifers worldwide. Our investigation is set within a probabilistic framework and relies on a numerical Monte Carlo approach targeting transient variable-density flow and solute transport in a three-dimensional randomly heterogeneous porous domain. The geological setting is patterned after the Argentona river basin, in the Maresme region of Catalonia (Spain). Our numerical study is concerned with exploring the effects of (a) random heterogeneity of the domain on SWI in combination with (b) a variety of groundwater withdrawal schemes. The latter have been designed by varying the screen location along the vertical direction and the distance of the wellbore from the coastline and from the location of the freshwater–saltwater mixing zone which is in place prior to pumping. For each random realization of the aquifer permeability field and for each pumping scheme, a quantitative depiction of SWI phenomena is inferred from an original set of metrics characterizing (a) the inland penetration of the saltwater wedge and (b) the width of the mixing zone across the whole three-dimensional system. Our results indicate that the stochastic nature of the system heterogeneity significantly affects the statistical description of the main features of the seawater wedge either in the presence or in the absence of pumping, yielding a general reduction of toe penetration and an increase of the width of the mixing zone. Simultaneous extraction of fresh and saltwater from two screens along the same wellbore located, prior to pumping, within the freshwater–saltwater mixing zone is effective in limiting SWI in the context of groundwater resources exploitation.

1 Introduction

Groundwater resources in coastal aquifers are seriously threatened by seawater intrusion (SWI), which can deteriorate the quality of freshwater aquifers, thus limiting their potential use. This situation is particularly exacerbated within areas with intense anthropogenic activities, which are associated with competitive uses of groundwater in connection, e.g., with agricultural processes, industrial processes and/or high urban water supply demand. Critical SWI scenarios are attained when seawater reaches extraction wells designed for urban freshwater supply, with severe environmental, social and economic implications (e.g., Custodio, 2010; Mas-Pla et al., 2014; Mazi et al., 2014).

The development of effective strategies for sustainable use of groundwater resources in coastal regions should be based on a comprehensive understanding of SWI phenomena. This challenging problem has been originally studied by assuming a static equilibrium between freshwater (FW) and seawater (SW) and a sharp FW–SW interface, where FW and SW are considered as immiscible fluids. Under these hypotheses, the vertical position of the FW–SW interface below the sea level, z, is given by the Ghyben–Herzberg solution, $z = h_F \rho_F / \Delta \rho$, where h_F is the FW head above sea level and $\Delta \rho = \rho_S - \rho_F$ is the density contrast, ρ_S and ρ_F, respectively, being SW and FW density. Starting from these works, several analytical and semi-analytical expressions have been developed to describe SWI under diverse flow configurations (e.g., Strack, 1976; Dagan and Zeitoun, 1998; Bruggeman, 1999; Cheng et al., 2000; Bakker, 2006; Nordbotten and Celia, 2006; Park et al., 2009). In this broad context, Strack (1976) evaluated the maximum (or critical) pumping rate to avoid encroachment of SW in FW pumping wells. All of these sharp-interface-based solutions neglect a key aspect of SWI phenomena, i.e., the formation of a transition zone where mixing between FW and SW takes place and fluid density varies with salt concentration. As a consequence, these models may significantly overestimate the actual penetration length of the SW wedge, leading to an excessively conservative evaluation of the critical pumping rate (Gingerich and Voss, 2005; Pool and Carrera, 2011; Llopis-Albert and Pulido-Velazquez, 2013).

A more realistic approach relies on the formulation and solution of a variable-density problem, in which SW and FW are considered as miscible fluids and groundwater density depends on salt concentration. The complexity of the problem typically prevents finding a solution via analytical or semi-analytical methods, with a few notable exceptions (Henry, 1964; Dentz et al., 2006; Bolster et al., 2007; Zidane et al., 2012; Fahs et al., 2014). Henry (1964) developed a semi-analytical solution for a variable-density diffusion problem in a (vertical) two-dimensional homogeneous and isotropic domain. Dentz et al. (2006) and Bolster et al. (2007) applied perturbation techniques to solve analytically the Henry problem for a range of (small and intermediate) values of the Péclet number, which characterizes the relative strength of convective and dispersive transport mechanisms. Zidane et al. (2012) solved the Henry problem for realistic (small) values of the diffusion coefficient. Finally, Fahs et al. (2014) presented a semi-analytical solution for a square porous cavity system subject to diverse salt concentrations at its vertical walls. Practical applicability of these solutions is quite limited, due to their markedly simplified characteristics. Various numerical codes have been proposed to solve variable-density flow and transport equations (e.g., Voss and Provost, 2002; Ackerer et al., 2004; Soto Meca et al., 2007; Ackerer and Younes, 2008; Albets-Chico and Kassinos, 2013). Numerical simulations can provide valuable insights into the effects of dispersion on SWI, a feature that is typically neglected in analytical and semi-analytical solutions. Abarca et al. (2007) introduced a modified Henry problem to account for dispersive solute transport and anisotropy in hydraulic conductivity. Kerrou and Renard (2010) analyzed the dispersive Henry problem within two- and three-dimensional randomly heterogeneous aquifer systems. These authors relied on computational analyses performed on a single three-dimensional realization, invoking ergodicity assumptions. Lu et al. (2013) performed a set of laboratory experiments and numerical simulations to investigate the effect of geological stratification on SW–FW mixing. Riva et al. (2015) considered the same setting as in Abarca et al. (2007) and studied the way quantification of uncertainty associated with SWI features is influenced by lack of knowledge of four key dimensionless parameters controlling the process, i.e., gravity number, permeability anisotropy ratio and transverse and longitudinal Péclet numbers. Enhancement of mixing in the presence of tidal fluctuations and/or FW table oscillations has been analyzed by Ataie-Ashtiani et al. (1999), Lu et al. (2009, 2015) and Pool et al. (2014) in homogeneous aquifers and by Pool et al. (2015) in randomly heterogeneous

three-dimensional systems (under ergodic conditions). Recent reviews on the topic are offered by Werner et al. (2013) and Ketabchi et al. (2016).

Several numerical modeling studies have been performed with the aim of identifying the most effective strategy for the exploitation of groundwater resources in domains mimicking the behavior of specific sites. Dausman and Langevin (2005) examined the influence of hydrologic stresses and water management practices on SWI in a superficial aquifer (Broward County, USA) by developing a variable-density model formed by two homogeneous hydrogeological units. Werner and Gallagher (2006) studied SWI in the coastal aquifer of the Pioneer Valley (Australia), illustrating the advantages of combining hydrogeological and hydrochemical analyses to understand salinization processes. Misut and Voss (2007) analyzed the impact of aquifer storage and recovery practices on the transition zone associated with the salt water wedge in the New York City aquifer, which was modeled as a perfectly stratified system. Cobaner et al. (2012) studied the effect of transient pumping rates from multiple wells on SWI in the Gosku deltaic plain (Turkey) by means of a three-dimensional heterogeneous model, calibrated using head and salinity data.

All the aforementioned field-scale contributions are framed within a deterministic approach, where the system attributes (e.g., permeability) are known (or determined via an inverse modeling procedure), so that the impact of uncertainty of hydrogeological properties on target environmental (or engineering) performance metrics is not considered. Exclusive reliance on a deterministic approach is in stark contrast with the widely documented and recognized observation that a complete knowledge of aquifer properties is unfeasible. This is due to a number of reasons, including observation uncertainties and data availability, i.e., available data are most often too scarce or too sparse to yield an accurate depiction of the subsurface system in all of its relevant details. Stochastic approaches enable us not only to provide predictions (in terms of best estimates) of quantities of interest, but also to quantify the uncertainty associated with such predictions. The latter can then be transferred, for example, into probabilistic risk assessment, management and protection protocols for environmental systems and water resources.

Only a few contributions studying SWI within a stochastic framework have been published to date. The vast majority of these works considers idealized synthetic showcases and/or simplified systems. To the best of our knowledge, only two studies (Lecca and Cau, 2009; Kerrou et al., 2013) have analyzed the transient behavior of a realistic costal aquifer within a probabilistic framework. Lecca and Cau (2009) evaluated SWI in the Oristano (Italy) aquifer by considering a stratified system where the aquitard is characterized by random heterogeneity. Kerrou et al. (2013) analyzed the effects of uncertainty in permeability and distribution of pumping rates on SWI in the Korba aquifer (Tunisia). Both works characterize the uncertainty of SWI phenomena in terms of the planar extent of the system characterized by a target probability of exceedance of a given threshold concentration.

Our stochastic numerical study has been designed to mimic the general behavior of the Argentona aquifer, in the Maresme region of Catalonia (Spain). This area, as well as other Mediterranean deltaic sites, is particularly vulnerable to SWI (Custodio, 2010). We note that the objective of this study is not the quantification of the SWI dynamics of a particular field site. Our emphasis is on the analysis of the impact of random aquifer heterogeneity and withdrawals on SWI patterns in a realistic scenario. Key elements of novelty in our work include the introduction and the study of a new set of metrics, aimed at investigating quantitatively the effects of random heterogeneity on the three-dimensional extent of the SW wedge penetration and of the FW–SW mixing zone. The impact of random heterogeneity is assessed in combination with three diverse withdrawal scenarios, designed by varying (i) the distance of the wellbore from the coastline and from the FW–SW mixing zone and (ii) the screen location along the vertical direction. We frame our analysis within a numerical Monte Carlo (MC) approach.

Section 2 provides a general description of the field site, the mathematical model adopted to simulate flow and transport phenomena in three-dimensional heterogeneous systems and the numerical settings. Section 3 illustrates the key results of our work and Sect. 4 contains our concluding remarks.

2 Materials and methods

2.1 Site description

To consider a realistic scenario that is relevant to SWI problems in highly exploited aquifers, we cast our analysis in a setting inspired from the Argentona river basin, located in the region of Maresme, in Catalonia, Spain (see Fig. 1a). This aquifer is a typical deltaic site, characterized by shallow sedimentary units and a flat topography. As such, it has a strategic value for anthropogenic (including agricultural, industrial and touristic) activities. The geological formation hosting the groundwater resource is mainly composed of a granitic Permian unit. A secondary unit of quaternary sediments is concentrated along the Argentona river.

Rodriguez Fernandez (2015) developed a conceptual and numerical model to simulate transient two-dimensional (horizontal) constant-density flow in the Argentona river basin across the area of about 35 km² depicted in Fig. 1b. The aquifer is heavily exploited, as shown by the large number of wells included in Fig. 1b, mainly located along the Argentona river. The latter is a torrential ephemeral stream, in which water flows only after heavy rain events. On these bases, it is assumed that the only water intake to the river comes from surface runoff of both granitic and quaternary

units. On the basis of transient hydraulic head measurements available at 21 observation wells (not shown in Fig. 1b) in the period 2006–2013, Rodriguez Fernandez (2015) characterized the site through a uniform permeability, whose value was estimated as $k_B = 1.77 \times 10^{-11}\,\mathrm{m}^2$, and provided estimates of temporally and spatially variable recharge rates, according to land use or cover (see Fig. 1c).

Our numerical analysis focuses on the coastal portion of the Argentona basin (Fig. 1c). This region extends for about 2.5 km along the coast (i.e., along the full width of the basin) and up to 750 m inland from the coast. The offshore extension of the aquifer is not considered. The size of the study area has been selected on the basis of the results of preliminary simulations on larger domains, which highlighted that salt concentration values were appreciable only within a narrow (less than 400 m wide) region close to the coast. The vertical thickness of the domain ranges from 50 m along the coast up to 60 m at the inland boundary, the underlying clay sequence being considered to represent the impermeable bottom of the aquifer. SWI is simulated by means of a three-dimensional variable-density flow and solute transport model based on the well-established finite element USGS SUTRA code (Voss and Provost, 2002) over the 8-year time window 2006–2013. Details of the mathematical and numerical model are discussed in the following sections.

2.2 Flow and transport equations

Fluid flow is governed by mass conservation and Darcy's law:

$$\frac{\partial (\phi \rho)}{\partial t} + \nabla^{\mathrm{T}} (\rho \boldsymbol{q}) = 0 \tag{1}$$

$$\frac{\partial (\phi \rho)}{\partial t} = \frac{1}{g} S_{\mathrm{S}} \frac{\partial p}{\partial t} + \phi \frac{\partial \rho}{\partial C} \frac{\partial C}{\partial t},$$

$$\boldsymbol{q} = -\frac{\boldsymbol{k}}{\mu} (\nabla p + \rho g \nabla z), \tag{2}$$

where $\boldsymbol{q}$ $(\mathrm{L\,T}^{-1})$ is the specific discharge vector, with components q_x, q_y and q_z, respectively, along x, y (see Fig. 1c) and z axes (z denoting the vertical direction); C (–) is solute concentration, or solute mass fraction, expressed as mass of solute per unit mass of fluid; $\boldsymbol{k}$ (L^2) is the diagonal permeability tensor, with components $k_{11} = k_x$, $k_{22} = k_y$ and $k_{33} = k_z$, respectively, along directions x, y and z; ϕ (–) is aquifer porosity; ρ $(\mathrm{M\,L}^{-3})$ and μ $(\mathrm{M\,L}^{-1}\,\mathrm{T}^{-1})$, respectively, are fluid density and dynamic viscosity; p $(\mathrm{M\,L}^{-1}\,\mathrm{T}^{-2})$ is fluid pressure; g $(\mathrm{L\,T}^{-2})$ is gravity; S_{S} (L^{-1}) is the specific storage coefficient, and the superscript T denotes transpose. Equations (1) and (2) must be solved jointly with the advection–dispersion equation:

$$\frac{\partial (\phi \rho C)}{\partial t} + \nabla^{\mathrm{T}} (\rho C \boldsymbol{q}) - \nabla^{\mathrm{T}} [\rho \mathbf{D} \nabla C] = 0. \tag{3}$$

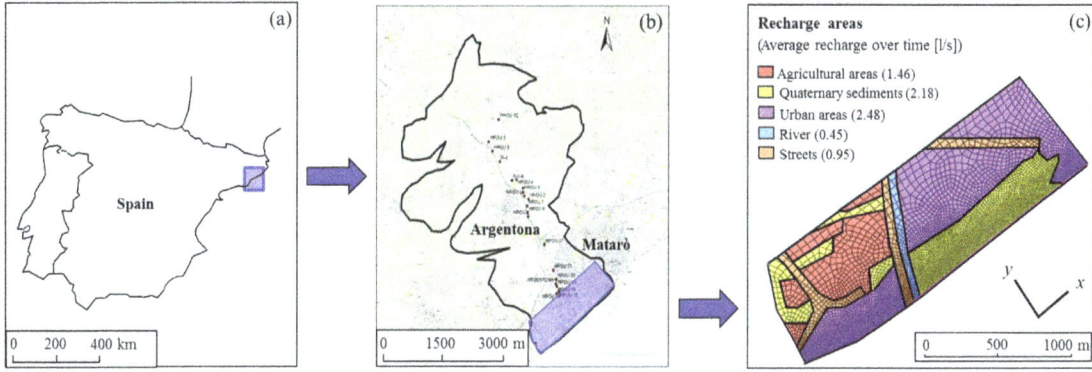

Figure 1. (a) Location of the Argentona river basin. **(b)** Two-dimensional, constant-density model (modified from Rodriguez Fernandez, 2015). Pumping wells (brown dots) and three-dimensional model location (blue shaded area) are also depicted. **(c)** Planar view of the three-dimensional variable-density flow and transport model. The unstructured grid and diverse recharge areas are also shown.

$\mathbf{D}$ ($L^2 T^{-1}$) is the dispersion tensor defined as

$$\mathbf{D} = (\phi D_m + \alpha_T |\mathbf{q}|)\mathbf{I} + (\alpha_L - \alpha_T)\frac{\mathbf{q}\mathbf{q}^T}{|\mathbf{q}|}, \quad (4)$$

where D_m ($L^2 T^{-1}$) is the molecular diffusion coefficient, and α_L and α_T (L), respectively, are the longitudinal and transverse dispersivity coefficients. Closure of the system of Eqs. (1)–(4) requires a relationship between fluid properties (ρ and μ) and solute concentration C. Fluid viscosity can be assumed constant in typical SWI settings and the following model has been shown to be accurate at describing the evolution of ρ with C (Kolditz et al., 1998):

$$\rho = \rho_F + (\rho_S - \rho_F)\frac{C}{C_S}, \quad (5)$$

where C_S is SW concentration. Table 1 lists aquifer and fluid parameters adopted.

2.3 Numerical model

The domain depicted in Fig. 1c is discretized through an unstructured three-dimensional grid formed by 101 632 hexahedral elements. The resolution of the mesh increases towards the sea, where our analysis requires the highest spatial detail. The element size along both horizontal directions x and y ranges from 10 m (in a 200 m wide region along the coast) to 60 m (close to the inland boundary). Element size along the vertical direction is 2 and 4 m, respectively, within the first 10 m from the top surface and across the remaining 40 m. These choices are consistent with constraints for numerical stability, namely $\Lambda_L \leq 4\alpha_L$, Λ_L being the distance between element sides measured along a flow line and $\alpha_L = 5$ m. In the absence of transverse dispersivity estimates and in analogy with main findings of previous works (e.g., Cobaner et al., 2012), we set $\alpha_T = \alpha_L/10$.

A sequential Gaussian simulation algorithm (Deutsch and Journel, 1998) is employed to generate unconditional log

Table 1. Parameters adopted in the numerical model.

Parameter	Value
Freshwater density, ρ_F (kg m^{-3})	1000
Seawater density, ρ_S (kg m^{-3})	1025
Seawater mass fraction, C_S (–)	0.035
Fluid viscosity, μ (kg m^{-1} s^{-1})	1×10^{-3}
Effective porosity, ϕ (–)	0.15
Specific storage, S_S (m^{-1})	0.01
Permeability, k_B (m^2)	1.77×10^{-11}
Molecular diffusion, D_m (m^2 s^{-1})	1×10^{-9}
Longitudinal dispersivity, α_L (m)	5.0
Transverse dispersivity, α_T (m)	0.5

permeability fields, $Y(\mathbf{x}) = \ln k(\mathbf{x})$, characterized by a given mean $\langle Y \rangle = \ln k_B$ and variogram structure. Since no Y data are available, for the purpose of our simulations we assume that the spatial structure of Y is described by a spherical variogram, with moderate variance, i.e., $\sigma_Y^2 = 1.0$, and isotropic correlation scale $\lambda = 100$ m. This value is consistent with the observation that the integral scale of log conductivity and transmissivity values inferred worldwide using traditional (such as exponential and spherical) variograms tends to increase with the length scale of the sampling window at a rate of about 1/10 (Neuman et al., 2008). The MC approach allows us to provide a probabilistic description of the SWI processes associated with space–time distributions of flux and solute concentration obtained in each random realization of the permeability field by solving the transient flow and transport Eqs. (1)–(5). Note also that $\lambda > 5\Delta$ in our model, Δ being the largest cell size within a 200 m wide region along the coast. A model satisfying this condition has the benefit of (a) yielding a proper reconstruction of the spatial correlation structure of Y and (b) limiting the occurrence of excessive variations between values of aquifer properties across neighboring cells (Kerrou and Renard, 2010) in the region where SWI takes place.

Equations (1)–(5) are solved jointly, adopting the following boundary conditions. The lateral boundaries perpendicular to the coastline and the base of the aquifer are impervious; along the inland boundary we set a time-dependent prescribed hydraulic head, h_F (with $h_F = z + p/(\rho_F g)$ ranging between 1.2 and 3.5 m), and $C = C_F$; along the coastal boundary we impose a prescribed head, $h_S = 0$ (with $h_S = z + p/(\rho_S g)$), and

$$(qC - \mathbf{D}\nabla C) \cdot n = \begin{cases} q \cdot \mathbf{n} \, C & \text{if } q \cdot n < 0 \\ q \cdot \mathbf{n} \, C_S & \text{if } q \cdot n > 0 \end{cases}, \qquad (6)$$

n being the normal vector pointing inward along the boundary; i.e., the concentration of the water entering and leaving the system across the coastal boundary is, respectively, equal to C_S and C. The area of interest is characterized by five recharge zones, identified on the basis of the land use (inferred from the SIGPAC, 2005, dataset). The total recharge varies slightly in time (on a monthly basis), with mean value equal to $7.6\,\mathrm{L\,s^{-1}}$ (see Fig. 1c). Initial conditions are set as $h = 0$ and $C = 0$, as commonly assumed in the literature (e.g., Jakovovic et al., 2016) and flow and transport are simulated for an 8-year period (2006–2013) with a uniform time step $\Delta t = 1$ day. Varying the initial conditions (e.g., setting, $h = 2.4$ m, corresponding to the mean value of h set along the inland boundary), did not lead to significantly different results at the end of the simulated 8-year period, $t = t_w$ (details not shown).

A pumping well is activated at time t_w and flow and transport are simulated for 8 additional months. We consider three diverse pumping schemes, reflecting realistic engineered operational settings. The borehole is located along the vertical cross-section B–B′, as shown in Fig. 2a. Scheme 1 (S1) and Scheme 2 (S2) consider the pumping well to be placed 180 m away from the shoreline, i.e., at $y' = y/\lambda = 1.8$, outside of the mixing zone which is in place prior to pumping in all MC realizations. In S1 (Fig. 2b) the well is screened in the upper part of the aquifer, the screen starting from 2 m below the top and extending across a total thickness of 15 m. The double-negative hydraulic barrier is a technology employed to manage (or possibly prevent) SWI. It relies on the joint use of an FW (*production* well) and an SW (*scavenger* well) pumping well. We employ this setting in S2 (Fig. 2c), which is designed by adding an additional screen in the lower part of the aquifer to S1, starting from 34 m below the ground surface and extending across a total thickness of 12 m. In S2, production and scavenger wells are located along the same borehole, complying with technical and economic requirements typically associated with field applications (Pool and Carrera, 2009; Saravanan et al., 2014; Mas-Pla et al., 2014). This scheme is also particularly appealing when applied to renewable energy resources, such as the pressure retarded osmosis, which allows conversion of the chemical energy of two fluids (FW and SW) into mechanical and electrical energy

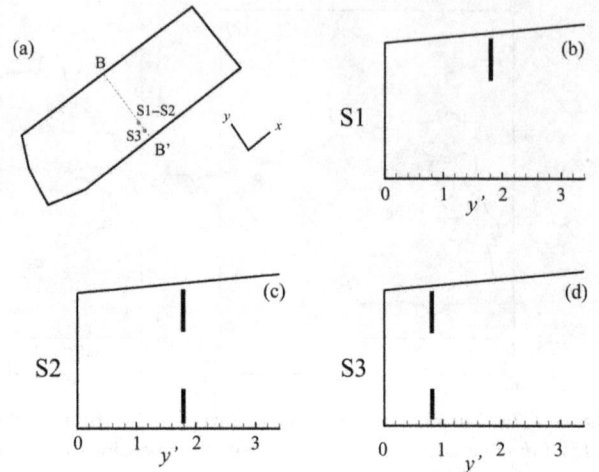

Figure 2. **(a)** Planar view of the location of the boreholes. Vertical cross section B–B′ and well-screen location for scheme **(b)** S1, **(c)** S2 and **(d)** S3.

(Panyor, 2006). Scheme 3 (S3, Fig. 2d) shares the same operational design of S2, but the borehole is moved seawards, at $y' = 0.8$, i.e., within the mixing zone in place before pumping. In all pumping schemes, the withdrawal rate is constant in time and uniformly distributed along the well screens, where pumping is implemented by setting a flux-type condition. We set a total withdrawal rate $Q = 5\,\mathrm{L\,s^{-1}}$ at the upper well screen in S1, S2 and S3; an additional extraction at the same rate Q is imposed along the lower screen in S2 and S3. During the pumping period, transport is simulated using a time step $\Delta t = 30$ min for the first month and with increasing Δt progressively up to a maximum value $\Delta t = 120$ min for the remaining 7 months, as the system showed progressively smoother variations while approaching steady state. In Sect. 3 all schemes are compared against a benchmark case, Scheme 0 (S0), where no pumping wells are active.

2.4 Local and global SWI metrics

To provide a comprehensive characterization of SWI phenomena across the whole three-dimensional domain, we quantify the extent of the SW wedge and of the associated mixing zone on the basis of seven metrics, as illustrated in Fig. 3.

For each MC realization and along each vertical cross-section perpendicular to the coast, we evaluate (i) toe penetration, L'_T, measured as the distance from the coast of the iso-concentration curve $C/C_S = 0.5$ at the bottom of the aquifer; (ii) solute spreading at the toe, L'_S, evaluated as the separation distance along the direction perpendicular to the coast (y axis in Fig. 1c) between the iso-concentration curves $C/C_S = 0.25$ and $C/C_S = 0.75$ at the bottom of the aquifer; (iii) mean width of the mixing zone, W'_{MZ}, i.e., the spatially averaged vertical distance between iso-concentration curves $C/C_S = 0.25$ and $C/C_S = 0.75$ within the region

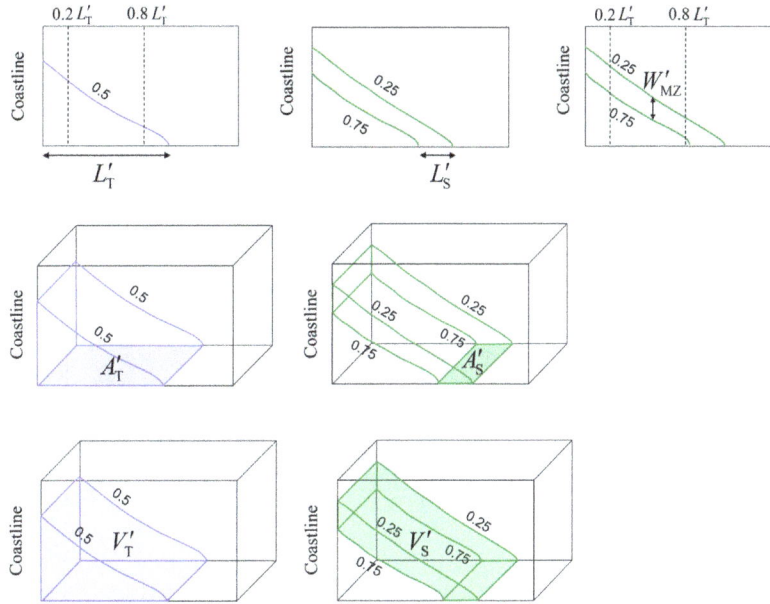

Figure 3. Graphical representation of the SWI metrics defined in Sect. 2.4.

$0.2 L'_T \leq y' \leq 0.8 L'_T$. All quantities L'_T, L'_S and W'_{MZ} are dimensionless, corresponding to distances rescaled by λ, the correlation scale of Y. We also analyze the (dimensionless) areal extent of SW penetration and solute spreading at the bottom of the aquifer by evaluating (iv) A'_T and (v) A'_S, as obtained by integrating L'_T and L'_S along the coastline. Finally, we quantify SWI across the whole thickness of the aquifer by evaluating the dimensionless volumes enclosed between (vi) the sea boundary and the iso-concentration surface $C/C_S = 0.5$, here termed V'_T, and (vii) the isosurfaces $C/C_S = 0.25$ and $C/C_S = 0.75$, here termed V'_S. First- and second-order statistical moments of each of these metrics are evaluated within the adopted MC framework. With the settings employed here, a flow and transport simulation on a single system realization is associated with a computational cost of about 12 h (on an Intel Core™ i7-6900K CPU@ 3.20 GHz processor). Results illustrated in the following sections are based on a suite of 400 MC simulations (i.e., 100 MC simulation for each scenario, S0–S3). Details concerning the analysis of convergence of the first and second statistical moments of the metrics here introduced are provided in Appendix A.

3 Results and discussion

3.1 Effects of three-dimensional heterogeneity on SWI

The effects of heterogeneity on SWI are inferred by comparing the results of our MC simulations against those obtained for an equivalent homogeneous aquifer, characterized by an effective permeability, k_{ef}. The latter is here evaluated as $k_{ef} = e^{\langle Y \rangle + \sigma_Y^2 / 6} = 2.09 \times 10^{-11} \, \mathrm{m}^2$ (Ababou, 1996). A first clear effect of heterogeneity is noted on the struc-

ture of the three-dimensional flow field. This is elucidated in Fig. 4, where we depict the permeability color map and streamlines (dashed curves) obtained at the end of the 8-year simulation period along the vertical cross-section B–B' perpendicular to the coastline. Note that henceforth, results are depicted in terms of dimensionless spatial coordinates, $y' = y/\lambda$ and $z' = z/\lambda$. Flow within the equivalent homogeneous aquifer (Fig. 4b) is essentially horizontal at locations far from the zone where SWI phenomena occur. Otherwise, the vertical flux component, q_z, is non-negligible throughout the whole domain for the heterogeneous system (Fig. 4c) and streamlines tend to focus towards regions characterized by large permeability values. Figure 4 also depicts iso-concentration curves $C/C_S = 0.25$, 0.5 and 0.75 within the transition zone (red curves). It can be noted that iso-concentration profiles tend to be sub-parallel to streamlines directed towards the seaside boundary. In the homogeneous domain the slope of these curves varies mildly and in a gradual manner from the top to the bottom of the aquifer. Their slope in the heterogeneous domain is irregular and markedly influenced by the spatial arrangement of permeability. In other words, the way streamlines are refracted at the boundary between two blocks of contrasting permeability drives the local pattern of concentration contour lines. As a consequence, iso-concentration curves tend to become sub-vertical when solute is transitioning from regions characterized by high k values to zones associated with moderate to small k, a sub-horizontal pattern being observed when transitioning from low to high k values.

Figure 5 depicts isolines $C/C_S = 0.5$ at the bottom of the aquifer (Fig. 5a) and along three vertical cross-sections, selected to exemplify the general pattern observed in the sys-

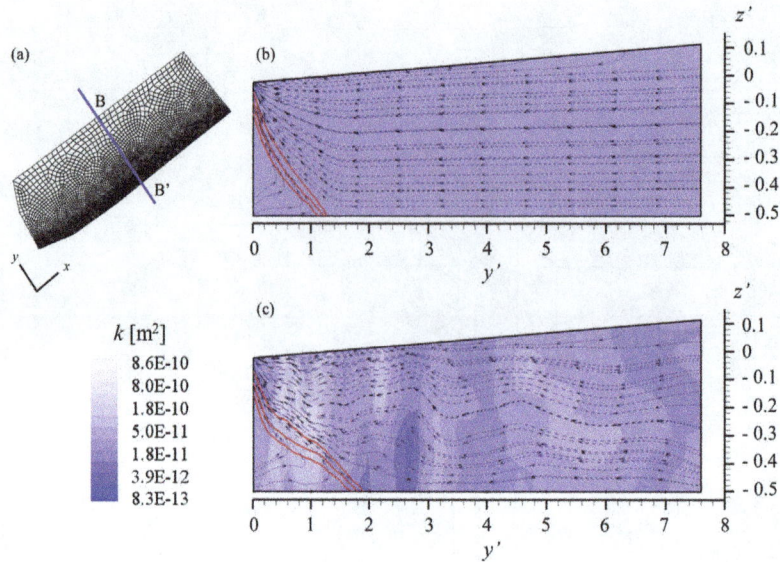

Figure 4. Permeability distribution (color contour plots), streamlines (black dashed curves) and iso-concentration curves $C/C_S = 0.25, 0.5$ and 0.75 (solid red curves) along the cross-section B–B$'$ evaluated at $t = t_w$ within **(b)** the equivalent homogeneous aquifer and **(c)** one random realization of k. Vertical exaggeration = 5.

tem (Fig. 5b–d), evaluated for (i) the 100 heterogeneous realizations analyzed (dotted blue curves), (ii) the equivalent homogeneous system (denoted as *Hom*; solid red curve) and (iii) the configuration obtained by averaging the concentration fields across all MC heterogeneous realizations (denoted as *Ens*; solid blue curve). These results suggest that iso-concentration curves exhibit considerably large spatial variations within a single realization. Comparison of the results obtained for *Ens* and *Hom* reveals that the mean wedge penetration at the bottom layer is slightly overestimated by the solution computed within the equivalent homogeneous system. Otherwise, along the coastal vertical boundary, the extent of the area with (mean) relative concentration larger than 0.5 is generally underestimated by *Hom*. These outcomes (hereafter called rotation effects) are consistent with previous literature findings associated with (a) deterministic models (Abarca et al., 2007) or (b) stochastic analyses performed under ergodicity assumptions (Kerrou and Renard, 2010, and Pool et al., 2015). Abarca et al. (2007) showed that a similar effect can be observed in a homogenous domain when considering increasing values of the dispersion coefficient. Kerrou and Renard (2010) and Pool et al. (2015) noted that the strength of the rotation effect increases with the variance of the log permeability field.

In the following, we analyze values of $\xi'' = \xi'/\xi'^{\text{Hom}}$, ξ' representing a given SWI metric (as listed in Sect. 2.4) and ξ'^{Hom} being the corresponding metric obtained on the equivalent homogeneous system. Figure 6 depicts the range of variability of ξ'' across the set of MC realizations (symbols). Each plot is complemented by the depiction of (i) the (ensemble) average of ξ'', as evaluated through all MC realizations,

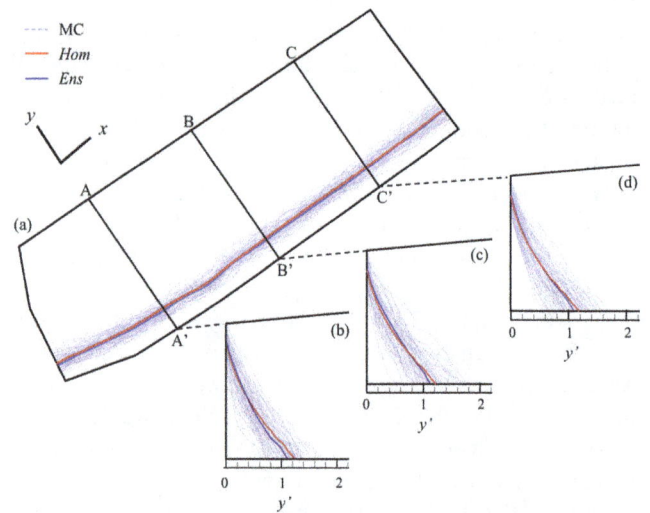

Figure 5. MC-based (dotted blue curves) iso-concentration lines $C/C_S = 0.5$ at $t = t_w$ along **(a)** the bottom of the aquifer and **(b–d)** three vertical cross-sections perpendicular to the coast. The ensemble-average concentration curves (solid blue curve) and the results obtained within the equivalent homogeneous system (solid red curve) are also reported. Vertical exaggeration = 5.

$\langle \xi'' \rangle$ (black solid line); (ii) the confidence intervals, $\langle \xi'' \rangle \pm \sigma_{\xi''}$ (black dashed lines), $\sigma_{\xi''}$ being the standard deviation of ξ''; (iii) ξ''^{Ens}, evaluated on the basis of the ensemble-average concentration field (blue line). Inspection of Fig. 6a–c complements the qualitative analysis of Fig. 5 and suggests that heterogeneity causes (on average) a slight reduction of the toe penetration, $\langle L_T'' \rangle$ and $L_T''^{\text{Ens}}$ being slightly less than 1, and

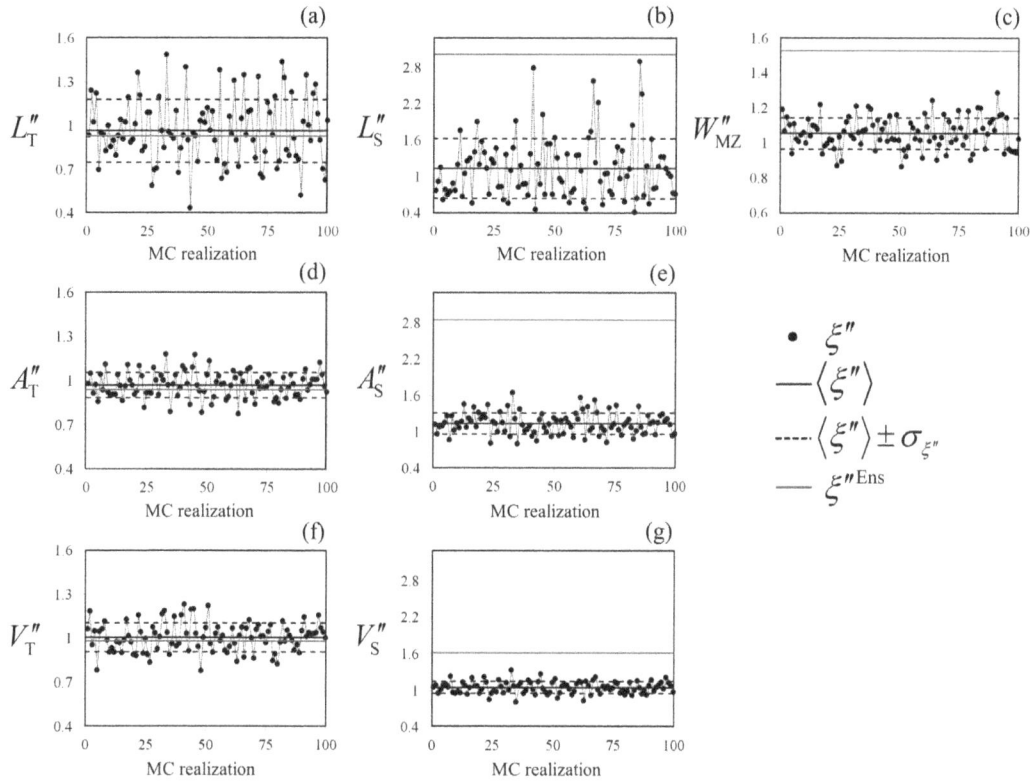

Figure 6. Metrics $\xi'' = \xi'/\xi'^{\text{Hom}}$ evaluated at $t = t_w$ within each MC heterogeneous realization (dots) and on the basis of the ensemble-average concentration field (blue line). Ensemble mean metrics (solid black line) and confidence intervals of width equal to ± 1 standard deviation of ξ'' about their mean (dashed black lines) are also depicted. Metrics in **(a–c)** are computed along the vertical cross-section B–B'.

an enlargement of the mixing zone, $\langle L_S'' \rangle$ and $\langle W_{\text{MZ}}'' \rangle$ being larger than 1. The results of Fig. 6a–c also emphasize that, while $L_T''^{\text{Ens}}$ calculated from the ensemble-average concentration distributions is virtually indistinguishable from $\langle L_T'' \rangle$, $L_S''^{\text{Ens}}$ and $W_{\text{MZ}}''^{\text{Ens}}$ markedly overestimate $\langle L_S'' \rangle$ and $\langle W_{\text{MZ}}'' \rangle$, respectively, as they visibly lie outside of the corresponding confidence intervals of width $\sigma_{\xi''}$. Note that even as Fig. 6a–c have been computed along cross-section B–B', qualitatively similar results have been obtained along all vertical cross-sections, as also suggested by the behavior of the dimensionless areal extent of toe penetration, A_T'' (Fig. 6d), and solute spreading, A_S'' (Fig. 6e), as well as of the dimensionless volumes V_T'' and V_S'' (Fig. 6f–g). Overall, the results in Fig. 6 clearly indicate that the ensemble-average concentration field can provide accurate estimates of the average wedge penetration while rendering biased estimates of quantities characterizing mixing. Our findings are consistent with previous studies (e.g., Dentz and Carrera, 2005; Pool et al., 2015) in showing that an analysis relying on the ensemble concentration field tends to overestimate significantly the degree of mixing and spreading of the solute because it combines the uncertainty associated with sample-to-sample variations of (a) the solute center of mass and (b) the actual spreading. Our results also highlight that the extent of this overestimation decreases whenever integral rather than local quantities

are considered: $V_S''^{\text{Ens}}$ in Fig. 6g is indeed about half of $L_S''^{\text{Ens}}$ and $A_S''^{\text{Ens}}$.

3.2 Effects of pumping on SWI in three-dimensional heterogeneous media

Here, we investigate SWI phenomena in heterogeneous systems when a pumping scheme is activated as described in Sect. 2.3. Figure 7 collects contour maps of relative concentration C/C_S obtained for all schemes along cross-section B–B' at the end of the 8-month pumping period, with reference to (i) *Hom* (the equivalent homogeneous system, left column) and (ii) *Ens* (the ensemble-average concentration field, right column). Contour lines $C/C_S = 0.25$, 0.50 and 0.75 are also highlighted. Extracting FW according to the engineered solution S1 (Fig. 7c–d) results in a slight landward displacement of the SWI wedge and in an enlargement of the transition zone, as compared to S0 (Fig. 7a–b), where no pumping is activated. This behavior can be observed in the homogeneous as well as (on average) in the heterogeneous settings, and is related to the general decrease of the piezometric head within the inland side caused by pumping which, in turn, favors SWI. One can also note that the partially penetrating well induces non-horizontal flow (in the proximity of the well) thus enhancing mixing along the vertical direction. Wedge

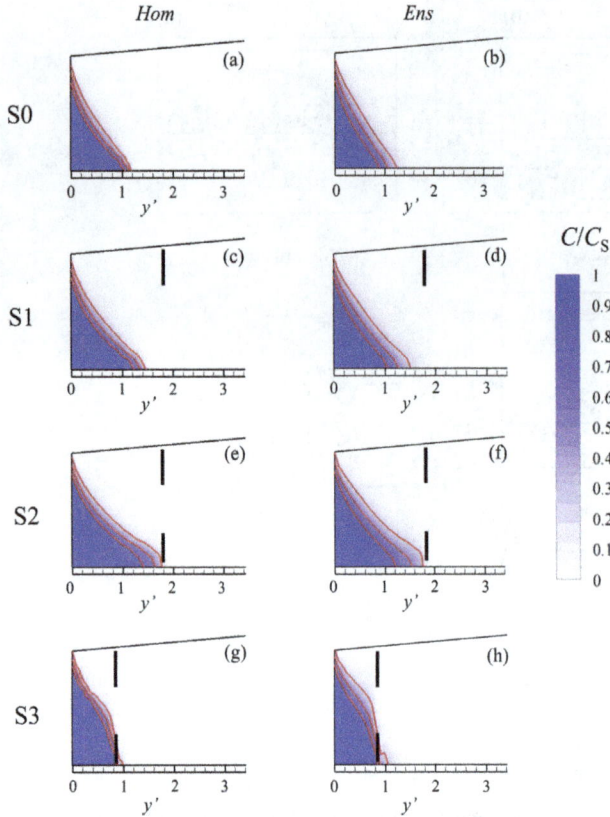

Figure 7. Concentration distribution (color contour plots) and iso-concentration lines $C/C_S = 0.25$, 0.5 and 0.75 (solid red curves) along the cross-section B–B$'$ shown in Fig. 2a evaluated at the end of the pumping period within the equivalent homogeneous aquifer (left column) and the ensemble-average concentration field (right column). Black vertical lines represent the location of the well screens. Vertical exaggeration = 5.

penetration and solute spreading are further enlarged in S2 (see Fig. 7e–f), where the total extracted volume is increased with respect to S1 through an additional pumping rate at the bottom of the aquifer. However, when the pumping well is operating within the transition zone (S3, Fig. 7g–h) the SW wedge tends to recede, approaching the pumping well location. In this case, the lower screen acts as a barrier limiting the extent of the SWI at the bottom of the aquifer.

The combined effects of groundwater withdrawal and stochastic heterogeneity on SWI are investigated quantitatively through the analysis of the temporal evolution of the seven metrics introduced in Sect. 2.4.

Figure 8 illustrates mean values of ξ' (with $\xi' = L_T'$, L_S' and W_{MZ}') computed for all pumping schemes considered across the collection of all heterogeneous realizations (black line) along the vertical cross-section B–B$'$ versus dimensionless time $t' = (t - t_w) Q/\lambda^3$. Confidence intervals $\langle\xi'\rangle \pm \sigma_{\xi'}$ are also depicted (dashed lines). As additional terms of comparison, Fig. 8 also includes ξ'^{Hom} (red line) and ξ'^{Ens} (blue line), respectively, evaluated in the equivalent homogeneous

domain and in the ensemble-average concentration field. Figure 8 shows that the toe penetration, L_T', and the solute spreading, as quantified by L_S' and W_{MZ}', do not vary significantly with time in the absence of pumping (S0) because stationary boundary conditions are imposed for $t' > 0$. Pumping schemes S1 and S2 cause the progressive inland displacement of the toe, together with an overall increase of spreading at the bottom of the aquifer. This phenomenon is more severe in S2, where SW and FW are simultaneously extracted. On the other hand, the vertical width of the mixing zone, e.g., W_{MZ}', is not significantly affected by the pumping within schemes S1–S2. Configuration S3, in which the pumping well is located within the mixing zone, leads to the most pronounced changes in the shape and position of the SW wedge. The toe penetration first decreases rapidly in time and then stabilizes around the well location. Quantities L_S' and W_{MZ}' show an early time increase, suggesting that a rapid displacement of the wedge enhances mixing and spreading. As the toe stabilizes over time, both L_S' and W_{MZ}' decrease, reaching values equal to (or slightly less than) those detected in the absence of pumping. Consistent with the observations of Sect. 3.1, Fig. 8 supports the findings that (i) an equivalent homogeneous domain is typically characterized by the largest toe penetration and the smallest vertical width of the mixing zone, the only exception being observed in S3 at late times, when $L_T'^{Hom} \cong L_T'^{Ens} < \langle L_T'\rangle$; (ii) basing the characterization of the mixing zone on ensemble-average concentrations enhances the actual effects of heterogeneity and yields overestimated mixing zone widths. Figure 8 also highlights that the extent of the confidence interval associated with toe penetration is approximately constant in time and does not depend significantly on the analyzed flow configuration. Otherwise, the confidence intervals associated with our mixing zone metrics depend on the pumping scenario and tend to increase with time if FW and SW are simultaneously extracted, especially in S2.

The fully three-dimensional nature of the analyzed problem is exemplified in Fig. 9, where we depict isolines $C/C_S = 0.5$ along the bottom of the aquifer at the end of the pumping period for the equivalent homogeneous system (Fig. 9a) and as a result of ensemble averaging across the collection of heterogeneous fields (Fig. 9b). The spatial pattern of iso-concentration contours associated with pumping schemes S1–S3 departs from the corresponding results associated with S0 within a range extending for about 10λ around the well, along the direction parallel to the coastline.

We quantify the global effect of pumping on the three-dimensional SW wedge by evaluating, for each pumping scenario and each MC realization, the temporal evolution of $\xi^* = (\xi' - \xi_{S0}')/\xi_{S0}'$, i.e., the relative percentage variation of areal extent, $\xi' = A_T'$, A_S', and volumetric extent, $\xi' = V_T'$, V_S', of wedge penetration and mixing zone with respect to their counterparts computed in the absence of pumping ξ_{S0}'. Figure 10 shows that the penetration area, A_T^*, and the solute spreading, as quantified through A_S^*, at the bottom

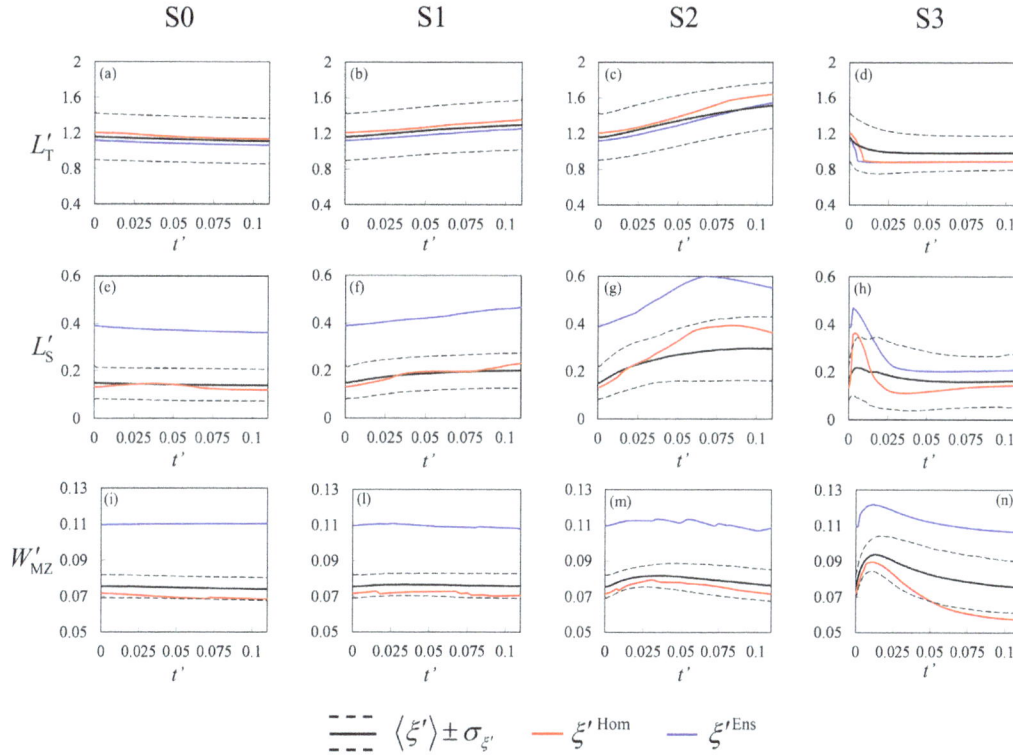

Figure 8. Temporal evolution of $\xi' = L'_T$, L'_S and W'_{MZ} evaluated along the vertical cross-section B–B' shown in Fig. 2a during the pumping period, for all pumping schemes within the equivalent homogeneous domain (red lines) and the ensemble-average concentration field (blue lines). MC-based mean values of ξ' (black lines) and confidence intervals of width equal to ± 1 standard deviation of ξ' about their mean are also depicted.

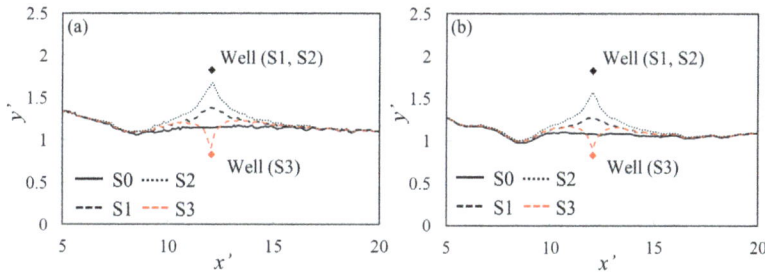

Figure 9. Isolines $C/C_S = 0.5$ at the end of the pumping period along the bottom of the aquifer for the four schemes S0–S3 evaluated **(a)** within the equivalent homogeneous domain and **(b)** from the ensemble-average concentration field.

of the aquifer tend to increase with pumping time. Corresponding results for V^*_T and V^*_S are depicted in Fig. 11. The scenario where these quantities are most affected by pumping is S2, where the FW and SW pumping wells are located outside the transition zone (in place prior to pumping). Placing both wells within the transition zone prior to pumping yields a significant decrease of the effect of pumping on both areal and volumetric metrics. We further note that the mean penetration area and volume can be accurately determined by the solution obtained within the equivalent homogeneous domain and is associated with a relatively small uncertainty, as quantified by the confidence intervals. Heterogeneity ef-

fects are clearly visible on spreading along the bottom of the aquifer. Figure 10d–f highlight that A^{*Hom}_S and A^{*Ens}_S are not accurate approximations of $\langle A^*_S \rangle$. Spreading uncertainty, as quantified by $\sigma_{A^*_S}$, is significantly larger than $\sigma_{A^*_T}$, especially in pumping scenario S2. On the other hand, the spreading of the SWI volume, as quantified by V^*_S (Fig. 11d–f), can be accurately estimated by the homogeneous equivalent system, as well as by the ensemble mean concentration field, and it is characterized by a relatively low uncertainty.

Our analyses document that for a stochastically heterogeneous aquifer an operational scheme of the kind engineered in S3 (i) is particularly efficient for the reduction of SWI

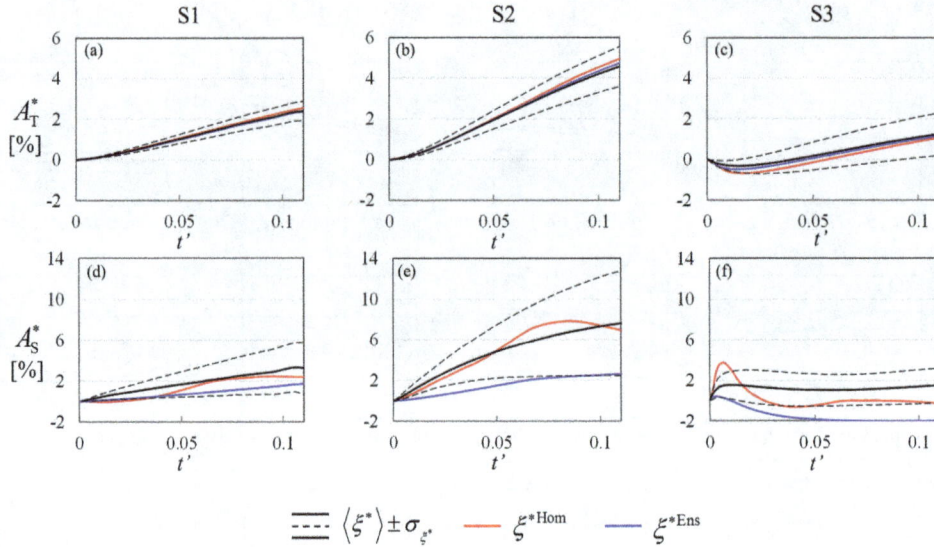

Figure 10. Temporal evolution of $\xi^* = \left(\xi' - \xi'_{S0}\right)/\xi'_{S0}$, where $\xi' = A'_T$, A'_S, evaluated for all pumping schemes within the equivalent homogeneous domain (red lines) and the ensemble-average concentration field (blue lines). MC-based mean values of ξ^* (black lines) and confidence intervals of width equal to ± 1 standard deviation of ξ^* about their mean are also depicted.

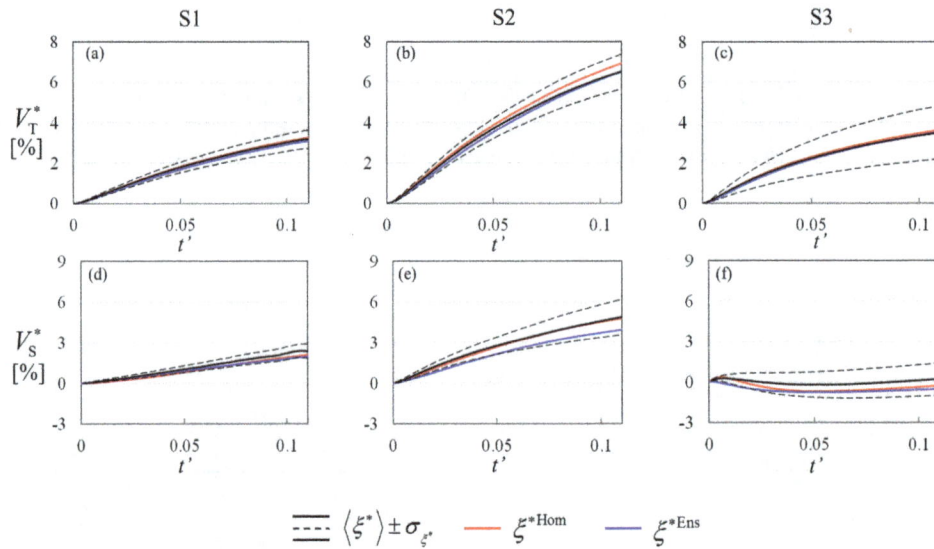

Figure 11. Temporal evolution of $\xi^* = \left(\xi' - \xi'_{S0}\right)/\xi'_{S0}$, where $\xi' = V'_T$, V'_S, evaluated for all pumping schemes within the equivalent homogeneous domain (red lines) and the ensemble-average concentration field (blue lines). MC-based mean values of ξ^* (black lines) and confidence intervals of width equal to ± 1 standard deviation of ξ^* about their mean are also depicted.

maximum penetration (localized at the bottom of the aquifer) and (ii) is advantageous in controlling the extent of the volume of the SW wedge, as compared to the double-negative barrier implemented in S2. However, this withdrawal system may lead to the salinization of the FW extracted from the upper screen due to upconing effects. This aspect is further analyzed in Fig. 12 where the temporal evolution of C_T/C_S is depicted, C_T being the salt concentration associated with the total mass of fluid extracted by the upper screen (production well) in S3. Results for the equivalent homogeneous

system (red curve), each of the heterogeneous MC realizations (grey dashed curves) and the ensuing ensemble-average concentration $\langle C_T \rangle/C_S$ (blue curve) are depicted. Vertical bars represent the 95 % confidence interval around the ensemble mean, evaluated as $\langle C_T \rangle/C_S \pm 1.96\, \sigma_{C_T/C_S}/\sqrt{n}$. Here σ_{C_T/C_S} is the standard deviation of C_T/C_S and $n = 100$ is the number of MC realizations forming the collection of samples. The width of these confidence intervals can serve as a metric to quantify the order of magnitude of the uncertainty associated with the estimated mean. The equivalent homoge-

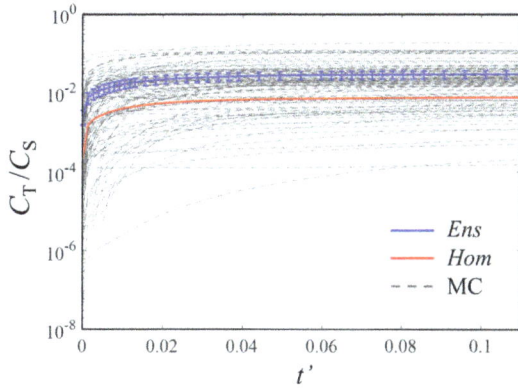

Figure 12. Temporal evolution of C_T/C_S for pumping scheme S3 within the equivalent homogeneous system (red curve) and for each heterogeneous realization (gray dashed curves). The ensemble-average (blue curve) dimensionless concentration and its 95 % confidence interval are also depicted.

neous system significantly underestimates the ratio $\langle C_T \rangle / C_S$. Moreover, Fig. 12 highlights the marked variability of the results across the MC space. As such, it reinforces the conclusion that the mean value $\langle C_T \rangle$ is an intrinsically weak indicator of the actual salt concentration at the producing well.

4   Conclusions

We investigate quantitatively the role of stochastic heterogeneity and groundwater withdrawal on seawater intrusion (SWI) in a randomly heterogeneous coastal aquifer through a suite of numerical Monte Carlo (MC) simulations of transient variable-density flow and transport in a three-dimensional domain. Our work attempts to include the effects of random heterogeneity of permeability and groundwater withdrawal within a realistic and relevant scenario. For this purpose, the numerical model has been tailored to the general hydrogeological setting of a coastal aquifer, i.e., the Argentona river basin (Spain), a region which is plagued by SWI. To account for the inherent uncertainty associated with aquifer hydrogeological properties, we conceptualize our target system as a heterogeneous medium whose permeability is a random function of space. The SWI phenomenon is studied through the analysis of (a) the general pattern of iso-concentration curves and (b) a set of seven dimensionless metrics describing the toe penetration and the extent of the mixing or transition zone. We compare results obtained across a collection of $n = 100$ MC realizations and for an equivalent homogeneous system. Our work leads to the following major conclusions.

Heterogeneity of the system affects the SW wedge along all directions both in the presence and absence of pumping. On average, our heterogeneous system is characterized by toe penetration and extent of the mixing zone that are, respec-

tively, smaller and larger than their counterparts computed in the equivalent homogeneous system.

Ensemble (i.e., across the MC realizations) mean values of linear, L_T', areal, A_T', and volumetric, V_T', metrics representing 50 % of SWI penetration virtually coincide with their counterparts evaluated in the ensemble-average concentration field.

Ensemble mean values of linear, L_S' and W_{MZ}', and areal, A_S', metrics representing the mixing and spreading of SWI penetration are markedly overestimated by their counterparts evaluated on the basis of the ensemble-average concentration field. Therefore, average concentration fields, typically estimated through interpolation of available concentration data, cannot be employed to provide reliable estimates of solute spreading and mixing.

All of the tested pumping schemes lead to an increased SW wedge volume compared to the scenario where pumping is absent. The key aspect controlling the effects of groundwater withdrawal on SWI is the position of the wellbore with respect to the location of the saltwater wedge in place prior to pumping. The toe penetration decreases or increases depending on whether the well is initially (i.e., before pumping) located within or outside the seawater intruded region, respectively. The water withdrawal scheme that is most efficient for the reduction of the maximum inland penetration of the seawater toe is the one according to which freshwater and saltwater are, respectively, extracted from the top and the bottom of the same borehole, initially located within the SW wedge. This result suggests the potential effectiveness of the so-called "negative barriers" in limiting intrusion, even when considering the uncertainty effects stemming from our incomplete knowledge of permeability spatial distributions.

Salt concentration, C_T, of water pumped from the producing well is strongly affected by permeability heterogeneity. Our MC simulations document that C_T can vary by more than 2 orders of magnitude amongst individual realizations, even in the moderately heterogeneous aquifer considered in this study. As such, relying solely on values of C_T obtained through an effective homogeneous system or on the ensemble-average estimates $\langle C_T \rangle$ does not yield reliable quantification of the actual salt concentration at the producing well.

Future development of our work includes the analysis of the influence of the degree of heterogeneity and of the functional format of the covariance structure of the permeability field on SWI, also in the presence of multiple pumping wells.

Appendix A

Here, we assess the stability of the MC-based first- and second-order (statistical) moments associated with the metrics introduced in Sect. 2.4. Figures A1 and A2, respectively, depict the sample mean, $\langle \xi'' \rangle$, and standard deviation, $\sigma(\xi'')$, of all metrics evaluated at the end of the 8-year period versus the number of MC simulations, n. The estimated 95% confidence intervals, computed according to Eqs. (3) and (8) of Ballio and Guadagnini (2004), are also depicted. Figures A1 and A2 show that the oscillations displayed by the quantities of interest are in general limited and do not hamper the strength of the main message of our work. As expected, moments of integral quantities (i.e., A_T'', A_S'', V_T'', V_S'', W_{MZ}'') tend to stabilize faster than their counterparts evaluated for local quantities (L_T'' and L_S'').

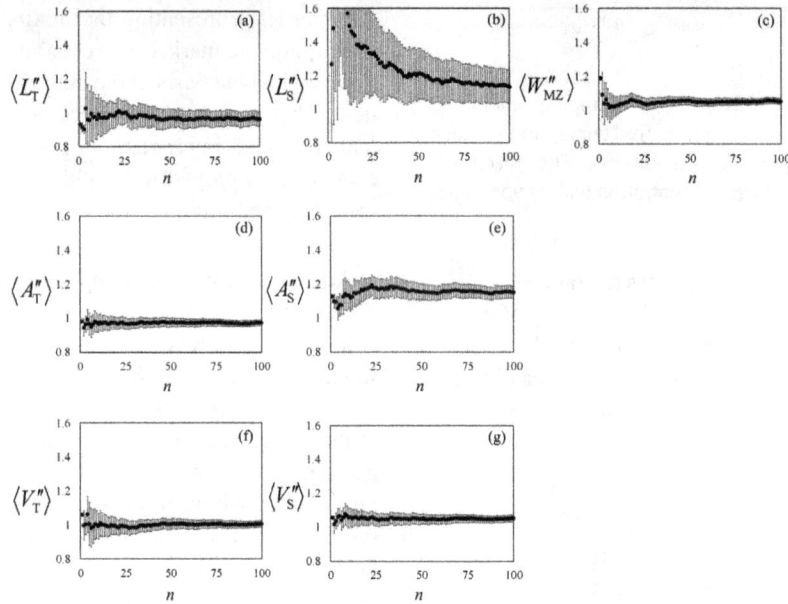

Figure A1. Sample mean of the seven dimensionless metrics at $t = t_w$ and associated 95% confidence intervals versus ensemble size, n.

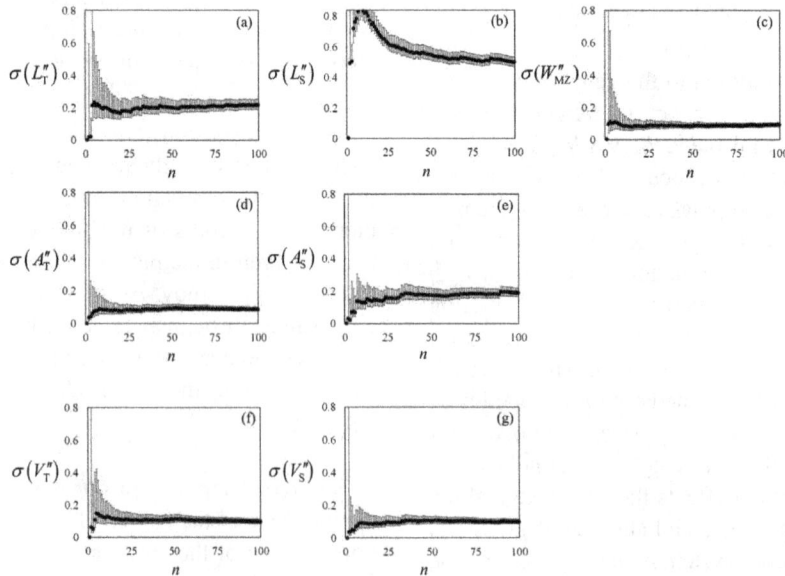

Figure A2. Sample standard deviation of the seven dimensionless metrics at $t = t_w$ and associated 95% confidence intervals versus ensemble size, n.

Competing interests. The authors declare that they have no conflict of interest.

Acknowledgements. The authors would like to thank the EU and MIUR for funding, in the frame of the collaborative international consortium (WE-NEED) financed under the ERA-NET WaterWorks2014 Cofunded Call. This ERA-NET is an integral part of the 2015 Joint Activities developed by the Water Challenges for a Changing World Joint Programme Initiative (Water JPI). We are grateful to Albert Folch, Xavier Sanchez-Vila and Laura del Val Alonso of the Universitat Politècnica de Catalunya and Jesus Carrera of the Spanish Council for Scientific Research for sharing data on the hydrogeological characterization of the Argentona site with us.

Edited by: Bill X. Hu

References

Ababou, R.: Random porous media flow on large 3D grids: numerics, performance and application to homogeneization, in: Environmental studies, mathematical, computational & statistical analysis, edited by: Wheeler, M. F., Springer, New York, 1–25, 1996.

Abarca, E., Carrera, J., Sánchez-Vila, X., and Dentz, M.: Anisotropic dispersive Henry problem, Adv. Water Resour., 30, 913–926, https://doi.org/10.1016/j.advwatres.2006.08.005, 2007.

Ackerer, P. and Younes, A.: Efficient approximations for the simulation of density driven flow in porous media, Adv. Water Resour., 31, 15–27, https://doi.org/10.1016/j.advwatres.2007.06.001, 2008.

Ackerer, P., Younes, A., and Mancip, M.: A new coupling algorithm for density-driven flow in porous media, Geophys. Res. Lett., 31, L12506, https://doi.org/10.1029/2004GL019496, 2004.

Albets-Chico, X. and Kassinos, S.: A consistent velocity approximation for variable-density flow and transport in porous media, J. Hydrol., 507, 33–51, https://doi.org/10.1016/j.jhydrol.2013.10.009, 2013.

Ataie-Ashtiani, B., Volker, R. E., and Lockington, D. A.: Tidal effects on sea water intrusion in unconfined aquifers, J. Hydrol., 216, 17–31, https://doi.org/10.1016/S0022-1694(98)00275-3, 1999.

Bakker, M.: Analytic solutions for interface flow in combined confined and semi-confined, coastal aquifers, Adv. Water Resour., 29, 417–425, https://doi.org/10.1016/j.advwatres.2005.05.009, 2006.

Ballio, F. and Guadagnini, A.: Convergence assessment of numerical Monte Carlo simulations in groundwater hydrology, Water Resour. Res., 40, W04603, https://doi.org/10.1029/2003WR002876, 2004.

Bolster, D. T., Tartakovsky, D. M., and Dentz, M.: Analytical models of contaminant transport in coastal aquifers, Adv. Water Resour., 30, 1962–1972, https://doi.org/10.1016/j.advwatres.2007.03.007, 2007.

Bruggeman, G. A.: Analytical Solutions of Geohydrological Problems, Developments in Water Science, 46, Elsevier Science, Amsterdam, 956 pp., 1999.

Cheng, A. H.-D., Halhal, D., Naji, A., and Ouazar, D.: Pumping optimization in saltwater-intruded coastal aquifers, Water Resour. Res., 36, 2155–2165, https://doi.org/10.1029/2000WR900149, 2000.

Cobaner, M., Yurtal, R., Dogan, A., and Motz, L. H.: Three dimensional simulation of seawater intrusion in coastal aquifers: A case study in the Goksu Deltaic Plain, J. Hydrol., 464–465, 262–280, https://doi.org/10.1016/j.jhydrol.2012.07.022, 2012.

Custodio, E.: Coastal aquifers of Europe: an overview, Hydrogeol. J., 18, 269–280, https://doi.org/10.1007/s10040-009-0496-1, 2010.

Dagan, G. and Zeitoun, D. G.: Seawater-freshwater interface in a stratified aquifer of random permeability distribution, J. Contam. Hydrol., 29, 185–203, https://doi.org/10.1016/S0169-7722(97)00013-2, 1998.

Dausman, A. and Langevin, C. D.: Movement of the Saltwater Interface in the Surficial Aquifer System in Response to Hydrologic Stresses and Water-Management Practices, Broward County, Florida, U.S. Geological Survey Scientific Investigations, Report 2004-5256, 73 pp., 2005.

Dentz, M. and Carrera, J.: Effective solute transport in temporally fluctuating flow through heterogeneous media, Water Resour. Res., 41, W08414, https://doi.org/10.1029/2004WR003571, 2005.

Dentz, M., Tartakovsky, D. M., Abarca, E., Guadagnini, A., Sanchez-Vila, X., and Carrera, J.: Variable density flow in porous media, J. Fluid Mech., 561, 209–235, https://doi.org/10.1017/S0022112006000668, 2006.

Deutsch, C. V. and Journel, A. G.: GSLIB: Geostatistical software library and user's guide, Oxford University Press, Oxford, UK, 348 pp., 1998.

Fahs, M., Younes, A., and Mara, T. A.: A new benchmark semi-analytical solution for density-driven flow in porous media, Adv. Water Resour., 70, 24–35, https://doi.org/10.1016/j.advwatres.2014.04.013, 2014.

Gingerich, S. B. and Voss, C. I.: Three-dimensional variable-density flow simulation of a coastal aquifer in southern Oahu, Hawaii, USA, Hydrogeol. J., 13, 436–450, https://doi.org/10.1007/s10040-004-0371-z, 2005.

Henry, H. R.: Effects of dispersion on salt encroachment in coastal aquifers, US Geological Survey Water-Supply Paper 1613-C, C71–C84, 1964.

Jakovovic, D., Werner, A. D., de Louw, P. G. B., Post, V. E. A., and Morgan, L. K.: Saltwater upconing zone of influence, Adv. Water Resour., 94, 75–86, https://doi.org/10.1016/j.advwatres.2016.05.003, 2016.

Kerrou, J. and Renard, P.: A numerical analysis of dimensionality and heterogeneity effects on advective dispersive seawater intrusion processes, Hydrogeol. J., 18, 55–72, https://doi.org/10.1007/s10040-009-0533-0, 2010.

Kerrou, J., Renard, P., Cornaton, F., and Perrochet, P.: Stochastic forecasts of seawater intrusion towards sustainable groundwater management: application to the Korba aquifer (Tunisia), Hydrogeol. J., 21, 425–440, https://doi.org/10.1007/s10040-012-0911-x, 2013.

Ketabchi, H., Mahmoodzadeh, D., Ataie-Ashtiani, B., and Simmons, C. T.: Sea-level rise impacts on seawater intrusion in coastal aquifers: Review and integration, J. Hydrol., 535, 235–255, https://doi.org/10.1016/j.jhydrol.2016.01.083, 2016.

Kolditz, O., Ratke, R., Diersch, H. J. G., and Zielke, W.: Coupled groundwater flow and transport: 1. Verification of variable density flow and transport models, Adv. Water Resour., 21, 27–46, https://doi.org/10.1016/S0309-1708(96)00034-6, 1998.

Lecca, G. and Cau, P.: Using a Monte Carlo approach to evaluate seawater intrusion in the Oristano coastal aquifer: A case study from the AQUAGRID collaborative computing platform, Phys. Chem. Earth, 34, 654–661, https://doi.org/10.1016/j.pce.2009.03.002, 2009.

Llopis-Albert, C. and Pulido-Velazquez, D.: Discussion about the validity of sharp-interface models to deal with seawater intrusion in coastal aquifers, Hydrol. Process., 28, 3642–3654, https://doi.org/10.1002/hyp.9908, 2013.

Lu, C. H., Kitanidis, P. K., and Luo, J.: Effects of kinetic mass transfer and transient flow conditions on widening mixing zones in coastal aquifers, Water Resour. Res., 45, W12402, https://doi.org/10.1029/2008WR007643, 2009.

Lu, C. H., Chen, Y. M., Zhang, C., and Luo, J.: Steady-state freshwater-seawater mixing zone in stratified coastal aquifers, J. Hydrol., 505, 24–34, https://doi.org/10.1016/j.jhydrol.2013.09.017, 2013.

Lu, C. H., Xin, P., Li, L., and Luo, J.: Seawater intrusion in response to sea-level rise in a coastal aquifer with a general-head inland boundary, J. Hydrol., 522, 135–140, https://doi.org/10.1016/j.jhydrol.2014.12.053, 2015.

Mas-Pla, J., Ghiglieri, G., and Uras, G.: Seawater intrusion and coastal groundwater resources management. Examples from two Mediterranean regions: Catalonia and Sardinia, Contrib. Sci., 10, 171–184, https://doi.org/10.2436/20.7010.01.201, 2014.

Mazi, K., Koussis, A. D., and Destouni, G.: Intensively exploited Mediterranean aquifers: resilience to seawater intrusion and proximity to critical thresholds, Hydrol. Earth Syst. Sci., 18, 1663–1677, https://doi.org/10.5194/hess-18-1663-2014, 2014.

Misut, P. E. and Voss, C. I.: Freshwater-saltwater transition zone movement during aquifer storage and recovery cycles in Brooklyn and Queens, New York City, USA, J. Hydrol., 337, 87–103, https://doi.org/10.1016/j.jhydrol.2007.01.035, 2007.

Neuman, S. P., Riva, M., and Guadagnini, A.: On the geostatistical characterization of hierarchical media, Water Resour. Res., 44, W02403, https://doi.org/10.1029/2007WR006228, 2008.

Nordbotten, J. M. and Celia, M. A.: An improved analytical solution for interface upconing around a well, Water Resour. Res., 42, W08433, https://doi.org/10.1029/2005WR004738, 2006.

Panyor, L.: Renewable energy from dilution of salt water with fresh water: pressure retarded osmosis, Desalination, 199, 408–410, https://doi.org/10.1016/j.desal.2006.03.092, 2006.

Park, N., Cui, L., and Shi, L.: Analytical design curves to maximize pumping or minimize injection in coastal aquifers, Ground Water, 47, 797–805, https://doi.org/10.1111/j.1745-6584.2009.00589.x, 2009.

Pool, M. and Carrera, J.: Dynamics of negative hydraulic barriers to prevent seawater intrusion, Hydrogeol. J., 18, 95–105, https://doi.org/10.1007/s10040-009-0516-1, 2009.

Pool, M. and Carrera, J.: A correction factor to account for mixing in Ghyben-Herzberg and critical pumping rate approximations of seawater intrusion in coastal aquifers, Water Resour. Res., 47, W05506, https://doi.org/10.1029/2010WR010256, 2011.

Pool, M., Post, V. E. A., and Simmons, C. T.: Effects of tidal fluctuations on mixing and spreading in coastal aquifers: Homogeneous case, Water Resour. Res., 50, 6910–6926, https://doi.org/10.1002/2014WR015534, 2014.

Pool, M., Post, V. E. A., and Simmons, C. T.: Effects of tidal fluctuations and spatial heterogeneity on mixing and spreading in spatially heterogeneous coastal aquifers, Water Resour. Res., 51, 1570–1585, https://doi.org/10.1002/2014WR016068, 2015.

Riva, M., Guadagnini, A., and Dell'Oca, A.: Probabilistic assessment of seawater intrusion under multiple sources of uncertainty, Adv. Water Resour., 75, 93–104, https://doi.org/10.1016/j.advwatres.2014.11.002, 2015.

Rodriguez Fernandez, H.: Actualizacion del modelo de flujo y transporte del acuifero de la riera de Argentona, MS Thesis, Universitat Politecnica de Catalunya, Spain, 2015.

Saravanan, K., Kashyap, D., and Sharma, A.: Model assisted design of scavenger well system, J. Hydrol., 510, 313–324, https://doi.org/10.1016/j.jhydrol.2013.12.031, 2014.

SIGPAC: Sistema de Información Geografica de la Plolitica Agraria Comunitaria, available at: http://sigpac.mapa.es/fega/visor/ (last access: 8 May 2018), 2005.

Soto Meca, A., Alhama, F., and Gonzalez Fernandez, C. F.: An efficient model for solving density driven groundwater flow problems based on the network simulation method, J. Hydrol., 339, 39–53, https://doi.org/10.1016/j.jhydrol.2007.03.003, 2007.

Strack, O. D. L.: A single-potential solution for regional interface problems in coastal aquifers, Water Resour. Res., 12, 1165–1174, https://doi.org/10.1029/WR012i006p01165, 1976.

Voss, C. I. and Provost, A. M.: SUTRA, A model for saturated-unsaturated variable-density ground-water flow with solute or energy transport, US Geological Survey Water-Resources Investigations Report 02-4231, 250 pp., 2002.

Werner, A. D. and Gallagher M. R.: Characterisation of sea-water intrusion in the Pioneer Valley, Australia using hydrochemistry and three-dimensional numerical modelling, Hydrogeol. J., 14, 1452–1469, https://doi.org/10.1007/s10040-006-0059-7, 2006.

Werner, A. D., Bakker, M., Post, V. E. A., Vandenbohede, A., Lu, C., Ataie-Ashtiani, B., Simmons, C. T., and Barry, D. A.: Seawater intrusion processes, investigation and management: recent advances and future challenges, Adv. Water Resour., 51, 3–26, https://doi.org/10.1016/j.advwatres.2012.03.004, 2013.

Zidane, A., Younes, A., Huggenberger, P., and Zechner, E.: The Henry semianalytical solution for saltwater intrusion with reduced dispersion, Water Resour. Res., 48, W06533, https://doi.org/10.1029/2011WR011157, 2012.

4

Groundwater origin, flow regime and geochemical evolution in arid endorheic watersheds: a case study from the Qaidam Basin, northwestern China

Yong Xiao[1], Jingli Shao[2], Shaun K. Frape[3], Yali Cui[2], Xueya Dang[4,5], Shengbin Wang[6,7], and Yonghong Ji[8]

[1]Faculty of Geosciences and Environmental Engineering, Southwest Jiaotong University, Chengdu, 611756, China
[2]School of Water Resources and Environment, China University of Geosciences (Beijing), Beijing, 100083, China
[3]Department of Earth and Environmental Sciences, University of Waterloo, Waterloo, N2L 3G1, Canada
[4]Xi'an Center of Geological Survey, China Geological Survey, Xi'an, 710054, China
[5]Key Laboratory of Groundwater and Ecology in Arid and Semi-arid Regions,
China Geological Survey, Xi'an, 710054, China
[6]Key Lab of Geo-environment of Qinghai Province, Xining, 810007, China
[7]Bureau of Qinghai Environmental Geological Prospecting, Xining, 810007, China
[8]Lunan Geo-Engineering Exploration Institute of Shandong Province, Yanzhou, 272100, China

Correspondence: Jingli Shao (jshao@cugb.edu.cn)

Abstract. Groundwater origin, flow and geochemical evolution in the Golmud River watershed of the Qaidam Basin was assessed using hydrogeochemical, isotopic and numerical approaches. The stable isotopic results show groundwater in the basin originates from precipitation and meltwater in the mountainous areas of the Tibetan Plateau. Modern water was found in the alluvial fan and shallow aquifers of the loess plain. Deep confined groundwater was recharged by paleowater during the late Pleistocene and Holocene under a cold climate. Groundwater in the low-lying depression of the central basin is composed of paleobrines migrated from the western part of the basin due to tectonic uplift in the geological past. Groundwater chemistry is controlled by mineral dissolution (halite, gypsum, anhydrite, mirabilite), silicate weathering, cation exchange, evaporation and mineral precipitation (halite, gypsum, anhydrite, aragonite, calcite, dolomite) and varies from fresh to brine with the water types evolving from $HCO_3 \cdot Cl\text{-}Ca \cdot Mg \cdot Na$ to Cl-Na, Cl-K-Na and Cl-Mg type along the flow path. Groundwater flow patterns are closely related to stratigraphic control and lithological distribution. Three hierarchical groundwater flow systems, namely local, intermediate and regional, were identified using numerical modeling. The quantity of water discharge from these three systems accounts for approximately 83 %, 14 % and 3 %, respectively, of the total groundwater quantity of the watershed. This study can enhance the understanding of groundwater origin, circulation and evolution in the Qaidam Basin as well as other arid endorheic watersheds in northwestern China and elsewhere worldwide.

1 Introduction

Closed basins in arid and semiarid areas (e.g., the Great Artesian Basin and Murray Basin in Australia, Minqin Basin and Qaidam Basin in China, Death Valley in United States) have been the focus of attention due to their water scarcity, fragile ecology and rich mineral resources related to salt lakes (Edmunds et al., 2006; Lowenstein and Risacher, 2009; Love et al., 2013, 2017; Shand et al., 2013; Stone and Edmunds, 2014; He et al., 2015; Cartwright et al., 2017; Priestley et al., 2017a; Xiao et al., 2017). Groundwater plays a vital role in water supply, ecology maintenance, transportation of chemical components, and the formation of oil, gas reservoirs and mineral resources in these basins (Toth, 1980; Jiang et al., 2014; Jiao et al., 2015; Xiao et al., 2017). Understanding the regimes of groundwater recharge, flow and hydrogeochemical evolution is essential to maintain proper management and implement sustainable utilization of groundwater and mineral resources, as well as maintain the ecological environment (Cartwright et al., 2010a; Herrera et al., 2016).

In the arid northwest of China there are many closed basins such as the Tarim, Qaidam, Junggar and Minqin basins, in which the low-lying discharge areas are occupied by saline lakes, salt playas and salt crusts. The Qaidam Basin (Fig. 1a, b), the largest closed basin of the Tibetan Plateau, has the most plentiful number of salt lakes and salt playas and almost all varieties of salt deposits (Zheng et al., 1993), as well as rich oil and gas reservoirs (Tan et al., 2011; Ye et al., 2014). Considerable research has been conducted to provide support for water supply and mineral resource exploitation in the basin (Chen and Bowler, 1986; Vengosh et al., 1995; Lowenstein and Risacher, 2009; Li et al., 2010; Tan et al., 2011; Hou et al., 2014; Ye et al., 2014; Chen et al., 2017). However, most of the previous studies focused on the groundwater in the piedmont areas (Wang and Ren, 1996; Wang et al., 2010; Zhang, 2013; Hou et al., 2014; Su et al., 2015; Xu et al., 2017), material source of salt lakes (Vengosh et al., 1995; Lowenstein and Risacher, 2009; Tan et al., 2011; Chen et al., 2015) and the evolution of salt lakes (Chen and Bowler, 1986; Chen et al., 2017). The systematic understanding of regional groundwater regimes is still inadequate. This would limit the comprehensive planning and management of groundwater and salt lake mineral resource exploitation, and finally make it difficult to safeguard the circulation of the groundwater system and maintain the eco-environmental balance. Therefore, several attempts have been made to understand the regional groundwater regimes (Tan et al., 2009; Gu et al., 2017; Xiao et al., 2017), but very little research reported the circulation and evolution of groundwater from the mountain pass area to the central terminal lake area due to the notable difficulties in moving through and access the swamps on the lacustrine plain. This would greatly limit the full understanding of the role of hydrogeological processes in the basin.

Hydrogeological survey efforts have been undertaken in the Golmud River watershed of the basin since 2015 and have developed a better understanding of regional hydrogeological conditions. The main objective of this study is to assess the regional hydrogeological regime of closed basins in the arid northwest of China, using the Golmud River watershed as a case study. To achieve this aim, a comprehensive approach using environmental isotopes (^{2}H, ^{18}O, ^{3}H, ^{13}C, ^{14}C) and hydrochemistry coupled with numerical simulation was performed. Stable hydrogen and oxygen can provide valuable information on the origin and recharge environment of groundwater, and radioactive isotopes such as ^{3}H and ^{14}C record the residence time of groundwater (Cartwright et al., 2007; Awaleh et al., 2017; Huang et al., 2017). Hydrochemical composition has recorded the recharge water characteristics, hydrostratigraphic information, geochemical interaction and other processes along the groundwater flow path (Redwan and Moneim, 2015; Verma et al., 2016; Love et

al., 2017) and thus can be used to track groundwater evolution. Numerical simulation of groundwater flow is an essential tool to synthesize hydrogeological information and reveal groundwater flow patterns (Bredehoeft and Konikow, 2012; Anderson et al., 2015; Tóth et al., 2016). The combination of these approaches is robust to reveal groundwater origin, flow regimes, renewability, hydrochemical evolution and interaquifer mixing, as well as surface water and groundwater interactions, etc., in basins with complex hydrogeology or sparse monitoring data, and has been successfully applied in many basins such as the Great Artesian Basin and Murray Basin in Australia, Michigan Basin in US, Minqin Basin and Ordos Plateau in China, Stampriet Basin in Africa (Edmunds et al., 2006; Banks et al., 2010; Love et al., 2013, 2017; Stone and Edmunds, 2014; Su et al., 2016; Cartwright and Morgenstern, 2017; Petts et al., 2017; Priestley et al., 2017b).

The specific aims of the present study are to: (1) identify the recharge source of groundwater, (2) assess the regional groundwater chemistry characteristics, (3) determine the controlling mechanisms of hydrogeochemistry, (4) delineate regional groundwater flow patterns and (5) ultimately establish systematic regional groundwater regimes from the mountain pass to the terminal lake in the typical Golmud watershed of Qaidam Basin. This study would provide insights into the origin, recharge environment, flow regime and geochemical evolution of regional groundwater in arid endorheic watersheds of Qaidam Basin and provide reference for other arid closed basins in northwestern China as well as similar endorheic watersheds worldwide.

2 Study area

The Qaidam Basin is a large closed basin located on the northeastern margin of the Tibetan plateau, surrounded by the Qilian Mountains to the north, the Kunlun Mountains to the south and the Altun Mountains to the west (Fig. 1b). The study area, Golmud River watershed (GRW), is located in the southern part of the Qaidam Basin hosting the second largest river, the Golmud River, running from the Kunlun Mountains in the south to the low-lying depression in the north central part of the Basin (Fig. 1c). The Qarhan salt lake is the largest salt lake in China, located at the northern margin of the GRW, adjacent to the Golmud River's terminal Lake Dabusun. The third largest city on the Tibetan plateau, Golmud City, is also located in the GRW.

The outcropping stratigraphy of the GRW ranges from Proterozoic to Quaternary in age. The Quaternary strata are found in the mountainous areas to the south. These strata have undergone magmatic activity, uplift and tectonic movements, as well as intense weathering, resulting in massive material sources of sediments to the basin. The Quaternary deposits have thicknesses ranging from hundreds of meters in the piedmont area to thousands of meters in the low-lying depression (basin center) (Zheng et al., 1993; Chen et al.,

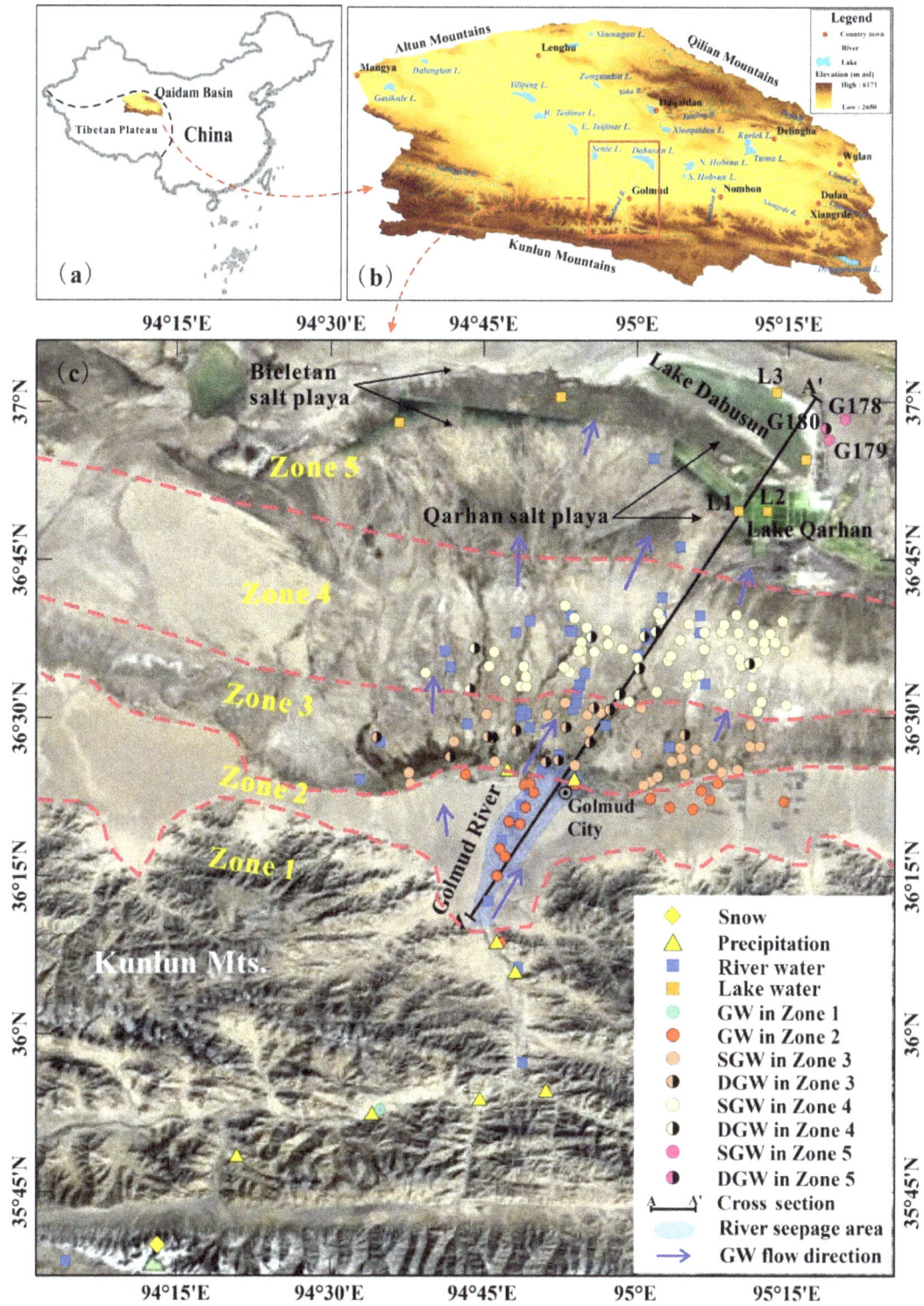

Figure 1. Location of the study area (**a**) within China, (**b**) within the Qaidam Basin, and (**c**) details of sampling location and groundwater and physiographic zones within the study area. Please note that the above figure contains disputed territories.

2017). Field surveys found that salt crusts are formed on the ground surface in locations near the terminal areas of streams. Core drilling records also show many salt-bearing deposits such as halite, calcium, sulfate and sodium sulfate were observed throughout the strata (Chen and Bowler, 1986). The regional Quaternary aquifers in the basin vary from single unconfined gravel and sand layers with hydraulic conductivity (K) greater than $50\,m\,d^{-1}$ in the alluvial fan to multilayers of silt and clay with hydraulic conductivity (K) ranging from 0.1 to $0.001\,m\,d^{-1}$ in the low-lying depression (basin center). Three continuous aquitards (clay layers) are found in the basin at depths of 60, 290 and 450 m (Fig. 8), which have significant influences on confining groundwater flow (Shao et al., 2017).

The climate in the GRW is extremely variable, both spatially and temporally. Precipitation in the Kunlun Mountains is more than 200 mm per year, but less than 50 mm in the basin, and also presents a gradual decreasing trend from the piedmont area to the low-lying central depression. The potential evaporation is extremely high (>2600 mm per year). This hyper-arid climate results in aquifers in the basin that do not obtain effective recharge from the local precipitation. Groundwater in the basin is mainly recharged by Golmud River seepage through the riverbed in the alluvial fan and bedrock lateral inflow at the southern mountain front, and flows from the alluvial fan in the south to the basin center in the north (Fig. 1c). Much of the groundwater overflows as springs at the front of the alluvial fan due to the fining of sediments in the aquifer's downdip. The depth to groundwater is less than 3 m in most areas from the front of alluvial fan to the basin center, resulting in significant potentially evaporate loss of groundwater. The regional groundwater finally discharges to the terminal lake and undergoes large evaporate loss (Shao et al., 2017).

Based on the terrain, sediments and hydrogeological condition, the study area can be divided into 5 zones. Zone 1 is the Kunlun mountainous area, and Zone 2 is the alluvial fan plain at the Kunlun piedmont. Zones 3, 4 and 5 occur on the loess plain, where Zone 3 is the main groundwater overflow (discharge) zone of the watershed, and Zone 5 is the terminal lake zone (low-lying depression of the watershed) with salt crusts and playa. Zone 4 is the transition zone (middle to lower stream area of the watershed) between Zone 3 and 5.

3 Materials and methods

3.1 Hydrochemical and isotopic sampling and analytical methods

A total of 228 water samples were collected from GRW in 2015 and 2016, including 180 groundwater samples and 48 surface water samples (42 river water and 6 lake water samples) (Fig. 1c). Groundwater samples were collected from both shallow phreatic aquifers and deep confined aquifers.

Surface water samples were obtained along the Golmud River, as well as from Lake Qarhan, Lake Dabusun and other small lakes in the low-lying depression area (basin center). In addition, 1 snow (snowmelt water) sample, 8 precipitation samples and 90 brine water samples (groundwater) with hydrogen and oxygen stable isotope data were obtained from China's stable isotope geochemistry database (http://210.73.59.163/isogeochem/, last access: 31 May 2018). The location of the snow and precipitation samples are shown in Fig. 1c. The detailed locations of the 90 brine water samples are not known, but it is known that all of these samples were collected from the Qarhan salt playa and Bieletan salt playa (Fig. 1c).

For groundwater sampling, all wells and boreholes, except those that were artesian, were pumped for several well volumes to remove the stagnant water in the wells and boreholes and monitored until the electrical conductivity (EC) of the pumping water was stable. The sampling procedure followed is described in Huang et al. (2016) and Chen et al. (2011). Samples for major element analysis were collected in two 250 mL high-density polyethylene bottles after filtration using 0.45 μm filter membranes (Huang et al., 2016). Samples for tritium (^{3}H) and stable isotope (^{2}H, ^{18}O) analysis were collected in 500 and 50 mL glass bottles, respectively, that were filled to overflowing after rinsing and were sealed tightly. ^{13}C and ^{14}C samples were collected by adding $BaCl_2$ and CO_2-free NaOH to 120 L groundwater at pH = 12, obtaining $BaCO_3$ for dissolved inorganic carbon (DIC) analysis (Chen et al., 2011). The method used eliminates contact with the atmosphere in order to avoid CO_2 atmospheric contamination.

Parameters such as the water temperature (T), pH and EC were measured in the field with an in situ multiparameter instrument (Multi 350i/SET, Munich, Germany), and redox potential (Eh) was also determined in situ using a portable ORP tester (CLEAN ORP30 Tester, California, US). Major chemistry and isotopes (^{2}H, ^{18}O, ^{3}H, ^{13}C and ^{14}C) of the sampled water were analyzed at the Laboratory of Groundwater Sciences and Engineering in the Institute of Hydrogeology and Environmental Geology, Chinese Academy of Geological Sciences (Shijiazhuang, Hebei Province, China). Major cations (K^+, Na^+, Ca^{2+}, Mg^{2+}) were measured by inductively coupled plasma–mass spectrometry (Agilent 7500ce ICP-MS, Tokyo, Japan). Total dissolved solids (TDSs) and HCO_3^- were determined by gravimetric analysis and acid-base titration, respectively. Cl^- and SO_4^{2-} were analyzed using spectrophotometry (PerkinElmer Lambda 35, Waltham, MA, USA). The ionic charge balance of all samples were within 5 % difference. $\delta^{18}O$, $\delta^{2}H$ and $\delta^{13}C$ were measured by isotope ratio mass spectrometry using a Finnigan MAT 253, and $\delta^{18}O$, $\delta^{2}H$ were reported relative to the Vienna Standard Mean Ocean Water (VSMOW) standard, and $\delta^{13}C$ was reported relative to the Vienna Pee Dee Belemnite (VPDB). The analytical errors are ±0.2 ‰ for $\delta^{18}O$, ±1.0 ‰ for $\delta^{2}H$

and $\pm0.5\%$ for $\delta^{13}C$. The tritium content was determined using electrolytic enrichment and a liquid scintillation technique (Chen et al., 2011) with the precision of ±0.3 TU. The activity of ^{14}C was analyzed by liquid scintillation counting (1220 Quantulus), and expressed as a percentage of modern carbon (pMC) with the precision of ±0.3 % (Su et al., 2018).

3.2　Two-dimensional groundwater flow numerical simulation

It is assumed that the variation of density and viscosity of waters could be neglected for calculations involving most of the flow system (Zone 1–4). For simplicity, groundwater in the terminal lake zone (Zone 5) is also regarded as being mainly driven by gravity. Thus the equation governing variably saturated groundwater flow is as follows (Richards, 1931):

$$\frac{\partial}{\partial t}\phi S = \text{div}[K\nabla h],$$

where ϕ is the porosity, S is the liquid saturation, K is the hydraulic conductivity (m d^{-1}), and h is the hydraulic head (m). The TOUGH2 code (Transport Of Unsaturated Groundwater and Heat), which has quite robust simulation capabilities, is used to numerically solve this equation (Pruess et al., 1999).

The cross section parallel to the main direction of groundwater flow in GRW (Figs. 1 and 8) was selected for the two-dimensional flow simulation. This section starts at the mountain pass and ends at the terminal Lake Dabusun, with an approximate length of 100 km. Boundaries were specified according to the hydrogeology condition. The southern lateral boundary and top boundary in the alluvial fan are defined as given flux boundaries, and the bottom boundary of the section is regarded as a zero flux boundary. The springs and evaporation are set as mixed boundaries. The lake boundary in the basin center is specified as a given head boundary.

An irregular discretization was conducted vertically to capture the variation of the water table near the ground surface and also implement an efficient simulation. Cells are presented with the minimum thickness of 0.1 m near the ground surface and gradual increasing thickness downward, with a maximum thickness of about 80 m. Equal discretization was applied in the horizontal direction with a horizontal size of 1000 m for one cell. The initial permeability of various lithologies is specified based on the borehole drilling records and pumping test results, with the K_h (horizontal hydraulic conductivity) in the range of 10^2–10^{-3} m d^{-1} and anisotropy ratio $K_h/K_v = 5$–10 (K_v is vertical hydraulic conductivity) (Shao et al., 2017). In this study, the model is used to present the flow pattern under equilibrium conditions, and thus the recharge rates and hydraulic heads are given according to the annual average values. Evaporation was modeled using a newly developed method described by Hao et al. (2016), and the initial potential evaporation specified is 2600 mm per year. Springs are simulated using the DELV

module in TOUGH2, and the productivity index (PI) specified in the DELV module is calculated using the following equation (Pruess et al., 1999):

$$\text{PI} = \frac{2\pi(k\,\Delta z)}{\ln(\sqrt{A/\pi}/r)+s-1/2},$$

where Δz is the layer thickness (m), A is the grid block area (m^2), r is the spring radius (m), and s is the skin factor. Annual average observed hydraulic heads are used as natural constraints for the model calibration.

4　Results

4.1　Hydrochemistry of surface waters and groundwaters

The statistical summaries of chemical analysis results for surface water and groundwater are presented in Table 1. River waters (RW) from the mountain pass (Zone 2) to the low-lying depression (Zone 5) are slightly alkaline, with a range in pH values from 7.94 to 9.45. Fresh lake water (FLW) L1, which was sampled from the freshwater lake (relatively fresh compared to other salt lakes) recharged directly by river water in the low-lying depression (Zone 5), is also slightly alkaline, with a pH value of 8.98. Samples from the salt lakes such as Lake Qarhan and Lake Dabusun are slightly acidic, with values range from 6.03 to 6.28. Groundwater in the study area is neutral to slightly alkaline. The shallow phreatic groundwater (SGW) shows an evolving trend from slightly alkaline to slightly acidic along the flow path with the pH varying from 9.34 to 6.03. However, deep confined groundwater (DGW) samples are all slightly alkaline, with pH values ranging between 7.83 and 8.69. The redox potential of SGW is in the range of 123–162 mV from alluvial fan to middle to lower stream areas (Zone 2, 3 and 4), suggesting an oxidation condition. The Eh values of DGW vary from 153 to 40 mV along the flow path (Zone 3 to 4), indicating the redox condition gradually evolves from a state of oxidation to reduction (Fig. 3e).

Surface water and groundwater present distinct major solute chemistry across the study area. As shown in Table 1, the concentration of ions in RW demonstrates an increase along river flow paths, with TDS values varying from 393 to 2319 mg L^{-1}. The TDS value of FLW (L1) is much higher than that of RW in the low-lying depression (Zone 5), with the TDS value of 10 937 mg L^{-1}, while the salt lake waters (SLWs) have extremely high TDS values ranging from 339 098 to 403 758 mg L^{-1}. The dominant ions of RW are HCO_3^- and Na^+ with the concentration range of 184–215 mg L^{-1} for HCO_3^- and 63–92 mg L^{-1} for Na^+, respectively, in the alluvial fan area (Zone 2). These dominant ions gradually evolve to Cl^- and Na^+ with the concentration range of 655–1 776 mg L^{-1} for Cl^- and 438–996 mg L^{-1} for

Table 1. Statistical summary of physical and chemical parameters of the surface water and groundwater in the Golmud Watershed, Qaidam Basin, China.

Place	Source		pH	TDS mg L^{-1}	Ca mg L^{-1}	Mg mg L^{-1}	Na mg L^{-1}	K mg L^{-1}	Cl mg L^{-1}	HCO$_3$ mg L^{-1}	SO$_4$ mg L^{-1}
Zone 2	RW	Min.	8.03	393	30.9	26.3	63.0	3.7	87	184.0	68.5
		Max.	8.41	523	38.9	36.4	92.1	5.3	144	214.6	100.5
		Mean	8.28	462	35.8	32.1	80.2	4.5	114	198.1	85.4
	SGW	Min.	7.62	236	14.8	18.9	78.5	2.0	90.3	89.1	28.8
		Max.	8.83	1171	86.2	80.0	232.2	10.7	436.7	309.0	235.0
		Mean	8.07	618	49.0	36.4	131.0	6.9	173.7	221.8	116.0
Zone 3	RW	Min.	8.25	368	37.7	24.4	56.1	3.5	77.4	178.1	68.7
		Max.	8.55	1266	64.2	83.9	276.0	13.3	382.0	335.0	256.0
		Mean	8.43	670	49.1	42.8	139.8	6.8	177.5	232.8	133.9
	SGW	Min.	7.42	443	20.6	16.9	98.0	2.0	83.1	132.0	116.2
		Max.	9.32	12 116	359.0	474	3385.0	187.0	4316	1018	3059
		Mean	8.09	1853	81.9	81	476.9	22.7	554.9	293.0	501.9
	DGW	Min.	7.89	404	30.0	18.6	65.9	4.1	84.0	204.8	41.4
		Max.	8.64	676	52.4	37.7	162.0	9.7	145.0	310.0	218.0
		Mean	8.19	547	41.3	29.9	95.3	6.2	93.7	248.2	88.3
Zone 4	RW	Min.	7.94	616	39.7	38.7	104.5	5.5	152.0	246.0	97.5
		Max.	8.64	1833	82.6	116.9	432.2	18.7	580.7	448.0	382.3
		Mean	8.29	1013	55.9	57.7	232.7	9.6	289.3	303.1	191.1
	SGW	Min.	7.08	528	10.4	15.5	102.0	5.2	83.1	48.4	97.9
		Max.	9.34	185 006	2048	4058	56 200	1709	67 063	1179	82 202
		Mean	8.01	32 029	322.7	635	10 464	248	11 550	375	8627
	DGW	Min.	7.83	514	6.7	8.0	68.7	4.0	84.0	39.9	63.9
		Max.	8.69	7184	456.0	48.6	2048	39.8	1292	445	3648
		Mean	8.25	1401	59.0	24.8	406.8	8.4	263.1	269.3	483.2
Zone 5	RW	Min.	8.73	1741	56.1	92.8	437.9	12.1	655.2	159.7	208.1
		Max.	9.45	3268	81.0	118.5	996.3	18.9	1776	448.3	396.8
		Mean	9.09	2319	70.7	106.3	628.3	15.3	1061	268.1	289.9
	FLW	Representative	8.98	10 937	113.7	696.3	2957	231.8	5912	314.9	660.7
	SLW	Min.	6.03	399 098	116.2	99 500	4137	3168	276 849	1941	7717
		Max.	6.28	403 758	177.2	100 240	4740	4122	285 780	3118	10 894
		Mean	6.16	401 428	146.7	99 870	4439	3645	281 315	2530	9306
	SGW	Min.	6.03	336 229	1541	53 480	12 388	11 798	215 561	506.0	984.5
		Max.	8.56	361 200	5871	64 860	35 712	22 351	239 340	874.3	8313
		Mean	7.30	348 715	3706	59 170	24 050	17 075	227 451	690	4649
	DGW	Representative	8.64	370 940	1927	57 079	32 378	19 372	222 404		7851

RW: river water; SGW: shallow phreatic groundwater; DGW: deep confined groundwater; FLW: relatively fresh lake water; SLW: salt lake water.

Na$^+$ in the low-lying depression (Zone 5). FLW (L1) has the same dominant ions with RW in the low-lying depression (Zone 5), but with a higher concentration of 5912 mg L^{-1} for Cl$^-$ and 2957 mg L^{-1} for Na$^+$. SLW is dominated by Cl$^-$ and Mg^{2+} with the concentration range of 276 849 to 285 780 mg L^{-1} for Cl$^-$ and 99 500 to 100 240 mg L^{-1} for Mg^{2+}, respectively. Overall, the surface water types evolve from HCO$_3$ · Cl-Ca · Mg · Na type in the alluvial fan area

(Zone 2) to Cl-Na, Cl-K-Na and Cl-Mg types in the low-lying central depression (Zone 5) (Fig. 2a).

Groundwater shows a similar hydrochemical evolution along the flow path. The average TDS values vary from 618 to 32 029 mg L^{-1} for SGW and from 547 to 1401 mg L^{-1} for DGW from the upstream area (Zone 2) to the middle to lower stream area (Zone 4). DGW is much fresher when contrasted with the SGW at the same location (Fig. 2c). There is essentially no difference in TDS between SGW and DGW from

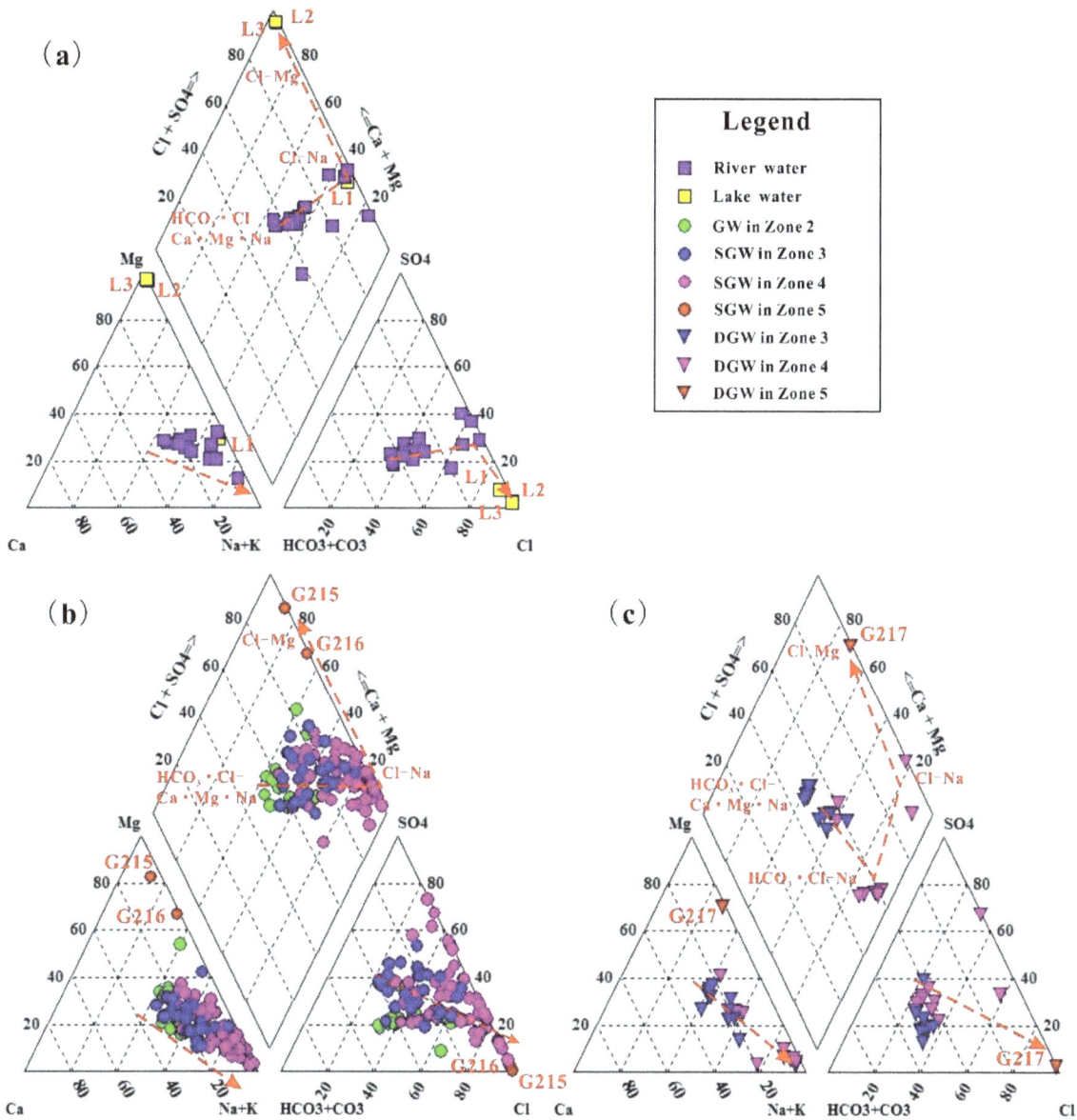

Figure 2. Piper diagrams of sampled surface water and groundwater. **(a)** Surface waters; **(b)** shallow phreatic groundwaters; **(c)** deep confined groundwaters from the Qaidam Basin, China (Red dashed lines and arrows indicate the direction of evolutionary flow systems).

the central depression (Zone 5), with the values ranging from 336 229 to 361 200 mg L^{-1} for SGW and 370 940 mg L^{-1} for representative DGW (Table 1). Groundwater in the alluvial fan area (Zone 2) is dominated by HCO$_3^-$, Cl$^-$ and Na$^+$ with the concentrations ranging from 89 to 309 mg L^{-1} for HCO$_3^-$, from 90 to 437 mg L^{-1} for Cl$^-$ and from 79 to 232 mg L^{-1} for Na$^+$, respectively. To the middle to lower stream area (Zone 4), the dominant ions vary to Cl$^-$ and Na$^+$ for both SGW and DGW. The mean concentration of Cl$^-$ is 11 550 mg L^{-1} for SGW and 263 mg L^{-1} for DGW, and the average concentration of Na$^+$ is 10 464 mg L^{-1} for SGW and 407 mg L^{-1} for DGW. All groundwaters in-

cluding SGW and DGW in the basin center (Zone 5) are dominated by Cl$^-$, Na$^+$ and Mg^{2+}. SGW has the concentration ranging from 215 561 to 227 451 mg L^{-1} for Cl$^-$, from 12 388 to 35 713 mg L^{-1} for Na$^+$ and from 53 480 to 64 860 mg L^{-1} for Mg^{2+}. The concentration of representative DGW is 222 404 mg L^{-1} for Cl$^-$, 32 378 mg L^{-1} for Na$^+$ and 57 079 mg L^{-1} for Mg^{2+}. Overall, the water types of both SGW and DGW evolve from HCO$_3$ · Cl-Ca · Mg · Na type in the upstream area (Zone 2) to Cl-Na type in the middle to lower stream area (Zone 4), and eventually to Cl-Mg type in the low-lying depression (Zone 5) (Fig. 2b).

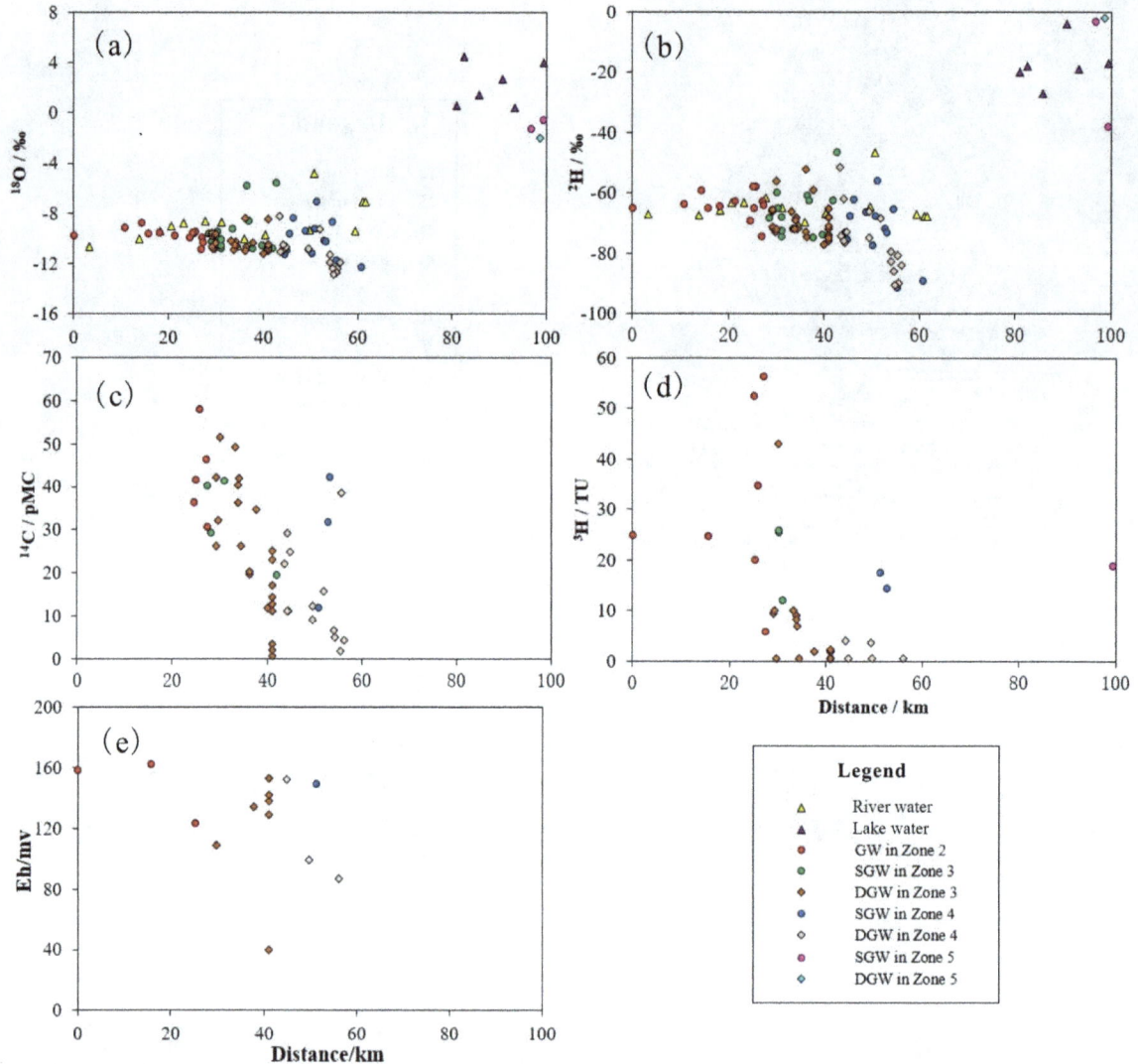

Figure 3. Isotopic data and Eh vs. distance from the mountain pass along the groundwater flow paths. **(a)** ^{18}O vs. distance, **(b)** ^{2}H vs. distance, **(c)** ^{14}C vs. distance, **(d)** ^{3}H vs. distance, **(e)** Eh vs. distance.

4.2   Stable and radio isotopes

The statistical summary of isotope results for precipitation, river water, lake water and groundwater can be found in Fig. 3 and Table 2. The representative snowmelt water in the Kunlun mountainous area (Zone 1) has a δD value of $-77.0\,‰$ and δ^{18}O value of $-11.9\,‰$. The δD and δ^{18}O values of precipitation in the mountainous area (Zone 1) are in the range of $-85.3\,‰$ to $-71.6\,‰$ and $-10.9\,‰$ to $-9.3\,‰$, with an average value of $-75.2\,‰$ and $-10.0\,‰$, respectively. The δD values of precipitation in the alluvial fan (Zone 2) range from $-68.1\,‰$ to $-66.2\,‰$, with the average value of $-67.2\,‰$ and δ^{18}O values ranging from $-10.1\,‰$ to $-9.7\,‰$ with an average value of $-9.9\,‰$. The enrichment of stable hydrogen and oxygen isotopes in precipitation from the mountainous area to the basin reflects the secondary

evaporation effect of precipitation in arid inland areas (Clark and Fritz, 1997). The δD and δ^{18}O values of river water vary from $-75.7\,‰$ to $-46.7\,‰$ and between $-11.1\,‰$ and $-4.8\,‰$, respectively, showing a gradual enrichment trend along the river flow path. Fresh and salt lake waters are all significantly enriched in heavy isotopes with values ranging from $-27.0\,‰$ to $-4.0\,‰$ for δD and from $0.4\,‰$ to $4.5\,‰$ for δ^{18}O. As shown in Fig. 3a and b, the SGW from the alluvial fan (Zone 2) to the middle to lower stream area (Zone 4) show an overall gradual enrichment trend along the flow path. In contrast, the DGW shows a significant depletion trend from the south to the north. Groundwater at different depths in the low-lying depression (Zone 5) are all brines with δD values ranging from $-66.0\,‰$ to $-2.0\,‰$ and the δ^{18}O values ranging between $-10.8\,‰$ and $-0.6\,‰$, demon-

Table 2. Statistical summary of isotopic analysis results of precipitation, surface water and groundwater in the Golmud Watershed, Qaidam Basin, China.

Place	Source	δD ‰ VSMOW			$\delta^{18}O$ ‰ VSMOW			^3H TU			^{14}C pMC			$\delta^{13}C$ ‰		
		Min.	Max.	Mean	Min.	Max.	Mean	Min.	Max.	Mean	Min.	Max.	Mean	Min.	Max.	Mean
Zone 1	SNW			−77.0			−11.9									
	PW	−85.3	−71.6	−75.2	−10.9	−9.3	−10.0									
	RW	−75.4	−64.8	−68.7	−11.1	−9.3	−10.1									
	GW			−65.0[a]			−9.7[a]									
Zone 2	PW	−68.1	−66.2	−67.2	−10.1	−9.7	−9.9									
	RW	−67.5	−63.2	−65.4	−10.7	−8.8	−9.6									
	SGW	−74.7	−58.0	−64.5	−10.9	−8.8	−9.8	20.0	56.3	35.5	30.6	57.9	42.5			
Zone 3	RW	−70.	−46.7	−63.6	−10.6	−4.8	−8.8									
	SGW	−74.8	−43.4	−63.5	−10.8	−4.3	−8.8	12.1	25.7	21.1	19.5	41.2	32.5			
	DGW	−82.9	−52.1	−71.6	−11.2	−8.4	−10.3	<1	10.1		0.7	49.2	23.4	−4.6	−1.7	−2.8
Zone 4	RW	−67.7	−67.2	−67.5	−9.4	−7.1	−8.2									
	SGW	−75.6	−56.0	−68.0	−11.3	−7.1	−9.6	14.4	17.5	16.0	11.9	42.2	27.3			
	DGW	−91.3	−51.4	−77.1	−12.8	−8.3	11.2	<1	4.1		1.9	38.6	14.8	−5.5	−3.9	4.8
Zone 5	RW			−51.3[a]			−6.6[a]									
	FLW			−20.0[a]			0.6[a]									
	SLW	−27.0	−4.0	−17.0	0.4	4.5	2.6									
	GW	−66.0	−2.0	−46.3	−10.8	−0.6	−8.2			18.9[a, b]						

SNW: snowmelt water; PW: precipitation water; RW: river water; SGW: shallow phreatic groundwater; DGW: deep confined groundwater; FLW: relatively fresh lake water; SLW: salt lake water. [a] Only one representative sample data. [b] Shallow phreatic water sample data.

strating relative enriched characteristics in contrast with the fresher groundwater in the upstream areas.

The ^3H values range from 56.3 to 12.1 TU in the SGW and from 25.7 to <1 TU in the DGW along the groundwater flow path (Fig. 3d). The ^{14}C activities in SGW vary from 57.9 to 11.9 pMC, and in DGW range from 49.15 to 0.7 pMC along the flow path (Fig. 3c). The spatial distributions of ^3H and ^{14}C results indicate increasing residence times for groundwaters in the aquifers from the south to north. While one shallow phreatic groundwater sample (G178) in the low-lying depression (Zone 5) was observed with relative high ^3H content (18.9 TU), this may be caused by the rapid infiltration of surface water in the flood period.

Groundwater in the alluvial fan (Zone 2) has a high tritium content ranging from 20.0 to 56.3 TU with the average value of 35.5 TU, indicating recharge by modern water less than 60 years old. Shallow groundwater in the overflow zone (Zone 3) and middle to lower stream area (Zone 4) also has a relative high tritium content in the range of 12.1–25.7 TU with an average value of 21.1 TU for Zone 3 and 14.4–17.5 TU with an average value of 16.0 TU for Zone 4, and the representative shallow phreatic water adjacent to the salt lake (Zone 5) also shows a high tritium content of 18.9 TU, presenting modern water isotopic signatures. This may be caused by the mixture with the infiltrating modern surface water. DGW in the overflow zone (Zone 3) and middle to lower stream area (Zone 4) is with tritium content ranging from <1 to 10.1 TU for Zone 3 and <1 to 4.1 TU for Zone 4. The elevated tritium determined from several deep

confined waters is most likely caused by mixtures with shallow phreatic water in the borehole; therefore, they cannot be used for groundwater age determination. Most DGW samples have a tritium content less than 1 TU, indicating they are not influenced by mixing with shallow groundwater in the boreholes. The age of this tritium-free DGW can be estimated using the radiocarbon activity.

Radiocarbon activity of groundwater can be significantly influenced by geochemical reactions (e.g., carbon minerals dissolution, isotopic exchange processes) during subsurface infiltration and in the aquifers (Cartwright et al., 2010b). It is therefore essential to correct the ^{14}C activity on the total dissolved inorganic carbon (TDIC) before using it for groundwater age estimation. Many models, such as statistical models, geochemical models, and mixing models, were proposed for ^{14}C activity correction. Most of the models are of limited interest due to the assumptions of fully closed systems or open systems, simplification, or even total ignorance of geochemical reactions beyond the recharge area. Using a model based on Carbon-13 is a good approach to correct the influence of geochemical reactions on ^{14}C activity of TDIC, and suitable for both open and closed systems. The measured apparent ^{14}C activity ($^{14}C_{uncorr}$) of TDIC was corrected using $\delta^{13}C$ as follows (Clark and Fritz, 1997):

$$^{14}C_{corr} = {}^{14}C_{uncorr} \frac{\delta^{13}C_{rech} - \delta^{13}C_{carb}}{\delta^{13}C_{TDIC} - \delta^{13}C_{carb}},$$

where $^{14}C_{corr}$ is the corrected ^{14}C activity of TDIC, $\delta^{13}C_{TDIC}$ is the measured $\delta^{13}C$ ratio of TDIC, $\delta^{13}C_{rech}$ is the assumed initial $\delta^{13}C$ ratio, and $\delta^{13}C_{carb}$ is the $\delta^{13}C$ ratio of carbonate being dissolved.

Groundwater in the study area is mainly recharged by Golmud River seepage in the upper alluvial fan located near parts of the Gobi desert where there is a lack of vegetation. The ^{14}C activity and $\delta^{13}C$ ratio of TDIC of the water would not be changed when infiltrating though the unsaturated zone. Thus, the $\delta^{13}C_{rech}$ ratio should be equal or close to the atmospheric value ($-6.4\,‰$). $\delta^{13}C_{carb}$ is close to $0\,‰$ (Clark and Fritz, 1997). Only some of the tritium-free DGW samples in Zone 3 and 4 have measured $\delta^{13}C$ data, and these were selected to calculate groundwater age using the aforementioned $\delta^{13}C$ correction approach. The age of DGW in Zone 3 and 4 ranges from 2264 to 20 754 years along the flow paths. Due to the absence of radiocarbon data, the age of paleogroundwater in Zone 5 cannot be calculated, but it is certain that the age is more than 20 000 years, which was deduced from the oldest age of groundwater in Zone 4 (20 754 years).

4.3 Two-dimensional groundwater flow modeling

The groundwater flow model was calibrated using annual average hydraulic heads from 63 different shallow wells measured in 2015 along the cross section (not shown in figure). The calibration shows a good match between simulated and observed hydraulic heads, as demonstrated in Fig. 4. The comparison results show that the fit to observed hydraulic heads is better in the loess plain (including Zone 3, 4, 5) with the maximum deviation less than 0.8 m, while relative poor in the alluvial fan (Zone 2) with the maximum deviation less than 5 m. The deviation in the loess plain is mainly caused by the heterogeneity and anisotropy in lithology (Gu et al., 2017). The relative large deviation in results within the alluvial fan is most likely attributed to the steep hydraulic gradient (Islam et al., 2017) and larger seasonal fluctuation of hydraulic heads. Over the whole study area, the RMSE (root mean squared error) is only 1.57 m; therefore, the calibrated model can be used to reveal the groundwater flow pattern.

The estimated hydraulic parameters are shown in Table 3. The estimated values of K_h are 56.3 m d^{-1} for gravel sand, 13.7 m d^{-1} for sand, 0.62 m d^{-1} for sandy silt, 0.13 m d^{-1} for silt and 0.001 m d^{-1} for clay. The anisotropy ratio of K_h / K_v was estimated as 10 for gravel sand and sand, and 5 for sandy silt, silt and clay. These parameters are effective values under the assumption of homogeneity in each layer. The water budget analysis indicates a dynamic equilibrium state with the equilibrium difference of 0.62 %. Springs are the dominant discharge form, followed by evaporation and lake discharge, accounting for 76.81 %, 22.44 % and 1.37 %, respectively.

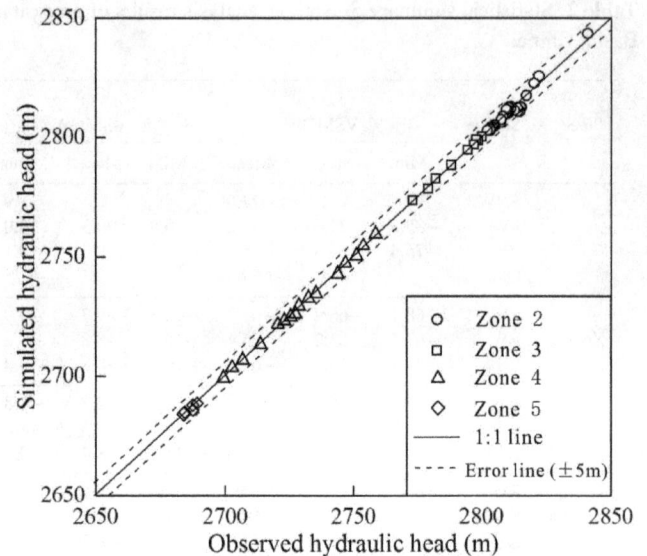

Figure 4. Comparison of observed and simulated hydraulic head values along the groundwater flow system, Golmud Watershed, China.

Table 3. Estimated parameters of different lithology from the Golmud Watershed, Qaidam Basin, China.

Lithology	K_h (m d^{-1})	Anisotropy ratio K_h / K_v	Porosity
Gravel sand	56.3	10	0.35
Sand	13.7	10	0.40
Sandy silt	0.62	5	0.5
Silt	0.13	5	0.6
Clay	0.001	5	0.65

5 Discussion

5.1 Water provenance and recharge characteristics

The δD and $\delta^{18}O$ isotope analysis results for different water types are shown in Fig. 5a in relation to the global meteoric water line (GMWL: $\delta D = 8 \times \delta^{18}O + 10$) (Craig, 1961). The Golmud watershed local meteoric water line (LMWL: $\delta D = 6.98 \times \delta^{18}O + 9.6$) (Wang, 2014) and Golmud watershed local evaporation line (LEL: $\delta D = 4.09 \times \delta^{18}O + 28.1$, $R^2 = 0.94$), which is the linear regression line of river and lake water in the study area, are also shown in Fig. 5a. The slope and intercept of the LMWL (6.98 and 9.6) are lower than those of the GMWL (8 and 10) as a result of secondary evaporation that occurred during precipitation, reflecting the arid climatic characteristics of the study area (Dogramaci et al., 2012; Wang et al., 2017).

As shown in Fig. 5a, most of the surface water and groundwaters in the study area are situated close to the GMWL and LMWL, indicating a meteoric origin. However, the spatial

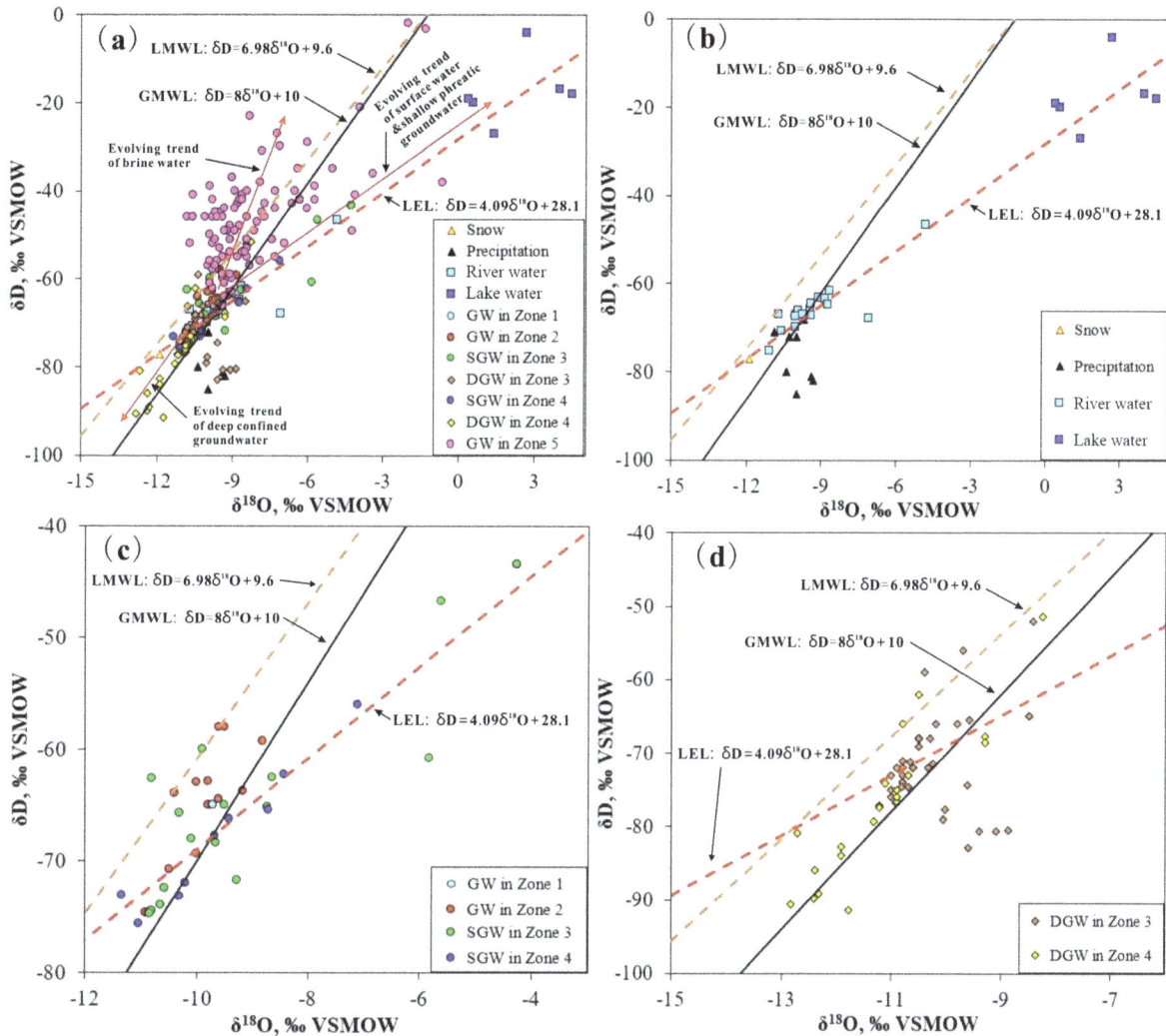

Figure 5. $\delta^{18}O$ vs. δD diagram of precipitation, surface water and groundwater for the Golmud study area of the Qaidam Basin, China. (**a**) All data; (**b**) snow, precipitation and surface waters; (**c**) shallow phreatic waters; (**d**) deep confined waters.

distribution of precipitation is extremely uneven. Most of the precipitation occurred in the Kunlun mountainous area to the south. Precipitation in the basin is very limited (annual rainfall less than 50 mm) and in this area there is little effective recharge to the aquifers (Xiao et al., 2017). Thus, surface water and groundwater in the study area mainly originates from meteoric water (including precipitation and snowmelt) in the mountainous areas. River water (δD: $-75.4‰$ to $-64.8‰$, $\delta^{18}O$: $-11.1‰$ to $-9.3‰$) and groundwater (δD: $-65.0‰$, $\delta^{18}O$: $-9.7‰$) in the mountainous area (Zone 1) have similar stable water isotopic signatures to precipitation (δD: $-85.3‰$ to $-71.6‰$, $\delta^{18}O$: $-10.9‰$ to $-9.3‰$) and snowmelt water (δD: $-77.0‰$, $\delta^{18}O$: $-11.9‰$) values from the Kunlun mountainous area (Zone 1), indicating their direct recharge relationship (Fig. 5a). River waters flow towards the northern low-lying depression of the central basin and show a gradual enrichment trend due to intensive evaporation. Lake

waters sampled from the low-lying depression (Zone 5) have the most enriched stable water isotope values and lie at the end of LEL defined by progressive evaporative enrichment of river water samples (Fig. 5b).

The δD and $\delta^{18}O$ values of the SGW and DGW demonstrate different varying trends along the groundwater flow path. The SGW shows a gradual positive enrichment trend in heavy isotopes along the LEL (Fig. 5c), implying the influence of evaporation. For the alluvial fan (Zone 2), the δD and $\delta^{18}O$ values of groundwater are very similar to that of river water in the alluvial fan (Zone 2) and groundwater in the mountainous area (Zone 1) (Table 2), indicating groundwater in the alluvial fan (Zone 2) is recharged directly by the seepage of river water and lateral inflow from the mountainous area, and out of the influence of evaporation. This is confirmed by similarities in major chemical composition (Fig. 2). The similar stable isotopic values also signify groundwater in the alluvial fan (Zone 2) has a short residence time, which is corroborated by elevated ^{3}H (20.0–56.3 TU, mean value of 35.5 TU) indicating the residence time is less than 60 years based on ^{3}H data (Xiao et al., 2017). SGW in the overflow zone (Zone 3) and the middle to lower stream area (Zone 4) has relative higher stable water isotope values compared with that in the alluvial fan and is plotted along the LEL, indicating SGW is influenced by evaporation from the overflow area (Zone 3) to the downstream. SGW in these two zones (Zone 3 and 4) also presents similar stable hydrogen and oxygen isotopic signatures as the river waters in the same area (Table 2), implying SGW has a very close hydraulic relationship with the rivers. The ^{3}H values of the SGW in Zone 3 and 4 range from 12.1 to 25.7 TU and from 14.4 to 17.5 TU, respectively, with the mean value of 21.1 and 16.0 TU, suggesting that SGW in these two zones contains a large component of modern water or mixtures of old and modern water.

DGW in the overflow zone (Zone 3) and the middle to lower stream area (Zone 4) is observed to have a completely opposite evolution trend in that the δD and $\delta^{18}O$ values become more depleted along the groundwater flow path (Fig. 5d). The depleted nature of the δD and $\delta^{18}O$ values may have two interpretations: (1) these aquifers have another recharge region where rainfall with low δD and $\delta^{18}O$ values occurs, or (2) the groundwater is ancient water recharged under colder climatic conditions (Chen et al., 2012; Awaleh et al., 2017). If (1) is the reason that groundwater would be more depleted in δD and $\delta^{18}O$ along the groundwater flow paths, it is difficult to construct a mixing model that would supply more depleted waters along a deep flow path. According to the groundwater age estimated using ^{14}C activity, the DGW in Zone 3 and 4 was recharged from 2264 BP to more than 20 754 BP (Holocene to late Pleistocene), which was a period when the climate changed from cold and wet condition (30 000 to 17 000 BP) to warm and dry conditions (14 000 BP to present) (Zhang et al., 2011). Consequently, it is believed that the depleted δD and $\delta^{18}O$ waters in the DGW were recharged by paleowater under a colder climate relative to the present day. Similar findings were reported in the adjoining Nomhon watershed of the Qaidam Basin (Xiao et al., 2017). Additionally, this is consistent with the paleoclimate findings recorded using groundwater data from other basins of northwestern China (He et al., 2015; Huang et al., 2017).

Groundwaters in the low-lying depression area (Zone 5), regardless of depth, are all brines with TDS values greater than 100 000 mg L^{-1}. Given the tectonic activity and depocenter migration within the Qaidam Basin over geological history (Chen and Bowler, 1986; Zhang, 1987), groundwater in the low-lying depression area (Zone 5) has a large component of paleobrines migrated from western Qaidam Basin due to the uplift in the past (Huang and Han, 2007). According to the ^{14}C age of DGW in Zone 4, the deduced age of DGW in the low-lying depression (Zone 5) is more than 20 000 years. SGW was observed with high ^{3}H content (18.9 TU) adjacent

to the terminal lake (G178) (Fig. 1 and Table 2), indicating mixing with leakage of modern surface water. As shown in Fig. 5a, most groundwaters in the basin center show a considerable deuterium excess, indicating that the original precipitation waters experienced considerable evaporation and vapor re-equilibration during recharge (Clark and Fritz, 1997).

5.2 Mechanisms controlling hydrochemistry

Generally, the composition of natural groundwater is primarily controlled by the chemical composition of recharge waters, water–aquifer matrix interaction and groundwater residence time (Redwan and Moneim, 2015; Verma et al., 2016). As exhibited in the diagrams between TDS vs. Na$^+$ / (Na$^+$ + Ca^{2+}) and Cl$^-$ / (Cl$^-$ + HCO$_3^-$) (Fig. 6), the major mechanisms controlling groundwater chemistry are water–rock interaction and evaporation–mineral precipitation processes (Gibbs, 1970). Water–rock interaction processes dominate the controls on groundwater chemistry at all depths in the alluvial fan (Zone 2) due to the great depth and the negligible impact of evaporation. For the overflow zone (Zone 3) and the middle to lower stream area (Zone 4), the governing mechanisms for SGW change from water–rock interaction to evaporation–mineral precipitation due to the gradual decrease in groundwater depth and recharge inputs from waters having undergone the influence of intensive evaporation in that part of the basin. Nearly all DGW in this part of the flow system are controlled by water–rock interaction. Two DGW samples are observed to plot in the evaporation–crystallization domain (Fig. 6). This is due to a high TDS and oversaturation of evaporative minerals (such as aragonite, calcite and dolomite) in the groundwater resulting in mineral precipitation (crystallization). For the low-lying depression (Zone 5), evaporation has a significant influence on the chemistry of SGW, and crystallization (precipitation) of many mineral phases is the primary geochemical process controlling both the SGW and DGW chemistry.

In order to further constrain the sources of solute in groundwater, the relationships between various ions are compared (Fig. 7). The relation of Na$^+$ vs. Cl$^-$ shows both SGW and DGW from the piedmont to the middle to lower stream area (Zone 2, 3, 4) are plotted along the 1 : 1 line (Fig. 7a), suggesting that halite dissolution is potentially a primary process or source of Na$^+$ and Cl$^-$ mineralization in groundwater. The calculated results of halite saturation index (SI$_{halite}$ <0) (Table 4) confirm that halite minerals of the aquifer matrix could be readily available to the groundwater. In addition, core drilling demonstrated that evaporated salts such as halite, calcium sulfate and sodium sulfate are widespread in the aquifer materials and can provide the solute source. Some of the SGW in Zone 4 is observed to have excess Na$^+$ relative to Cl$^-$ (Na / Cl ratios equal to 1.2–3.8), while groundwater in Zone 5, regardless of the depth, shows deficiency of Na$^+$ with respect to Cl$^-$ (Na / Cl ratios equal to 0.08–0.26), implying the existence of some other processes

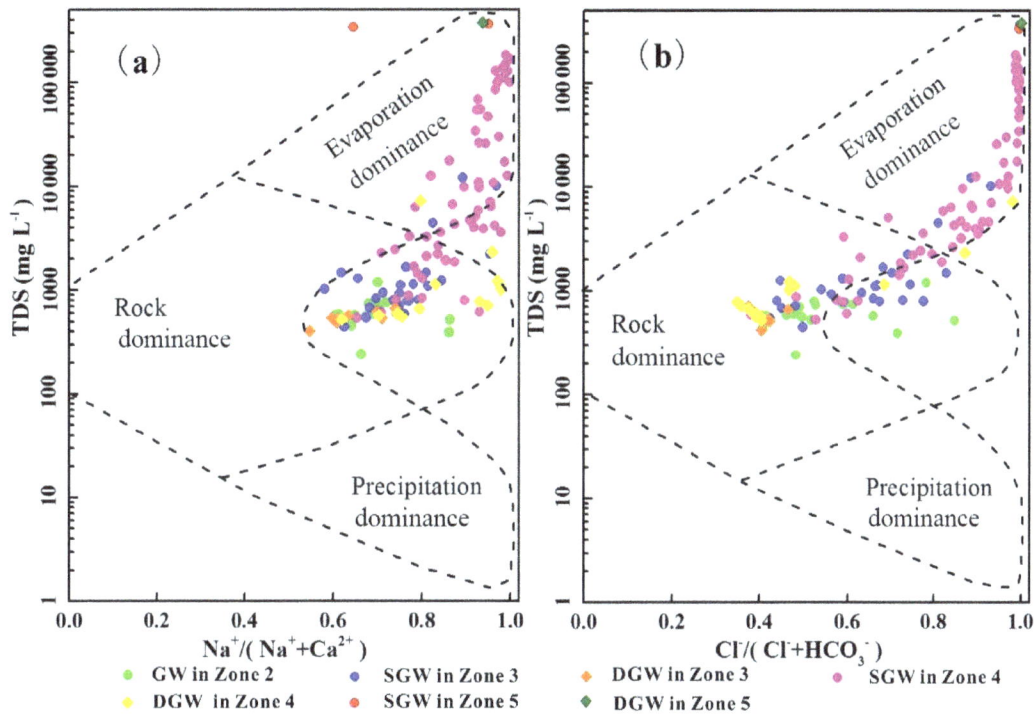

Figure 6. Diagrammatic representation showing the mechanisms controlling groundwater chemistry. (a) TDS vs. $Na^+ / (Na^+ + Ca^{2+})$; (b) TDS vs. $Cl^- / (Cl^- + HCO_3^-)$ (after Gibbs, 1970).

contributing Na^+ not Cl^- to groundwater and changing the ratio of Na^+ / Cl^-.

One explanation for the excess of Na^+ would be that the abundant Ca^{2+} and Mg^{2+} in fresh groundwater exchanges with the Na^+ on the surface of clay minerals, which results in an increase in Na^+ concentration and a decrease in Ca^{2+} and Mg^{2+} concentration in groundwater (Awaleh et al., 2017). The relationship of $[(Ca^{2+} + Mg^{2+})-(HCO_3^- + SO_4^{2-})]$ vs. $[(Na^+ + K^+)-Cl^-]$ (Fig. 7f) shows a regression line of $y = 1.0016x + 4.9078$ ($R^2 = 0.9966$) and corroborates the contribution of cation exchange (Ca^{2+} or $Mg^{2+} + 2NaX$ (solid) $\rightarrow$ $2Na^+ + CaX_2$ or MgX_2 (solid)) (Verma et al., 2016). In addition, silicate weathering (e.g., $2NaAlSi_3O_8$ (Albite) $+ 2CO_2 + 11H_2O \rightarrow 2Na^+ + Al_2Si_2O_5(OH)_5$ (Kaolinite) $+ 3H_4SiO_4 + 2HCO_3^-$) in the aquifers could also contribute Na^+, not Cl^-, to groundwater (Guo et al., 2015). The ratio of $Na^+ / (Cl^- + SO_4^{2-})$ is around 1 (Fig. 7b), demonstrating mirabilite ($Na_2SO_4 \cdot 10H_2O$) dissolution ($Na_2SO_4 \cdot 10H_2O \rightarrow 2Na^+ + SO_4^{2-}$) is an additional strong possible process that could also be responsible for the excess of Na^+ compared to Cl^- in groundwater (Jia et al., 2017). Groundwater in the basin with extremely high TDS concentration (more than $1\,000\,000\,mg\,L^{-1}$) has very low ratios of Na^+ / Cl^- (0.08–0.26) as a result of suspected reverse cation exchange ($Na^+ + CaX_2$ or MgX_2 (solid) $\rightarrow 2Na^+ + CaX_2$ or MgX_2 (solid)) (Fig. 7f).

The relationship between $(Ca^{2+} + Mg^{2+})$ and $(HCO_3^- + SO_4^{2-})$ shows that almost all groundwater from the piedmont to the middle to lower stream area (Zone 2, 3, 4) are plotted along the 1 : 1 line (Fig. 7c), implying the dissolution of minerals such as gypsum, anhydrite, aragonite, calcite and dolomite are the potential ion sources to groundwater in the mineralization process (Dogramaci et al., 2012). As shown in Fig. 7d, nearly all groundwater data plot away from the equiline of $(Ca^{2+} + Mg^{2+})$ vs. HCO_3^- (only three samples with the $(Ca^{2+} + Mg^{2+}) / HCO_3^-$ ratio in the range of 0.8–1.2, 6 samples with the ratio range of 0.2–0.5, and others with the ratio range of 1.2–604.2), indicating that the Ca^{2+}, Mg^{2+} and HCO_3^- are not primarily derived from the dissolution of aragonite, calcite and dolomite. The saturation index values of aragonite, calcite and dolomite are all almost greater than 0 in all samples (Table 4), suggesting the dissolution of these three minerals must be minimal. The saturation index values of gypsum and anhydrite for groundwater in these areas are all below zero (Table 4), corroborating the contribution of gypsum and anhydrite dissolution for groundwater mineralization. The deficiency of Ca^{2+} compared to SO_4^{2-} of groundwater (73.5 % of samples with the Ca / SO_4 ratio less than 0.8) presented in Fig. 7e is most likely as a result of the aforementioned mirabilite ($Na_2SO_4 \cdot 10H_2O$) dissolution and cation exchange. As mentioned earlier, the redox conditions of the deep confined aquifers in Zone 4 have evolved to a reduced environment, but due to the extremely low organic

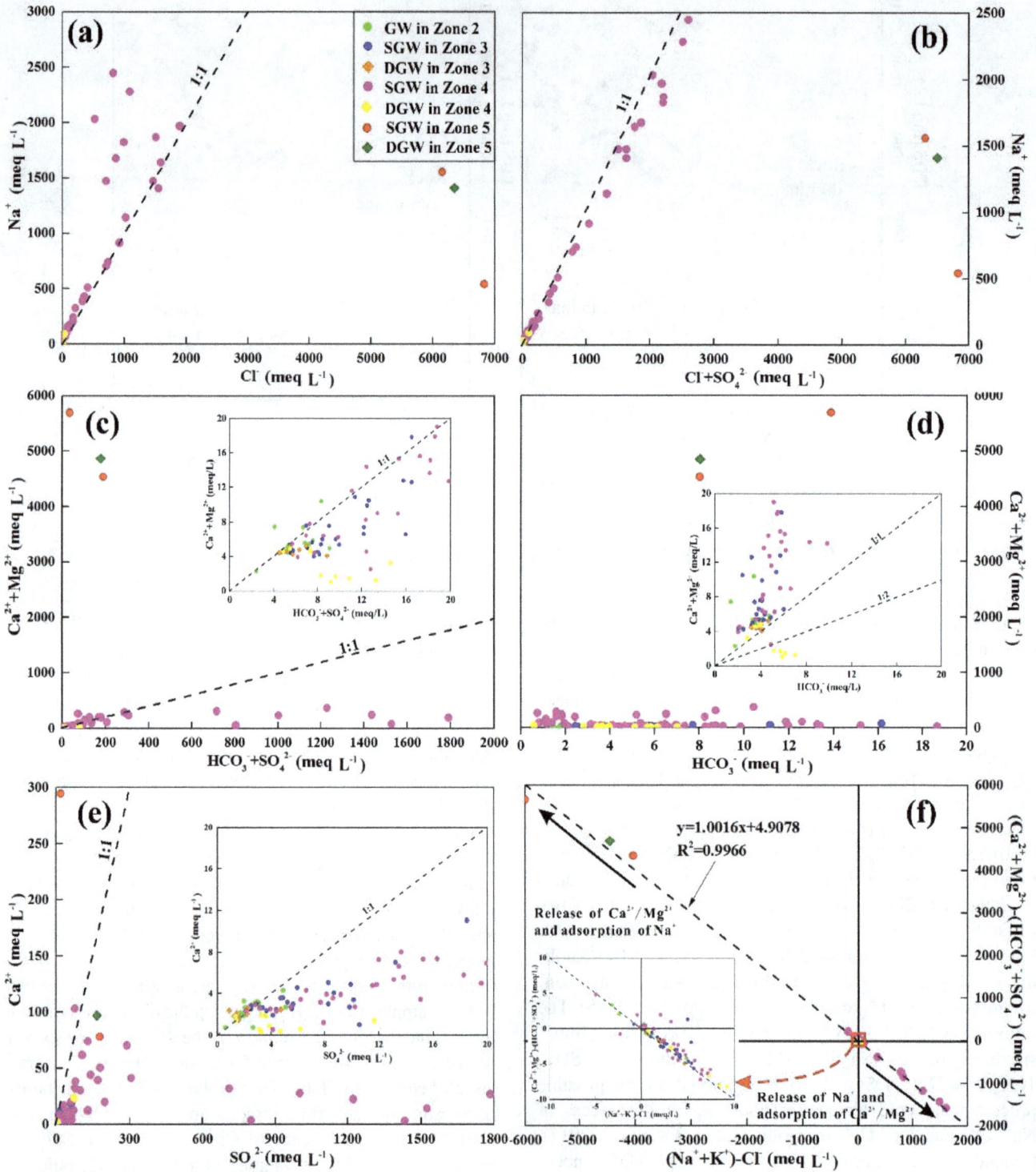

Figure 7. Bivariate plots ($meq\,L^{-1}$) of various ions in shallow phreatic and deep confined groundwater showed state **(a)** Na vs. Cl, **(b)** Na vs. $(Cl + SO_4)$, **(c)** $(Ca + Mg)$ vs. $(HCO_3 + SO_4)$, **(d)** $(Ca + Mg)$ vs. HCO_3, **(e)** Ca vs. SO_4, **(f)** $(Ca + Mg)$-$(HCO_3 + SO_4)$ vs. $[HCO_3 + SO_4]$.

Table 4. Saturation index of selected minerals from the Golmud Watershed, Qaidam Basin, China.

Place		Halite	Gypsum	Anhydrite	Aragonite	Calcite	Dolomite	Sylvite
	Min.	−6.87	−3.27	−3.75	−0.08	0.08	0.45	−7.59
SGW of Zone 2	Max.	−5.72	−1.88	−2.36	2.03	2.18	4.29	−6.90
	Mean	−6.43	−2.55	−3.02	1.29	1.45	2.85	−7.25
	Min.	−6.70	−3.12	−3.60	−0.16	−0.01	−0.78	−7.82
SGW of Zone 3	Max.	−3.70	−0.90	−1.36	2.24	2.40	4.97	−4.41
	Mean	−5.84	−2.04	−2.52	1.07	1.23	2.25	−6.66
	Min.	−6.98	−3.06	−3.52	−0.42	−0.26	−1.35	−7.65
DGW of Zone 3	Max.	−6.40	−2.67	−3.13	1.73	1.88	3.98	−7.07
	Mean	−6.82	−2.87	−3.33	0.89	1.05	1.91	−7.46
	Min.	−6.67	−3.18	−3.64	−0.89	−0.74	−2.00	−7.54
SGW of Zone 4	Max.	−1.16	0.32	−0.06	2.43	2.58	5.65	−2.44
	Mean	−3.93	−1.27	−1.71	1.47	1.63	3.45	−5.00
	Min.	−6.96	−3.80	−4.28	−1.02	−0.87	−2.49	−7.57
DGW of Zone 4	Max.	−4.30	−0.16	−0.65	1.39	1.54	3.41	−5.52
	Mean	−6.06	−2.89	−3.36	0.05	0.20	0.06	−7.16
	Min.	0.04	0.03	−0.16	2.29	2.44	6.78	−0.35
SGW of Zone 5	Max.	0.34	0.31	0.11	2.82	2.98	7.36	−0.17
	Mean	0.19	0.17	−0.03	2.56	2.71	7.07	−0.26
DGW of Zone 5		0.39	0.34	0.14	2.18	2.34	6.64	−0.21

carbon content in the sediments (Bowler et al., 1986; Chen and Bowler, 1986), sulfate reduction has a very limited influence on groundwater chemical evolution. This is also the reason that groundwater in the downstream area (Zone 4 and 5) has an abundant content of SO_4^{2-} in contrast to Ca^{2+}.

Groundwater in the low-lying depression (Zone 5) has extremely high TDS values ($>300\,000\,mg\,L^{-1}$) (Table 1) and almost all minerals are oversaturated (SI > 0) (Table 4); therefore, precipitation (crystallization) of minerals is the primary geochemical process in this part of the aquifer (Li et al., 2010). In addition, reverse cation exchange interaction and evaporation, which can be confirmed by the relationship of $[(Ca^{2+} + Mg^{2+})\text{-}(HCO_3^- + SO_4^{2-})]$ vs. $[(Na^+ + K^+)\text{-}Cl^-]$ (Fig. 7f) and the relation of stable water isotopes (Fig. 4), respectively, are also important mechanisms governing groundwater chemistry. Surface water has significant influences on the geochemical processes that occurred in the shallow aquifers. In the wet season, a large amount of relatively freshwater can reach the low-lying depression area (Zone 5) and infiltrate to the shallow aquifers. This would dilute the groundwater and dissolve the evaporated salts in the aquifers.

5.3 Groundwater flow and hydrogeochemical evolution

Theoretically, three types of groundwater flow systems, namely local, intermediate and regional, may occur in a large basin, and each flow system has its own characteristics based on aspects of flow path, recharge origin, cycle depth, cy-

cle amount, residence time, discharge position, hydrochemistry and controlling mechanisms (Toth, 1963). The cross-sectional groundwater flow modeling results demonstrated the groundwater flow paths in the study area are strictly controlled by distribution of the lithology (Fig. 8). Groundwater flow lines are shown to be upward convex in shape at the front of the alluvial fan and the middle-stream area due to an increase in relatively poor permeability due to the addition of finer, less permeable stratigraphic material. Based on the distribution of flow lines, three groundwater flow systems including local, intermediate and regional system were identified in the study area (Fig. 8).

The local groundwater flow system occurs in the shallow part of the alluvial fan (Zone 2) and overflow zone (Zone 3) with the deepest cycle depth within 250 m of the surface. This flow system obtains recharge water along the Golmud River flow path and discharges at the overflow zone. The water cycle quantity of the local system estimated by modeling accounts for approximately 83 % of the total quantity of groundwater in the watershed. Groundwater chemistry is mainly controlled by water–rock interaction and there appears to be very little evaporation. Groundwater has a rapid velocity in this part of the system, with a residence time of less than 60 years. As a result, groundwater here is fresh, with TDS values less than $1000\,mg\,L^{-1}$, and the water type is mainly $HCO_3 \cdot Cl\text{-}Ca \cdot Mg \cdot Na$. This system is the main source of water supply for Golmud city.

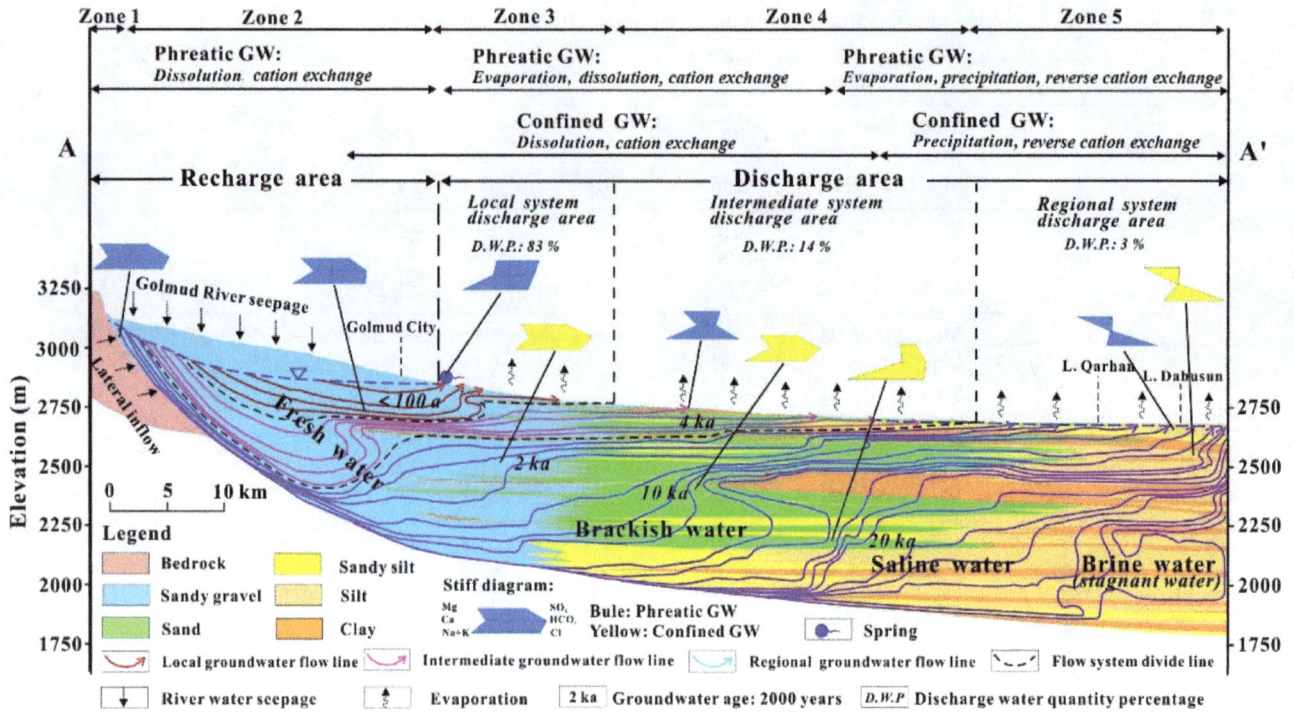

Figure 8. Conceptual model of groundwater flow and hydrochemical evolution in the Golmud watershed, China.

The intermediate flow system occurs below the local system and is recharged by river water seepage near the upper part of the alluvial fan. Groundwater flows to lower elevations towards the north and reaches its deepest cycle depth near 600 m at the middle part of the alluvial fan (Zone 2). Due to the increase in aquitards, water flow lines are presented as upward convex shapes at the middle to lower part of alluvial fan (Zone 2). Groundwater flow is constrained by two continuous aquitards (clay layers) at depths of 60 and 290 m (Fig. 8), respectively, at the front of the alluvial fan and overflow zone. The intermediate flow system discharges between the lower overflow zone (Zone 3) and the middle to lower stream area (Zone 4), as evidenced by springs and surface evaporation. The total cycle water quantity of the intermediate system accounts for approximately 14% of the total cycle groundwater amount in the watershed. Aquifers of this system in the alluvial fan have higher renewal rates due to their increased permeability, compared to those in the lower overflow zone (Zone 3) and the middle to lower stream area (Zone 4) that have relative low renewal rates as a result of a lithology dominated by finer sediments, with groundwater residence times of about 4000 years. Hydrochemistry is dominantly controlled by water–rock interaction, and also strongly influenced by evaporation within the discharge area. Because of the short residence time and a shortage of chemical solutes in the aquifer material, groundwater in the alluvial fan (Zone 2) generally maintains recharge water chemical characteristics that are fresh and of $HCO_3 \cdot Cl\text{-}Ca \cdot Mg \cdot Na$

type. Sufficient solutes in the aquifer medium and intensive evaporation in the fine soil plain results in the groundwaters gradually evolving to be brackish water and in some cases saline waters.

The regional groundwater flow system occurs under the intermediate system and is recharged at the upper part of the alluvial fan by river water seepage and lateral flow within the mountainous area, and it discharges at the basin center into terminal lakes resulting in evaporation. Groundwater flow paths are significantly controlled by the lithology (Fig. 8), and divided from the intermediate system by a continuous aquitard at a depth of 290 m. Aquifers of this system have very low water renewal rates with residence times up to and greater than 20 000 years. The modeled cycle water quantity of the regional system is only approximately 3%. Groundwater chemistry is mainly influenced by water–rock interaction in this system, except for shallow aquifers in the discharge area (Zone 5), which are strongly influenced by evaporation. Due to the substantial difference in residence time, water–rock interaction results in much different hydrochemical characteristics from the other aquifer systems. Groundwater that was presented as freshwater with a dominant water type of $HCO_3 \cdot Cl\text{-}Ca \cdot Mg \cdot Na$ in the alluvial fan (Zone 2) and overflow zone (Zone 3) become a brackish water type ($HCO_3 \cdot Cl\text{-}Na$) and a saline water type (Cl-Na) in the middle to lower stream area (Zone 4) and has evolved to be a brine water type mainly composed of Cl-Mg in the low-lying discharge area (Zone 5).

6   Conclusions

Previous studies on arid closed basins such as the Great Artesian Basin, Murray Basin, Death Valley and Minqin Basin have established a lot of typical groundwater circulation and evolution regimes. However, the Qaidam basin, a typical arid sedimentary closed basin formed with the uplift of the Tibetan plateau, has groundwater circulation patterns characterized by the complex tectonic activities, paleoclimate variation, arid climate characteristics, sedimentary lithology, and systematic evolution from fresh- to saltwater. Studies of this basin can enhance the understanding of groundwater origin, flow regime and hydrogeochemical evolution in such complex tectonic influenced arid sedimentary closed basins worldwide. Integration of hydrogeochemistry, isotopes and two-dimensional groundwater flow modeling was used to obtain insight into the hydrogeology in a typical arid endorheic watershed represented by the Qaidam Basin, Tibetan plateau. A number of key findings have come out of this study.

The groundwater in the basin originates from precipitation and meltwater in the mountainous areas to the south. Groundwater in the alluvial fan is recharged directly as a result of modern river water seepage and mountainous lateral inflow and has a rapid flow rate. Shallow phreatic waters in the overflow zone and the middle to lower stream area are supported by local and intermediate groundwater flow systems and have a close chemical and isotopic relationship with surface water. Deep confined groundwater in the overflow zone and the middle to lower stream area is recharged from paleometeoric water during the late Pleistocene and Holocene under a cold climate based on the results for stable water isotopic analyses. Groundwater in the low-lying depression (basin center) is made up of ancient brines that have possibly migrated from the western Qaidam Basin due to the uplift of the western basin in the geological past. Shallow phreatic aquifers in the low-lying depression (basin center) are also seasonally recharged by modern surface water during flooding periods.

Groundwater in the study area evolves from freshwater to brine water along the flow path. The hydrochemistry of groundwater in the alluvial fan is dominantly controlled by mineral dissolution and cation exchange, and occurs as slightly alkaline water with TDS values less than $1000\,\mathrm{mg\,L^{-1}}$ and a water type with a composition of $HCO_3 \cdot Cl\text{-}Ca \cdot Mg \cdot Na$. Deep confined groundwater chemistry in the overflow zone and middle to lower stream area is also controlled by mineral dissolution and cation exchange, as a result of longer residence times in the aquifers and shows a trend evolving from freshwater to brackish water and finally saline water with increasing solute inputs along the flow paths. As well as water–rock interaction, shallow phreatic water is also affected by intensive evaporation, and therefore, these waters can be much saltier than deep confined water. Groundwater in the low-lying depression (basin center) is composed of brine water, and the mineral precipitation coupled with reverse cation exchange is the dominant geochemical process controlling water chemistry. The impact of evaporation is also one of the important geochemical processes in the shallow phreatic aquifers, which can accelerate evaporated mineral precipitation. Salt dissolution occasionally occurred in the low-lying depression (basin center) during flood periods due to the infiltration of large amounts of fresh surface water.

Three different hierarchical groundwater flow systems were identified using the cross-sectional model. The continuous aquitards at depths of 60, 290 and 450 m have significant constraints on groundwater flow. The local flow system occurs in the shallow part of the alluvial fan and overflow zone and discharges in the overflow zone, with the deepest cycle depth within 250 m of surface. The intermediate system occurs below the local system and discharges in the lower overflow zone and middle to lower stream area with the deepest cycle depth reaching 600 m below surface. The regional system was separated from the intermediate system by a continuous aquitard at a depth of 290 m and discharges in the low-lying depression (basin center). Our calculation shows that the discharge water quantity of these three systems accounts for approximately 83 %, 14 % and 3 %, respectively.

This study enhanced the understanding of the origin, flow pattern, hydrochemical evolution and controlling mechanisms of the regional groundwater systems in the Qaidam Basin. These results can provide fundamental information for coping with future issues such as water conflicts, salt lake exploitation and climate warming in the basin and also provide references for understanding the hydrogeological processes in other similar endorheic watersheds of northwestern China and elsewhere in the world.

Author contributions.   This research was conceived by JS, YC and XD. SKF contributed ideas for analyses. YX, XD, SW and YJ carried out the field work. YX analyzed the data, carried out the model simulations, and wrote the paper with input from all the authors.

Competing interests.   The authors declare that they have no conflict of interest.

Acknowledgements.   This work was supported by the National Key R&D Program of China [2017YFC0406106] and the China Geological Survey [DD20160291]. We appreciate the help of Ge Zhang and Xiangzhi You at the Xi'an Center of Geological Survey, China Geological Survey; Zongyu Chen and Qichen Hao at the Institute of Hydrogeology and Environmental Geology, Chinese Academy of Geological Sciences; Xiaomin Gu at Nantong University; and Jingxing Liu and Dong Wang at the China University of Geosciences (Beijing). We are grateful to Editor Graham Fogg and the two anonymous reviewers whose insightful comments were very helpful in improving the paper.

Edited by: Graham Fogg

References

Anderson, M. P., Woessner, W. W., and Hunt, R. J.: Applied groundwater modeling: simulation of flow and advective transport, 2nd edn., Academic Press, Salt Lake City, 2015.

Awaleh, M. O., Baudron, P., Soubaneh, Y. D., Boschetti, T., Hoch, F. B., Egueh, N. M., Mohamed, J., Dabar, O. A., Masse-Dufresne, J., and Gassani, J.: Recharge, groundwater flow pattern and contamination processes in an arid volcanic area: Insights from isotopic and geochemical tracers (Bara aquifer system, Republic of Djibouti), J. Geochem. Explor., 175, 82–98, https://doi.org/10.1016/j.gexplo.2017.01.005, 2017.

Banks, E. W., Love, A. J., Simmons, C. T., and Shand, P.: Assessing surface water – Groundwater connectivity using hydraulic and hydrochemical approaches in fractured rock catchments, South Australia, Taylor & Francis, London, 2010.

Bowler, J. M., Qi, H., Kezao, C., Head, M. J., and Baoyin, Y.: Radiocarbon dating of playa-lake hydrologic changes: Examples from northwestern China and central Australia, Palaeogeogr. Palaeocl., 54, 241–260, https://doi.org/10.1016/0031-0182(86)90127-6, 1986.

Bredehoeft, J. D. and Konikow, L. F.: Ground-water models: validate or invalidate, Groundwater, 50, 493, 2012.

Cartwright, I. and Morgenstern, U.: Contrasting Transit Times and Water-rock Interaction in Australian Upland Catchments Draining Peatland and Eucalypt Forest, Proced. Earth. Plan. Sc., 17, 140–143, https://doi.org/10.1016/j.proeps.2016.12.032, 2017.

Cartwright, I., Weaver, T. R., Stone, D., and Reid, M.: Constraining modern and historical recharge from bore hydrographs, 3H, 14C, and chloride concentrations: Applications to dual-porosity aquifers in dryland salinity areas, Murray Basin, Australia, J. Hydrol., 332, 69–92, 2007.

Cartwright, I., Weaver, T., Cendón, D. I., and Swane, I.: Environmental isotopes as indicators of inter-aquifer mixing, Wimmera region, Murray Basin, Southeast Australia, Chem. Geol., 277, 214–226, https://doi.org/10.1016/j.chemgeo.2010.08.002, 2010a.

Cartwright, I., Weaver, T. R., Simmons, C. T., Fifield, L. K., Lawrence, C. R., Chisari, R., and Varley, S.: Physical hydrogeology and environmental isotopes to constrain the age, origins, and stability of a low-salinity groundwater lens formed by periodic river recharge: Murray Basin, Australia, J. Hydrol., 380, 203–221, 2010b.

Cartwright, I., Hofmann, H., Currell, M. J., and Fifield, L. K.: Decoupling of solutes and water in regional groundwater systems: The Murray Basin, Australia, Chem. Geol., 466, 466–478, https://doi.org/10.1016/j.chemgeo.2017.06.035, 2017.

Chen, A., Zheng, M., Shi, L., Wang, H., and Xu, J.: Magnetostratigraphy of deep drilling core 15YZK01 in the northwestern Qaidam Basin (NE Tibetan Plateau): Tectonic movement, salt deposits and their link to Quaternary glaciation, Quatern. Int., 436, 201–211, https://doi.org/10.1016/j.quaint.2017.01.026, 2017.

Chen, J., Liu, X., Wang, C., Rao, W., Tan, H., Dong, H., Sun, X., Wang, Y., and Su, Z.: Isotopic constraints on the origin of groundwater in the Ordos Basin of northern China, Environ. Earth Sci., 66, 505–517, https://doi.org/10.1007/s12665-011-1259-6, 2012.

Chen, K. and Bowler, J. M.: Late pleistocene evolution of salt lakes in the Qaidam basin, Qinghai province, China, Palaeogeogr. Palaeocl., 54, 87–104, 1986.

Chen, L., Ma, T., Ma, J., Du, Y., and Xiao, C.: Identification of material source for the salt lakes in the Qaidam Basin, Hydrogeology & Engineering Geology, 42, 101–107, 2015.

Chen, Z., Wei, W., Liu, J., Wang, Y., and Chen, J.: Identifying the recharge sources and age of groundwater in the Songnen Plain (Northeast China) using environmental isotopes, Hydrogeol. J., 19, 163–176, 2011.

Clark, I. D. and Fritz, P.: Environmental Isotopes in Hydrogeology, CRC press, New York, USA, 1997.

Craig, H.: Isotopic Variation in Meteoric Waters, Science, 133, 1702–1703, 1961.

Dogramaci, S., Skrzypek, G., Dodson, W., and Grierson, P. F.: Stable isotope and hydrochemical evolution of groundwater in the semi-arid Hamersley Basin of sub-tropical northwest Australia, J. Hydrol., 475, 281–293, 2012.

Edmunds, W. M., Ma, J., Aeschbach-Hertig, W., Kipfer, R., and Darbyshire, D. P. F.: Groundwater recharge history and hydrogeochemical evolution in the Minqin Basin, North West China, Appl. Geochem., 21, 2148–2170, https://doi.org/10.1016/j.apgeochem.2006.07.016, 2006.

Gibbs, R. J.: Mechanisms Controlling World Water Chemistry, Science, 170, 1088–1090, https://doi.org/10.1126/science.170.3962.1088, 1970.

Gu, X., Shao, J., Cui, Y., and Hao, Q.: Calibration of two-dimensional variably saturated numerical model for groundwater flow in arid inland basin, China, Curr. Sci., 113, 403–412, 2017.

Guo, X., Feng, Q., Liu, W., Li, Z., Wen, X., Si, J., Xi, H., Guo, R., and Jia, B.: Stable isotopic and geochemical identification of groundwater evolution and recharge sources in the arid Shule River Basin of Northwestern China, Hydrol. Process., 29, 4703–4718, 2015.

Hao, Q., Shao, J., Cui, Y., and Zhang, Q.: Development of a new method for efficiently calculating of evaporation from the phreatic aquifer in variably saturated flow modeling, Journal of Groundwater Science and Engineering, 4, 26–34, 2016.

He, J., Ma, J., Zhao, W., and Sun, S.: Groundwater evolution and recharge determination of the Quaternary aquifer in the Shule River basin, Northwest China, Hydrogeol. J., 23, 1745–1759, https://doi.org/10.1007/s10040-015-1311-9, 2015.

Herrera, C., Custodio, E., Chong, G., Lambán, L. J., Riquelme, R., Wilke, H., Jódar, J., Urrutia, J., Urqueta, H., Sarmiento, A., Gamboa, C., and Lictevout, E.: Groundwater flow in a closed basin with a saline shallow lake in a volcanic area: Laguna Tuyajto, northern Chilean Altiplano of the Andes, Sci. Total Environ., 541, 303–318, https://doi.org/10.1016/j.scitotenv.2015.09.060, 2016.

Hou, X., Zhang, J., and Liu, J.: Assessment of Groundwater Source of Piedmont Plain Area of China Northwest Arid Region Based On Numerical Modeling, Acta. Geol. Sin.-Engl., 88, 419–420, 2014.

Huang, G., Chen, Z., Sun, J., Wang, J., and Hou, Q.: Groundwater quality in aquifers affected by the anthropogenic and natural processes in an urbanized area, south China, Environ. Forensics, 17, 107–119, 2016.

Huang, L. and Han, F.: Evolution of salt lakes and palaeoclimate fluctuation in Qaidam Basin, Science Press, Beijing, 2007.

Huang, T., Pang, Z., Li, J., Xiang, Y., and Zhao, Z.: Mapping groundwater renewability using age data in the Baiyang alluvial fan, NW China, Hydrogeol. J., 25, 743–755, https://doi.org/10.1007/s10040-017-1534-z, 2017.

Islam, M. B., Firoz, A. B. M., Foglia, L., Marandi, A., Khan, A. R., Schüth, C., and Ribbe, L.: A regional groundwater-flow model for sustainable groundwater-resource management in the south Asian megacity of Dhaka, Bangladesh, Hydrogeol. J., 25, 617–637, https://doi.org/10.1007/s10040-016-1526-4, 2017.

Jia, Y., Guo, H., Xi, B., Jiang, Y., Zhang, Z., Yuan, R., Yi, W., and Xue, X.: Sources of groundwater salinity and potential impact on arsenic mobility in the western Hetao Basin, Inner Mongolia, Sci. Total Environ., 691, 601–602, 2017.

Jiang, X. W., Wan, L., Wang, J. Z., Yin, B. X., Fu, W. X., and Lin, C. H.: Field identification of groundwater flow systems and hydraulic traps in drainage basins using a geophysical method, Geophys. Res. Lett., 41, 2812–2819, 2014.

Jiao, J. J., Zhang, X., Yi, L., and Kuang, X.: Increased Water Storage in the Qaidam Basin, the North Tibet Plateau from GRACE Gravity Data, Plos One, 10, e0141442, https://doi.org/10.1371/journal.pone.0141442, 2015.

Li, M., Fang, X., Yi, C., Gao, S., Zhang, W., and Galy, A.: Evaporite minerals and geochemistry of the upper 400 m sediments in a core from the Western Qaidam Basin, Tibet, Quatern. Int., 218, 176–189, 2010.

Love, A. J., Shand, P., Karlstrom, K., Crossey, L., Rousseau-Gueutin, P., Priestley, S., Wholing, D., Fulton, S., and Keppel, M.: Geochemistry and Travertine Dating Provide New Insights into the Hydrogeology of the Great Artesian Basin, South Australia, Proced. Earth Plan. Sc., 7, 521–524, 2013.

Love, A. J., Shand, P., Fulton, S., Wohling, D., Karlstrom, K. E., Crossey, L., Rousseau-Gueutin, P., and Priestley, S. C.: A Reappraisal of the Hydrogeology of the Western Margin of the Great Artesian Basin: Chemistry, Isotopes and Groundwater Flow, Proced. Earth Plan. Sc., 17, 428–431, 2017.

Lowenstein, T. K. and Risacher, F.: Closed Basin Brine Evolution and the Influence of Ca–Cl Inflow Waters: Death Valley and Bristol Dry Lake California, Qaidam Basin, China, and Salar de Atacama, Chile, Aquat. Geochem., 15, 71–94, 2009.

Petts, D. C., Saso, J. K., Diamond, L. W., Aschwanden, L., Al, T. A., and Jensen, M.: The source and evolution of paleofluids responsible for secondary minerals in low-permeability Ordovician limestones of the Michigan Basin, Appl. Geochem., 86, 121–137, https://doi.org/10.1016/j.apgeochem.2017.09.011, 2017.

Priestley, S. C., Kleinig, T., Love, A. J., Post, V. E. A., Shand, P., Stute, M., Wallis, I., and Wohling, D. L.: Palaeohydrogeology and Transport Parameters Derived from 4 He and Cl Profiles in Aquitard Pore Waters in a Large Multilayer Aquifer System, Central Australia, Geofluids, 2017, 1–17, 2017a.

Priestley, S. C., Love, A. J., Post, V., Shand, P., Wohling, D., Kipfer, R., Payne, T. E., Stute, M., and Tyroller, L.: Environmental Tracers in Groundwaters and Porewaters to Understand Groundwater Movement Through an Argillaceous Aquitard, Proced. Earth Plan. Sc., 17, 420–423, 2017b.

Pruess, K., Oldenburg, C., and Moridis, G.: TOUGH2 user's guide version 2, Lawrence Berkeley National Laboratory, Berkeley, California, 1999.

Redwan, M. and Moneim, A. A. A.: Factors controlling groundwater hydrogeochemistry in the area west of Tahta, Sohag, Upper Egypt, J. Afr. Earth Sci., 118, 328–338, 2015.

Richards, L. A.: Capillary conduction of liquids through porous mediums, Physics, 1, 318–333, 1931.

Shand, P., Love, A. J., Gotch, T., Raven, M. D., Kirby, J., and Scheiderich, K.: Extreme Acidic Environments Associated with Carbonate Mound Springs in the Great Artesian Basin, South Australia, Proced. Earth Plan. Sc., 7, 794–797, 2013.

Shao, J., Cui, Y., Xiao, Y., Li, Y., and Zhao, D.: Groundwater cycle pattern and groundwater resource evaluation in Golmud watershed of Qaidam Basin, China University of Geosciences, Beijing, 2017.

Stone, A. E. C. and Edmunds, W. M.: Naturally-high nitrate in unsaturated zone sand dunes above the Stampriet Basin, Namibia, J. Arid. Environ., 105, 41–51, 2014.

Su, C., Cheng, Z., Wei, W., and Chen, Z.: Assessing groundwater availability and the response of the groundwater system to intensive exploitation in the North China Plain by analysis of long-term isotopic tracer data, Hydrogeol. J., 26, 1401–1415, https://doi.org/10.1007/s10040-018-1761-y, 2018.

Su, X., Xu, W., Yang, F., and Zhu, P.: Using new mass balance methods to estimate gross surface water and groundwater exchange with naturally occurring tracer 222Rn in data poor regions: a case study in northwest China, Hydrol. Process., 29, 979–990, 2015.

Su, X., Cui, G., Du, S., Yuan, W., and Wang, H.: Using multiple environmental methods to estimate groundwater discharge into an arid lake (Dakebo Lake, Inner Mongolia, China), Hydrogeol. J., 24, 1–16, 2016.

Tan, H., Rao, W., Chen, J., Su, Z., Sun, X., and Liu, X.: Chemical and Isotopic Approach to Groundwater Cycle in Western Qaidam Basin, China, Chinese Geogr. Sci., 19, 357–364, 2009.

Tan, H., Rao, W., Ma, H., Chen, J., and Li, T.: Hydrogen, oxygen, helium and strontium isotopic constraints on the formation of oilfield waters in the western Qaidam Basin, China, J. Asian Earth Sci., 40, 651–660, 2011.

Toth, J.: A theoretical analysis of groundwater flow in small drainage basins, J. Geophys. Res., 68, 4795-4812, https://doi.org/10.1029/JZ068i008p02354, 1963.

Toth, J.: Cross-formational gravity-flow of groundwater: a mechanism of the transport and accumulation of petroleum (the generalized hydraulic theory of petroleum migration), in: AAPG Studies in Geology No. 10: Problems of Petroleum Migration, The American Association of Petroleum Geologists, Tulsa, Oklahoma, 121–167, 1980.

Vengosh, A., Chivas, A. R., Starinsky, A., Kolodny, Y., Zhang, B., and Zhang, P.: Chemical and boron isotope compositions of nonmarine brines from the Qaidam Basin, Qinghai, China, Chem. Geol., 120, 135–154, 1995.

Verma, S., Mukherjee, A., Mahanta, C., Choudhury, R., and Mitra, K.: Influence of geology on groundwater–sediment interactions in arsenic enriched tectono-morphic aquifers of the Himalayan Brahmaputra river basin, J. Hydrol., 540, 176–195, 2016.

Wang, D. and Ren, F.: Abnormal Groundwater Chemistry in Ge'ermu Allubium Aquifer and its Origin, Journal of Changchun University of Earth Sciences, 26, 191–195, 1996.

Wang, L., Dong, Y., Xu, Z., and Qiao, X.: Hydrochemical and isotopic characteristics of groundwater in the northeastern Tennger Desert, northern China, Hydrogeol. J., 25, 2363–3275, https://doi.org/10.1007/s10040-017-1620-2, 2017.

Wang, Y.: Geochemistry Evolution and Water Cycle Patterns of Groundwater in Golmud River Basin, Master Degree, Chang'an University, Xi'an, China, 2014.

Wang, Y., Li, H., Ma, X., Jia, X., Kang, Q., Ma, G., Huang, Y., and Jia, J.: Use of Rushed Deep Lacustrine Freshwater in the Plain Area of Qaidam Basin, Northwestern Geology, 43, 113–119, 2010.

Xiao, Y., Shao, J., Cui, Y., Zhang, G., and Zhang, Q.: Groundwater circulation and hydrogeochemical evolution in Nomhon of Qaidam Basin, northwest China, J. Earth Syst. Sci., 126, 1–15, https://doi.org/10.1007/s12040-017-0800-8, 2017.

Xu, W., Su, X., Dai, Z., Yang, F., Zhu, P., and Huang, Y.: Multi-tracer investigation of river and groundwater interactions: a case study in Nalenggele River basin, northwest China, Hydrogeol. J., 25, 2015–2029, https://doi.org/10.1007/s10040-017-1606-0, 2017.

Ye, C., Zheng, M., Wang, Z., HAOWeilin, Lin, X., and Han, J.: Hydrochemistry of the Gasikule Salt Lake,Western Qaidam Basin of China, Acta. Geol. Sin.-Engl., 88, 170–172, 2014.

Zhang, J.: Groundwater resource evoluation of Xiangride-Nuomuhong Piedmont Plain, Master, China University of Geosciences, Beijing, China, 2013.

Zhang, M., Chen, Y., Yin, S., Zhang, J., Li, C., and Liu, G.: Sedimental features and paleo-environment reconstruction of the slope deposit at Xiaogangou of the Golmud River, Arid Land Geography, 34, 890–903, 2011.

Zhang, P.: Salt lakes in Qaidam Basin, Science Press, Beijing, 1987.

Zheng, M., Tang, J., Liu, J., and Zhang, F.: Chinese saline lakes, Hydrobiologia, 267, 23–36, 10.1007/bf00018789, 1993.

5

Speculations on the application of foliar ^{13}C discrimination to reveal groundwater dependency of vegetation and provide estimates of root depth and rates of groundwater use

Rizwana Rumman, James Cleverly, Rachael H. Nolan, Tonantzin Tarin, and Derek Eamus

Terrestrial Ecohydrology Research Group, School of Life Sciences, University of Technology Sydney, P.O. Box 123, Broadway, NSW 2007, Australia

Correspondence: Derek Eamus (derek.eamus@uts.edu.au)

Abstract. Groundwater-dependent vegetation is globally distributed, having important ecological, social, and economic value. Along with the groundwater resources upon which it depends, this vegetation is under increasing threat through excessive rates of groundwater extraction.

In this study we examined one shallow-rooted and two deep-rooted tree species at multiple sites along a naturally occurring gradient in depth-to-groundwater. We measured (i) stable isotope ratios of leaves (δ^{13}C), xylem, and groundwater (δ^2H and δ^{18}O); and (ii) leaf-vein density. We established that foliar discrimination of ^{13}C (Δ^{13}C) is a reliable indicator of groundwater use by vegetation and can also be used to estimate rooting depth. Through comparison with a continental-scale assessment of foliar Δ^{13}C, we also estimated the upper limits to annual rates of groundwater use. We conclude that maximum rooting depth for both deep-rooted species ranged between 9.4 and 11.2 m and that annual rates of groundwater use ranged from ca. 1400 to 1700 mm for *Eucalyptus camaldulensis* and from 600 to 900 mm for *Corymbia opaca*. Several predictions about hydraulic and leaf traits arising from the conclusion that these two species made extensive use of groundwater were supported by additional independent studies of these species in central Australia.

1 Introduction

Drylands cover 41 % of the earth's total land area (Reynolds et al., 2007) and are sub-categorized as hyper-arid, arid, semi-arid, and dry sub-humid areas. Hyper-arid, arid, and semi-arid regions are characterized by chronic water shortage with unpredictable rainfall (Clarke, 1991). Approximately 40 % of the world's population reside in drylands and groundwater represents a major water resource not only for human consumptive use, but also for groundwater-dependent ecosystems (GDEs, Eamus et al., 2006). Sustainable management of both groundwater and GDEs requires identification of the location of GDEs, rooting depth of vegetation, and rates of groundwater use, but attaining such information presents significant technical and cost challenges (Eamus et al., 2015).

Approximately 70 % of Australia is classified as semi-arid or arid (Eamus et al., 2006; O'Grady et al., 2011). Furthermore, annual potential evaporation exceeds annual rainfall across most of the continent; thus, most Australian biomes are water-limited according to the Budyko (1974) framework (Donohue et al., 2009). On average, central Australia receives less than 350 mm year^{-1} of rainfall, making water a primary limiting resource (Eamus et al., 2006). Because surface water bodies in this region are mostly ephemeral (NRE-TAS, 2009, although see Box et al., 2008, regarding the small number of permanent water bodies), groundwater plays an important role in maintaining ecosystem structure and function of terrestrial (especially riparian) vegetation (Eamus et al., 2006). Owing to the remoteness of much of Australia's interior, few studies have investigated groundwater use by vegetation communities in these semi-arid regions.

Stomatal conductance is regulated to maximize carbon gain whilst simultaneously minimizing transpiration (Cowan and Farquhar, 1977; Medlyn et al., 2011) and is sensitive to both soil and atmospheric water content (Prior et al., 1997; Thomas and Eamus, 1999). Intrinsic water-use efficiency (WUE_i), defined by the ratio of carbon gain to stomatal conductance, provides valuable insights into how vegetation responds to variation in water availability (Beer et al., 2009). Declining water supply results in increased WUE_i as stomatal conductance declines (Eamus et al., 2013). Discrimination against the ^{13}C isotope ($\Delta^{13}C$) is commonly used to calculate WUE_i. $\Delta^{13}C$ provides a time-integrated measure of WUE_i (Cernusak et al., 2011); in this study we examined spatial and seasonal patterns in $\Delta^{13}C$ across three tree species in the Ti Tree basin.

The present study was undertaken in the Ti Tree basin, which is the location of an important groundwater resource in central Australia (Cook et al., 2008a). Rainfall occurs mostly in large events during the austral summer (December–March); thus, there is minimal rainfall available for vegetation use over prolonged periods. The dry season in this region is characterized by declining soil water availability and high vapour pressure deficits (Eamus et al., 2013). Previous studies have documented several surprising attributes for a number of tree species in Ti Tree. O'Grady et al. (2009) observed that, despite living in an extremely water-limited environment, the specific leaf area (SLA) of *Corymbia opaca* and *Eucalytptus camaldulensis* was similar more to those from highly mesic environments than to species from arid environments. Similarly, Santini et al. (2016) observed that xylem wall thickness and vessel implosion resistance were significantly smaller in *E. camaldulensis* and *C. opaca* than in shallow-rooted *Acacia aneura*. Finally, differences in rates of water use and changes in midday water potential between the end of the wet season and the end of the dry season were minimal for *E. camaldulensis* and *C. opaca*, but were very large for *Acacia aptaneura* (which was previously classified as *Acacia aneura*; Maslin and Reid, 2012; Nolan et al., 2017). *A. aneura* and *A. aptaneura* often intermix with other members of the large Mulga complex of closely related *Acacia* species (Wright et al., 2016); thus, we will refer to *Acacia* spp. in the Mulga complex by the primary type, *A. aneura*. *A. aneura* is shallow-rooted and associated with shallow hard pans in this catchment (ca. 1 m below ground surface; Cleverly et al., 2016a, b), which prevents access to the groundwater below. *E. camaldulensis* is a riparian species, confined to narrow corridors along the ephemeral streams in Ti Tree (the Woodforde River and Allungra Creek) where groundwater depth is shallow (< 3 m); *C. opaca* is deep-rooted and may access groundwater to depths of 8 m or more (O'Grady et al., 2009). These observations lead to the first hypothesis tested in the present study: that WUE_i of *A. aneura* would be significantly larger than that of *E. camaldulensis* and *C. opaca*

because reliance on shallow stores of water by *Acacia* spp. imposes severe restrictions on water use, thus resulting in a large WUE_i.

Vertical (i.e. elevation) and horizontal distance from rivers receiving groundwater inflows in arid zones influences the degree to which trees access groundwater (O'Grady et al., 2006a; Thorburn et al., 1994). Trees closest to the river (i.e. elevationally and horizontally) have xylem water deuterium and ^{18}O isotope ratios (δ^2H and ^{18}O, respectively) that are close to the ratios of river and groundwater; trees further from the river have xylem water ratios increasingly different from those of groundwater and the river (O'Grady et al., 2006a). In endorheic basins like Ti Tree, evaporation of near-surface soil water imposes additional fractionation of δ^2H and ^{18}O (Craig, 1961) relative to groundwater; thus, δ^2H and $\delta^{18}O$ in xylem provide information on plant water source and climate (Cullen and Grierson, 2007). Variation in plant water sources with distance from the river can affect stomatal conductance and WUE_i. In this study we tested the hypothesis that WUE_i would increase with distance from the creek.

Differential access to water among co-occurring species within a biome results in variation of several morphological traits, including SLA (Warren et al., 2005), Huber value (Eamus et al., 2000; Sperry, 2000), and wood density (Bucci et al., 2004; Hacke et al., 2000). Leaf-vein density (LVD) is a trait that influences whole-plant performance. From a resource investment perspective, leaves are composed primarily of two components: mesophyll that undertakes photosynthesis and a leaf-vein network which delivers water and nutrients to the leaf. Investment in leaf veins is underpinned by resource allocation strategies (Niinemets et al., 2006, 2007; Niklas et al., 2007). Leaf-vein density is responsive to several environmental variables, but especially aridity (Uhl and Mosbrugger, 1999). Furthermore, LVD is positively correlated with leaf hydraulic conductance (K_{leaf}), maximum photosynthetic rate, and leaf-level gas-exchange rates (Brodribb et al., 2007; Sack et al., 2003; Sack and Frole, 2006; Sack and Holbrook, 2006). Consequently we investigate whether investment in LVD of three dominant overstorey tree species was affected by increasing depth-to-groundwater (DTGW).

The propensity for leaves to lose water matches the capacity of xylem to deliver the same volume of water (Brodribb and Holbrook, 2007; Meinzer and Grantz, 1990; Sperry, 2000), and positive correlations consequently occur between leaf-hydraulic conductance (K_{leaf}) and LVD (Brodribb et al., 2007; Sack and Holbrook, 2006). LVD provides a direct estimate of K_{leaf} because it correlates with the distance water must traverse from termini of the xylem network to sites of evaporation (Brodribb et al., 2010). Since transpiration is directly linked to availability of water to roots, we hypothesized that LVD will be correlated with depth-to-groundwater in plants for which groundwater is accessible; this correlation should be absent in species with shallow roots which cannot access groundwater. Whilst a number of studies have demonstrated increased LVD with increasing aridity along rainfall gradients (Brodribb et al., 2010; Brodribb and Holbrook, 2003; Sack and Holbrook, 2006), this relationship has not, to our knowledge, been examined in relation to DTGW. Finally, because LVD is strongly correlated with K_{leaf} and rates of leaf-scale gas exchange (Brodribb et al., 2007; Sack et al., 2003; Sack and Frole, 2006), we hypothesize that LVD will be significantly correlated with $\Delta^{13}C$ (and hence WUE$_i$).

To summarize, we address the following questions.

- Does access to groundwater by *E. camaldulensis* and *C. opaca* result in significantly smaller WUE$_i$ compared to *A. aneura*?

- Does LVD correlate with DTGW in the three species examined?

- Is there a correlation between LVD and $\Delta^{13}C$ (and hence WUE$_i$) for the three species examined?

- Does horizontal and vertical (i.e. elevational) distance from a known river flood-out zone influence foliar $\Delta^{13}C$ and WUE$_i$ of co-occurring species?

- Can foliar $\Delta^{13}C$ be used as an indicator of utilization of groundwater by vegetation of arid regions?

- Can foliar $\Delta^{13}C$ be used to estimate rooting depth and upper and lower bounds of rates of groundwater use?

2 Materials and methods

2.1 Site description

The study was conducted in the Ti Tree basin, a 5500 km^2 basin located approximately 200 km north of Alice Springs (NT) and 180 km north of the Tropic of Capricorn (22.28° S, 1933.25° E, 549 m a.s.l.). Climate is characterized as tropical and arid with hot summers and warm winters. The nearest Bureau of Meteorology station (Territory Grape Farm; Met Station 015643; within 25 km of all study sites) recorded mean and median annual precipitation of 319.9 and 29 mm, respectively (1987–May 2016; http://www.bom.gov.au/, last access: April 2018). Of the annual median rainfall, 72 % falls during the summer months (December–February) and 86 % falls during the monsoon season (November–April). Mean minimum and maximum monthly temperatures range from 5 and 22.6 °C in July to 22 and 37.5 °C in January.

The soil is a "red kandosol" (74 : 11 : 15, sand : silt : clay; Eamus et al., 2013), typical of large portions of semi-arid Australia, and has a high potential for drainage (Morton et al., 2011; Schmidt et al., 2010). Patches of hard siliceous soil are often observed and are likely surface expressions of the underlying hardpan (Cleverly et al., 2013), a common formation in the top 1–1.5 m in this type of soil (Cleverly et al., 2013, 2016a, b; Morton et al., 2011). The major potable source of water for this region is a large underground reservoir, recharged mainly by seepage from creek/river channels

and their flood-out zones, "mountain" front recharge, and occasional very heavy rainfall events (NRETAS, 2009; Calf et al., 1991). The Ti Tree basin has a natural gradient in DTGW. The depth of the water table below ground level is shallow (< 2 m) in the northern part and groundwater is lost through evapotranspiration (Shanafield et al., 2015), whereas DTGW reaches 60 m in the southern and western parts of the basin and 20–40 m in the eastern region (NRETA, 2007).

All study sites were characterized as being in one of three distinct vegetation types (Nolan et al., 2017; Cleverly et al., 2016a): (1) riparian, predominantly consisting of *Eucalyptus camaldulensis var. obtusa*, which line the banks of the ephemeral streams in the Ti Tree basin (Woodforde River and Allungra Creek); (2) low mixed woodland (*A. aneura* F. Muell. ex Benth., *A. aptaneura* Maslin & J. E. Reid, *A. kempeana* F. Muell.) with an understorey of shrubs, herbs, and C$_3$ and C$_4$ grasses; (3) tall, open *Corymbia* savanna with extensive Spinifex grass (*Triodia spp.*), sparse *Corymbia opaca* (D. J. Carr & S. G. M.Carr) K. D. Hill & L. A. S. Johnson trees, and occasional *Acacia* spp. trees. The Woodforde River and Allungra Creek are ephemeral streams that flow only after large extensive rainfall events. Nonetheless, perched aquifers beneath their riparian corridors are recharged by large storms (Villeneuve et al., 2015), providing long-term access to groundwater near ephemeral streams. Allungra Creek and its flood-out zone represent zones of local recharge (NRETAS, 2009). Overbank flooding and sheet flow occur in flood-outs where the Woodforde River and Allungra Creek enter the basin and split into a network of smaller channels (NRETA, 2007), resulting in an estimated 1.8 ML (megaliters) of groundwater recharge per year (NRETAS, 2009).

Four sampling plots (see the map in the Supplement) were established for determination of foliar $\Delta^{13}C$ and of $\delta^{18}O$ and δ^2H values for xylem water and groundwater from a nearby bore. DTGW in each of the four plots was 4.4, 8.3, 8.8, and 13.9 m, respectively. One of the four plots was located on the banks of the Woodforde River (plot 1, DTGW = 4.4 m). *E. camaldulensis* is the dominant tree species in plot 1, and *C. opaca* is also present. In the second plot, *A. aneura* is the dominant species (plot 2, DTGW = 8.3 m). *C. opaca* is the dominant tree species in the two remaining plots (plot 3, DTGW = 8.8 m; plot 4, 13.9 m).

$\Delta^{13}C$ of *E. camaldulensis*, *C. opaca*, and *A. aneura* were examined along three additional transects to investigate the influence of topography and of distance from a creek on WUE$_i$. The three transects were established perpendicularly to the banks of Allungra Creek, which is in an area of known groundwater recharge at the base of the hills that bound the southern extent of the basin (NRETAS, 2009). One of the three transects was located near a permanent water hole near the flood-out and the bottom of Allungra Creek. Transects two and three were 1–2 km upstream of the water hole. Transects extended from the creek bank within 1 m of the creek up to a maximum of 1800 m from the creek (vertical, that

Table 1. A summary of plots, transects, and spot measurements undertaken in the present study.

Plot/transect number	Depth to groundwater (m)	Species sampled	Replication	Isotopes analysed
Plots 1–4	4.4–13.9	*Eucalyptus camaldulensis, Acacia aneura (Mulga), Corymbia opaca*	2 or 3 trees per species, 3–5 samples per tree per isotope	^{18}O, deuterium of ground-water and xylem water; and ^{13}C of leaves/phyllodes
Transects 1–3 (Allungra Creek study)	Regional aquifer > 40 m but shallow ephemeral GW present after flood-outs	*Eucalyptus camaldulensis Acacia aneura*	2 or 3 trees per species, 3 or more samples per tree	^{13}C of leaves/ phyllodes
Additional spot measurements across the basin	20, 36, and 49.5	*Acacia aneura Corymbia opaca*	2 or 3 trees per species, 3 or more samples per tree	^{13}C of leaves/phyllodes

is, elevational, distance from the creek bed ranged from 0 to 4 m), across which vegetation graded from riparian forest to *Corymbia* open savanna, with occasional *Acacia* spp. trees interspersed throughout.

In addition to the four plots and three transects, "spot sampling" for foliar $\Delta^{13}C$ alone was performed at three sites to extend the examination of variation in WUE_i to 20, 36, and 49.5 m DTGW for *A. aneura* or 20 and 36 m for *C. opaca*. Finally, continental sampling of foliar $\Delta^{13}C$ of multiple dominant tree species was undertaken at seven sites distributed across Australia (Table S1 in the Supplement; Karan et al., 2016; Rumman et al., 2017) in order to allow comparison of foliar $\Delta^{13}C$ of our three Ti Tree species with a continental-scale regression of $\Delta^{13}C$ with rainfall. A summary of the isotopes analysed for each species and for each groundwater sampling undertaken, for each site, is given in Table 1.

2.2 Leaf sampling protocols and meteorology

Sampling was undertaken in April 2014 (end of the wet season) and September 2013 (end of the dry season). Three mature, healthy leaves on each of three branches from two or three replicate trees were sampled for $\Delta^{13}C$ in all plots, transects, and spot sampling sites. In addition, terminal branches of the trees in the four plots were collected for deuterium and $\delta^{18}O$ analyses of xylem water. Bore water samples were also collected in the four plots and the three spot sampling sites using groundwater samples were also collected from the bores located at each site using a HydraSleeve no-purge groundwater sampler (Cordry, 2003). Sampling along the three transects occurred at three or four points along each transect. The three trees of each of the dominant species at each location were located within 50 m of the bore. Leaves for leaf-vein analysis (see below) were collected during September 2013.

Climate conditions preceding and during the sampling periods were obtained from two eddy-covariance towers (Fluxnet sites AU-ASM and AU-TTE; Cleverly et al., 2016a; Cleverly, 2011, 2013). AU-ASM is located to the west of the Ti Tree at the spot sampling site where DTGW is 49.5 m.

AU-TTE is located near the eastern edge of the Ti Tree in plot 3 where DTGW is 8.8 m.

2.3 Stable isotope analyses: deuterium, $\delta^{18}O$ and foliar Delta^{13}C

Branch xylem water was extracted by cryogenic vacuum distillation (whereby samples are subject to a vacuum and water vapour is frozen using liquid nitrogen, as described in Ingraham and Shadel, 1992; West et al., 2006). Water from each branch sample was extracted for a minimum of 60–75 min (West et al., 2006). δ^2H and $\delta^{18}O$ analyses of branch water and groundwater were performed using a Picarro L2120-i Analyser for Isotopic H_2O. Five laboratory standards were calibrated against IAEA VSMOW2 – SLAP2 scale (Vienna Standard Mean Ocean Water 2, VSMOW2: $\delta^{18}O = 0$‰ and $\delta^2H = 0$‰; Standard Light Antarctic Precipitation 2, SLAP2: $\delta^{18}O = -55.5$‰ and $\delta^2H = -427.5$‰) and Greenland Ice Sheet Precipitation (GISP, $\delta^{18}O = -24.8$‰ and $\delta^2H = -189.5$,‰) as quality control references (IAEA, 2009). The standard deviation of the residuals between the VSMOW2 – SLAP2 value of the internal standards and the calculated values based on best linear fits was ca. 0.2‰ for $\delta^{18}O$ and ca. 1.0‰ for δ^2H.

2.4 Carbon isotope ratios of leaves

Leaf samples stored in paper bags were completely dried in an oven at 60 °C for 5 days. After drying, each leaf sample was finely ground to powder with a Retsch MM300 bead grinding mill (Verder Group, Netherlands) until homogeneous. Between 1 and 2 mg of ground material was sub-sampled in 3.5 mm × 5 mm tin capsules for analysis of the stable carbon isotope ratio ($\delta^{13}C$), generating three representative independent values per tree. All $\delta^{13}C$ analyses were performed in a Picarro G2121-i Analyser (Picarro, Santa Clara, CA, USA) for isotopic CO_2. Atropine and acetanilide were used as laboratory standard references. Results were normalized with the international standards sucrose (IAEA-CH-6, $\delta^{13}C_{VPDB} = -10.45$‰), cellulose (IAEA-CH-3, $\delta^{13}C_{VPDB} = -24.72$‰) and graphite

(USGS24, $\delta^{13}C_{VPDB} = -16.05\%$). The standard deviation of the residuals between IAEA standards and calculated values of $\delta^{13}C$ based on best linear fit was ca. 0.5‰.

2.5 Calculation of WUE$_i$ from $\delta^{13}C$ and hence $\Delta^{13}C$

WUE$_i$ was determined from ^{13}C discrimination ($\Delta^{13}C$), which is calculated from the bulk-leaf carbon isotope ratio ($\delta^{13}C$), using the following equations (Werner et al., 2012):

$$\Delta^{13}C\,(\%o) = \frac{R_a - R_p}{R_p} = \frac{\delta^{13}C_a - \delta^{13}C_p}{1 + \frac{\delta^{13}C_p}{1000}}, \tag{1}$$

$$WUE_i = \frac{c_a(b - \Delta^{13}C)}{1.6(b - a)}. \tag{2}$$

2.6 Leaf-vein density

A small sub-section (approximately $1\,cm^2$) of all leaves sampled for ^{13}C were used for LVD analysis, providing three leaf sub-sections per tree (nine samples per species). Due to the small size of *A. aneura* phyllodes, several phyllodes were combined and ground for ^{13}C analysis and one whole phyllode was used for LVD analysis. Each leaf section was cleared and stained following the approach described in Gardner (1975). A 5% (w/v) NaOH solution was used as the principal clearing agent. Leaf sections were immersed in the NaOH solution and placed in an oven at 40 °C overnight (Gardner, 1975). Phyllodes of *Acacia* proved difficult to clear effectively and were kept in the oven longer than overnight to aid clearing. Once cleared, the partially translucent leaf sections were stained with a 1% (w/v) safranin solution. Most leaf sections were stained for up to 3 min and then soaked with a 95% (w/v) ethanol solution until the vein network was sufficiently stained and the majority of colour was removed from the lamina. After staining, the cuticle was removed to aid in identifying the vein network. Following cuticle removal, leaf sub-sections were photographed using a Nikon microscope (model: SMZ800) at 40× magnification. Finally, minor veins were traced by hand and LVD was calculated as total vein length per unit area ($mm\,mm^{-2}$) using ImageJ version 1.48 (National Institutes of Health, USA).

2.7 Data and statistical analysis

Species-mean values ($n = 9$) for the dominant overstorey species at each location were calculated for δ^2H, $\delta^{18}O$, $\delta^{13}C$, and LVD. Relationships between $\Delta^{13}C$ and WUE$_i$ with DTGW were tested using regression analysis after testing for non-normality (Shapiro–Wilk test, $\alpha = 0.05$) and homogeneity of variances (Bartlett test). A Tukey post hoc test for multiple comparisons across sites was used to test for significance of variation as a function of DTGW. Breakpoints in functions with DTGW were determined using segmented regression analyses whereby the best fitting function is obtained by maximizing the statistical coefficient of explanation. The least squares method was applied to each of the two

segments while minimizing the sum of squares of the differences between observed and calculated values of the dependent variables. Next, one-way ANOVA was applied to determine significance of regressions and breakpoint estimates within a given season. We fully acknowledge that the number of plots with shallow DTGW was sub-optimal, but constraints arising from species distribution across the Ti Tree precluded additional sampling.

3 Results

3.1 Meteorological conditions during the study period

Mean daily temperature and mean daily vapour pressure deficit (VPD) were largest in summer (December–February) and smallest in winter (June–August) (Fig. 1). Daily sums of rainfall showed that the DTW 8.8 and 49.4 m sites received 265 and 228 mm rainfall, respectively, between the September 2013 and April 2014 sampling dates, representing more than 73% and 60% of the total respective rainfall received from January 2013 to June 2014 at these sites.

3.2 Variation in source-water uptake

Xylem water isotope ratios for *A. aneura* were widely divergent from the bore water stable isotope ratios (Fig. 2a) in both wet and dry seasons, indicative of a lack of access to groundwater. In contrast, δ^2H and $\delta^{18}O$ of xylem water in *E. camaldulensis* and *C. opaca* were predominantly (with two exceptions) tightly clustered around δ^2H and $\delta^{18}O$ of groundwater in the bores located within 50 m of the trees (Fig. 2b). There was little variation in xylem water composition between the end of the dry season and end of the wet season for either species, reflecting the consistent use of groundwater by these two species.

$\Delta^{13}C$ of *A. aneura*, sampled in the *Corymbia* savanna and *Acacia* spp. plots, was not significantly correlated with DTGW, in either season or across all values of DTGW (ANOVA F = 1.78; $p > 0.05$; Fig. 3a, b). As a consequence of these patterns in $\Delta^{13}C$, WUE$_i$ did not vary significantly for *A. aneura* across sites differing in DTGW (Fig. 4a) despite the large variability in WUE$_i$ (ranging from 62 to 92 $\mu mol\,mol^{-1}$) across sites and seasons. By contrast, foliar $\Delta^{13}C$ of *E. camaldulensis* and *C. opaca* declined significantly with increasing DTGW in both seasons at the sites where DTGW was relatively shallow (DTGW < ca. 12 m). Segmented regression analysis shown in Fig. 3c and d yielded breakpoints at 11.17 ± 0.54 m in September (ANOVA F = 11.548; df = 2, 38; $P < 0.01$) and 9.81 ± 0.4 m in April (ANOVA F = 14.67; df = 2, 47; $P < 0.01$), which represent the seasonal maximum depths from which groundwater can be extracted by these species. Thus, WUE$_i$ of *E. camaldulensis* and *C. opaca* was significantly smaller at the shallowest site than at sites with DTGW > 13.9 m (Fig. 4b), but did

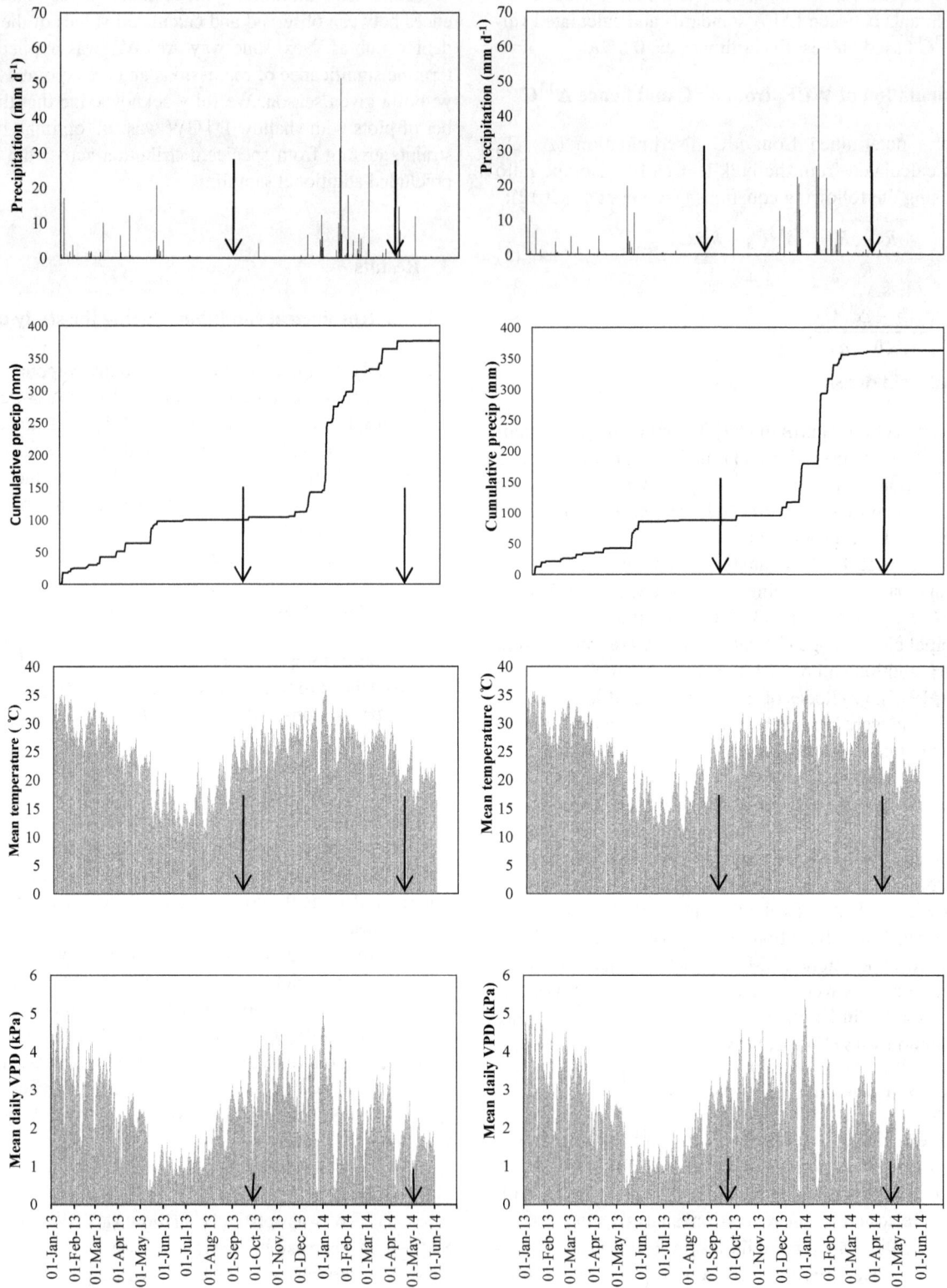

Figure 1. Mean daily meteorological conditions of daily precipitation, cumulative precipitation, mean air temperature, and vapour pressure deficit in January 2013–June 2014. Red lines show sampling periods in September 2013 (late dry season) and April 2014 (late wet season). On the left are data from the western EC tower (DTGW 49.4 m), and the other side shows data for the eastern EC tower (DTGW 8.8 m).

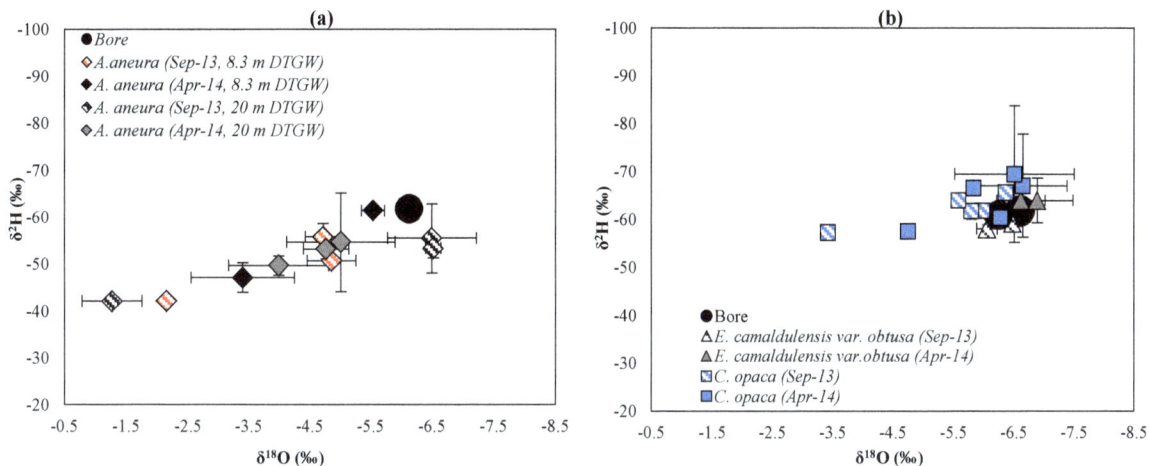

Figure 2. Comparison of xylem and bore water δ^2H–δ^{18}O plots for *Acacia aneura* sampled from 8.3 and 20 m DTGW **(a)** and *Eucalyptus camaldulensis* and *Corymbia opaca* sampled from 4.4, 8.8, and 13.9 m DTGW **(b)**. Error bars represent ± 1 standard error.

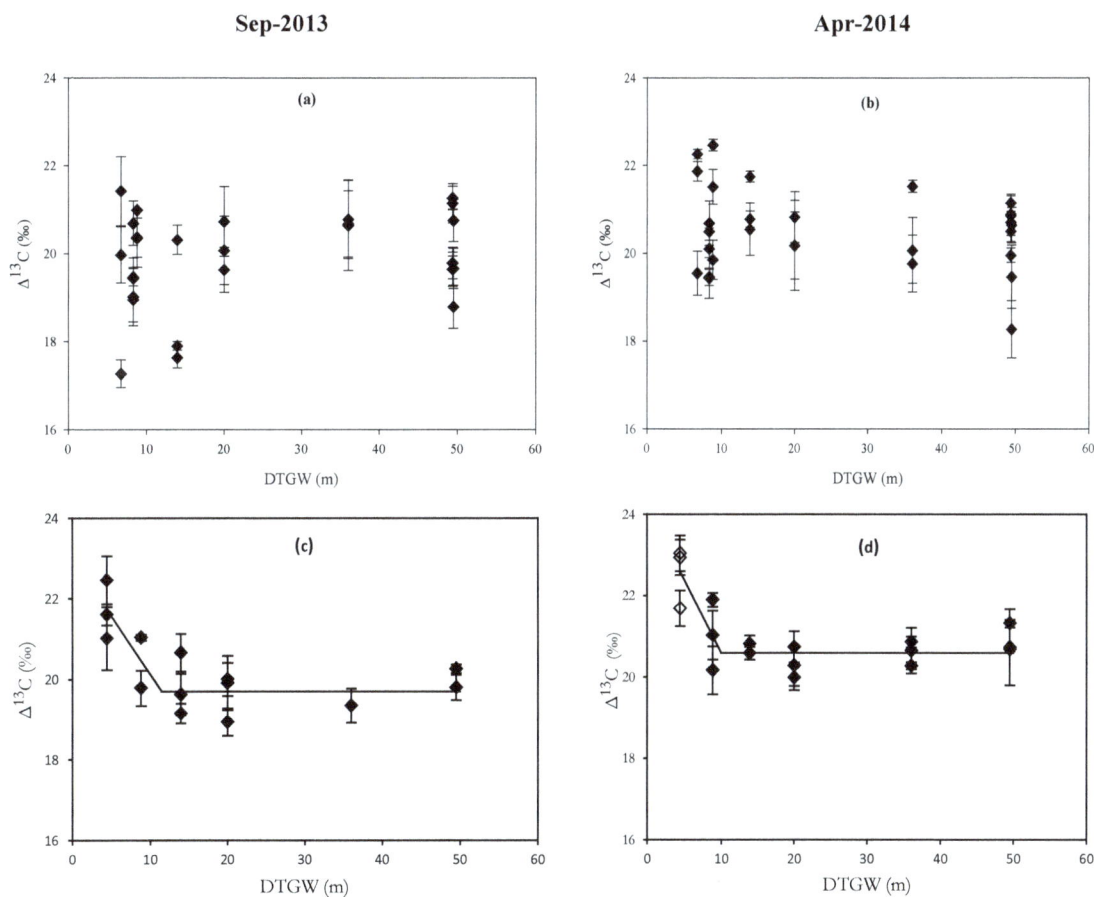

Figure 3. Carbon isotope discrimination in leaf dry matter (Δ^{13}C) plotted as a function of depth-to-groundwater (DTGW) in the Ti Tree basin. Panels **(a)** and **(b)** are data for *A. aneura* only; in panels **(c)** and **(d)** *C. opaca* (black symbols) and *E. camaldulensis* (red symbols) are presented. Left and right panels show September 2013 and April 2014 sampling, respectively. Lines in panels **(c)** and **(d)** are from segmented regression of the combined data. Error bars represent ± 1 standard error.

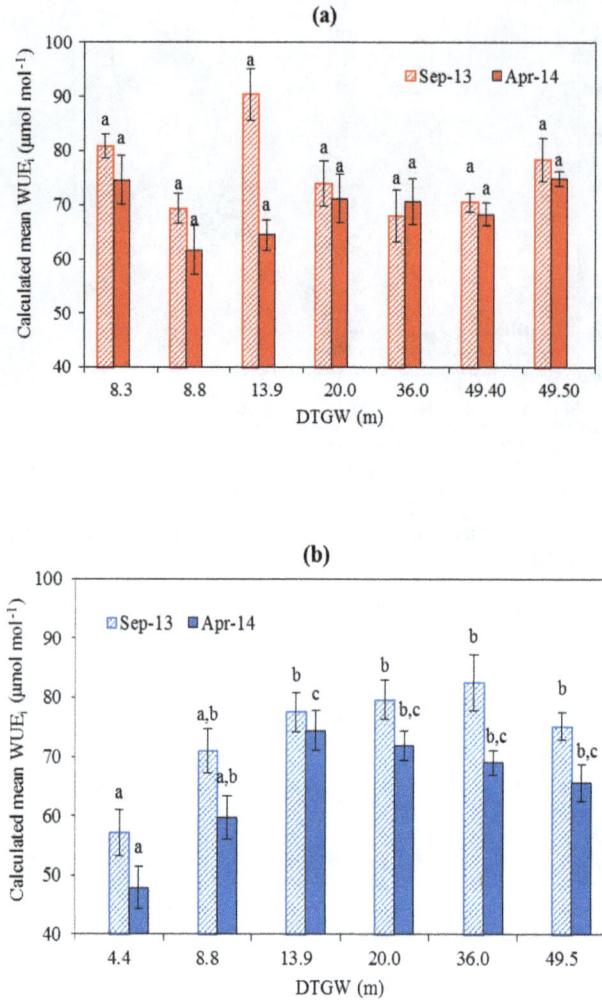

(a)

(b)

Figure 4. Leaf intrinsic water-use efficiency (WUE$_i$) calculated from Δ^{13}C in shallow-rooted *A. aneura* **(a)** and deep-rooted *E. camaldulensis* and *C. opaca* **(b)** across study sites for September 2013 (patterned column) and April 2014 (filled column). Bars within a season with the same letter are not significantly different across the depth-to-groundwater gradient (Tukey HSD, $p < 0.05$). Error bars represent ± 1 standard error.

not differ significantly across the deeper DTGW range (13.9–49.5 m; Fig. 4b).

As distance from Allungra Creek increased, foliar Δ^{13}C for both *E. camaldulensis* and *C. opaca* declined significantly (Fig. 5a), with concomitant increases in WUE$_i$ (Fig. 5b). There was no significant relationship for Δ^{13}C or WUE$_i$ with distance from the creek for *A. aneura* (data not shown).

3.3 Leaf-vein density

LVD did not vary significantly with increasing DTGW for *A. aneura* (Fig. 6a), but a significant increase in LVD with increasing DTGW was observed for DTGW < ca. 10 m in the two deep-rooted species (Fig. 6b). The breakpoint for *E. camaldulensis* and *C. opaca* (Fig. 6b; 9.36 m $\pm$ 0.6 m;

ANOVA F $= 6.38$; $P < 0.05$) agreed well with the two previous estimates (cf. Figs. 3 and 6), although again, constraints imposed by species distributions severely limited the number of samplings available at shallow DTGW sites. As with LVD and DTGW, no relationship was observed between LVD and Δ^{13}C for *A. aneura* (Fig. 7a), but a significant linear decline in Δ^{13}C with increasing LVD was observed for *E. camaldulensis* and *C. opaca* (Fig. 7b).

4   Discussion

Analyses of stable isotopes of bore water (i.e. groundwater) and xylem water across a DTGW gradient established that *A. aneura* adopted an "opportunistic" strategy of water use and was dependent on rainfall stored within the soil profile. This is consistent with previous studies, where *Acacia* spp. was shown to be very responsive to changes in upper soil moisture content, as expected given their shallow rooting depth and the presence of a shallow (< 1.5 m) hardpan below stands of *Acacia* spp. (Eamus et al., 2013; Pressland, 1975). Furthermore, very low predawn leaf-water potentials (< −7.2 MPa; Eamus et al., 2016) and very high sapwood density (0.95 g cm^{-3}; Eamus et al., 2016) in *A. aneura* of Ti Tree, and which are strongly correlated with aridity, confirm that they rely on soil water without access to groundwater, consistent with the findings of Cleverly et al. (2016b). By contrast, analyses of stable isotopes in groundwater and xylem water of *E. camaldulensis* and *C. opaca* established their access to groundwater, as has been inferred previously because of their large rates of transpiration in the dry season and consistently high (close to zero) predawn water potentials (Howe et al., 2007; O'Grady et al., 2006a, b). Importantly, we observed no significant change in xylem isotope composition for these two deep-rooted species (*E. camaldulensis* and *C. opaca*) between the end of the wet season and the end of the dry season, further evidence of year-round access to groundwater at the shallowest DTGW sites.

One specific aim of the present study was to determine whether discrimination against ^{13}C (Δ^{13}C) and resultant intrinsic water-use efficiency (WUE$_i$) could be used to identify access to groundwater. An increase in foliar Δ^{13}C represents decreased access to water and increasing WUE$_i$ (Leffler and Evans, 1999; Zolfaghar et al., 2014, 2017). The shallow-rooted *A. aneura* did not show any significant relationship of Δ^{13}C with DTGW during either season. Consequently mean WUE$_i$ showed no significant trend with increasing DTGW, consistent with the conclusion that *A. aneura* only accessed soil water during either season. In contrast, a breakpoint in the relationship between Δ^{13}C and DTGW was apparent between 9.4 m (derived from LVD results) and 11.2 m (derived from Δ^{13}C analyses) when data for the two species were combined. Where DTGW was larger than a threshold (DTGW > ca. 12 m), Δ^{13}C became independent of DTGW. Consequently, WUE$_i$ increased significantly as DTGW in-

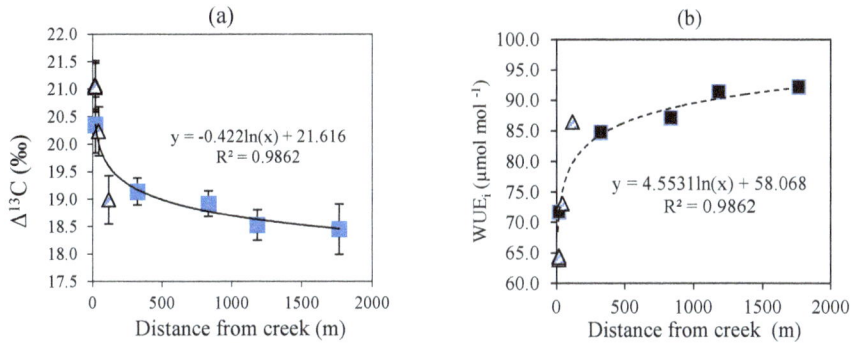

Figure 5. Δ^{13}C **(a)** and WUE$_i$ **(b)** of deep-rooted species sampled across the Ti Tree basin plotted as functions of distance from Allungra Creek bed. Striped triangles represent *E. camaldulensis* and blue squares represent *C. opaca*. Error bars represent ± 1 standard error. The regression is fitted only to the *E. camuldulensis* data. Note that the largest value of Δ^{13}C and the lowest value of WUE for *C. opaca* are three overlapping samples.

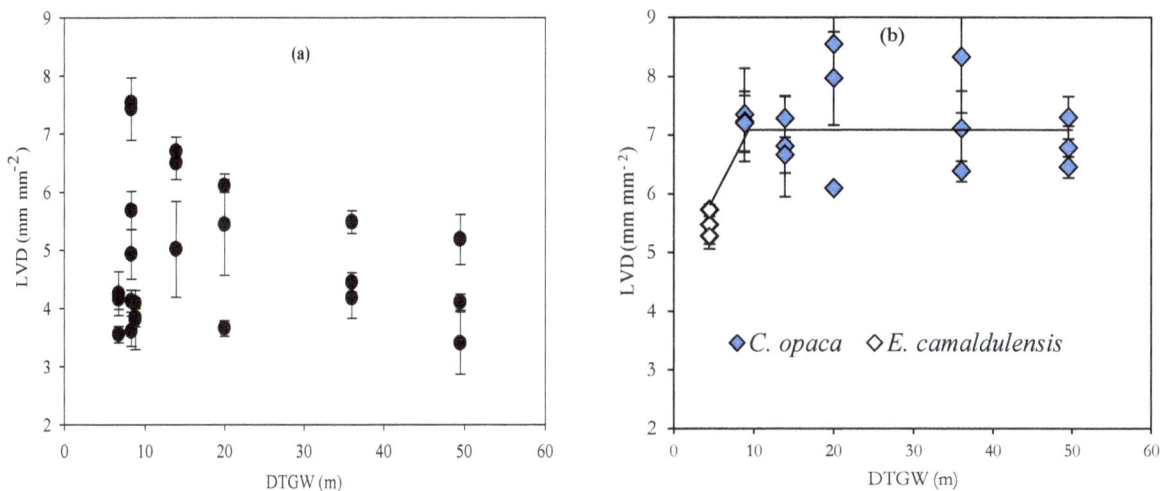

Figure 6. Leaf-vein density (LVD) of *A. aneura* **(a)** or *E. camaldulensis* (red symbols) and *C. opaca* (blue symbols) **(b)** as a function of depth-to-groundwater. Each symbol represents mean LVD calculated from three individual leaves. Error bars represent ± 1 standard error. A statistically significant correlation derived from segmented linear regression of leaf-vein density, for *E. camaldulensis* and *C. opaca* data combined, with depth-to-groundwater (DTGW) is shown in panel **(b)**. The r^2 and standard deviation slope of the regression below the breakpoint in **(b)** are 0.976 and 0.0031, respectively.

creased to these thresholds ($p < 0.001$), but did not vary with further increases in DTGW. We therefore suggest that foliar Δ^{13}C (or WUE$_i$) can be used as an indicator of groundwater access by vegetation. Δ^{13}C is less expensive and easier to measure than stable isotope ratios of water (δ^2H and δ^{18}O) in groundwater, soil water, and xylem water. Furthermore, canopies are generally more accessible than groundwater. Globally, identification of groundwater-dependent ecosystems has been hindered by the lack of a relatively cheap and easy methodology (Eamus et al., 2015); thus, Δ^{13}C shows great promise for identifying groundwater-dependent vegetation and ecosystems.

Whilst acknowledging the sub-optimal distribution of samples across the shallow DTGW range (< 10 m) from

which breakpoints in regressions were calculated (Figs. 3 and 6), which arose because of the natural distribution of trees across the basin, we can ask the question: are the speculated depths beyond which groundwater appears to become inaccessible supported by other independent studies of Australian trees? Several analyses support our suggestion of a lower limit of approximately 12 m beyond which groundwater is inaccessible to vegetation in this basin. Eamus et al. (2015) present the results of a seven-site (seven sites across the range 2.4–37.5 m DTGW), 18-trait, five-species study and identify a breakpoint between 7 and 9 m. Similarly, two recent reviews identify lower limits to root extraction of groundwater of 7.5 m (Benyon et al., 2006) and 8–10 m (O'Grady et al., 2010), while Cook et al. (1998b) established

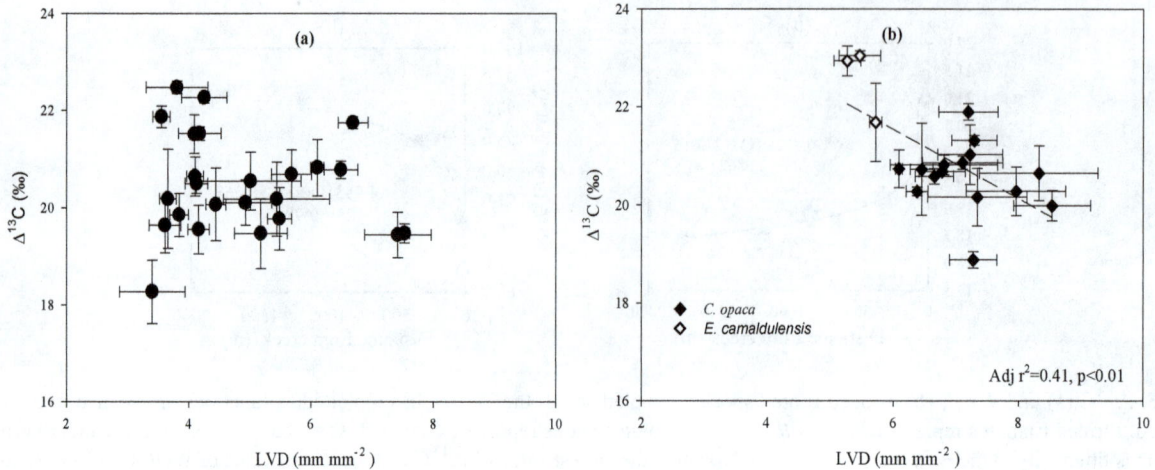

Figure 7. Relationships of leaf-vein density of **(a)** *A. aneura* and **(b)** *E. camaldulensis* (red symbols) and *C. opaca* (blue symbols) with bulk-leaf Δ^{13}C. Each symbol represents mean LVD and Δ^{13}C, with both variables measured on the same leaf. Error bars represent ± 1 standard error. A statistically significant correlation of LVD and Δ^{13}C of *E. camaldulensis* and *C. opaca* is plotted with a dashed line.

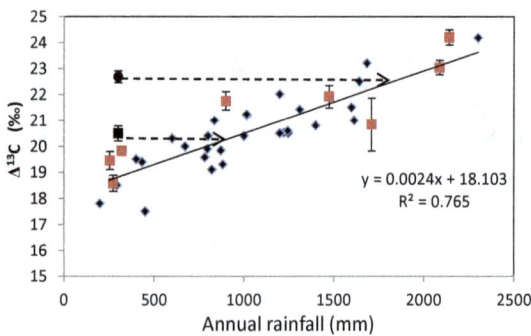

Figure 8. Relationships of discrimination against carbon-13 (Δ^{13}C) with annual rainfall observed in different studies across Australia. The diamonds represent observations made in eastern Australia (Stewart et al., 1995), northern Australia (Miller et al., 2001), and sites in New South Wales (Taylor, 2008). The red squares are data from a continental-scale assessment of foliar Δ^{13}C (Rumman et al., 2017). The black circle is the mean Δ^{13}C of *E. camaldulensis* and the black square is the mean Δ^{13}C *C. opaca*, both of which were measured in the current study. The 95 % CI for the mean Δ^{13}C is ± 0.403 and the s.e. of the slope is 0.000231. The black dashed arrows indicate the rainfall that would be required to account for the Δ^{13}C for *E. camaldulensis* and *C. opaca* if these two species relied only upon rainfall.

a limit of 8–9 m for a Eucalypt savanna. We therefore conclude that our estimates of the limits to groundwater accessibility appear reasonable.

Figure 8 shows combined Δ^{13}C from four Australian studies (Miller et al., 2001; Stewart et al., 1995; Taylor, 2008; Rumman et al., 2017), including one continental-scale study of foliar Δ^{13}C (Rumman et al., 2017). A single regression describes the data of all four independent studies. Thus, when rainfall is the sole source of water for vegetation, Δ^{13}C is strongly correlated with annual rainfall. In contrast, the mean Δ^{13}C for *E. camaldulensis* and *C. opaca* do not conform to the regression (Fig. 8). It appears that *E. camaldulensis* in Ti Tree "behaves" as though it were receiving approximately 1700 mm of rainfall, despite growing at a semi-arid site (ca. 320 mm average annual rainfall). This represents the upper limit to groundwater use by this species, assuming a zero contribution from rainfall (which is clearly very unlikely). The upper limit to annual groundwater use for *C. opaca* was similarly estimated to be 837 mm (Fig. 8). If all of the water from rainfall is used by these two species (which is also very unlikely), then the lower limit to groundwater use is the difference between rainfall and the estimates derived from Fig. 8, about 1380 mm for *E. camaldulensis* and 517 mm for *C. opaca*.

There are several independent estimates of annual tree water use for these two species which provide a valuable comparison to the estimates made above. O'Grady et al. (2009) showed that annual water use by riparian *E. camaldulensis* in Ti Tree was approximately 1642.5 m^3 m^{-2} sapwood year^{-1}. Assuming an average tree radius of 20 cm, a sapwood depth of 2 cm, and an average canopy ground cover of 25 m^2 per tree yields an annual water use of 1568 mm year^{-1}, encouragingly close to the estimate (1700 mm) derived from Fig. 8. The estimate for annual water use by *C. opaca* from O'Grady et al. (2009) is 837 mm, in reasonable agreement with the estimate from the average Δ^{13}C of *C. opaca* and the regression in Fig. 8 (ca. 900 mm). Because depth-to-groundwater for *C. opaca* is significantly larger (ca. 8–10 m) than that for *E. camaldulensis* (ca. 2–4 m), the resistance to water flow imposed by the xylem's path length is larger for *C. opaca* than *E. camaldulensis* and therefore water use may be expected to be smaller in the former than the latter, as observed.

Using an entirely different methodology from that used here, O'Grady et al. (2006c) estimated annual groundwater use by riparian vegetation on the Daly River in northern Australia to be between 694 and 876 mm, while O'Grady and Holland (2010) showed annual groundwater use to range from 2 to > 700 mm in their continental-scale review of Australian vegetation. Therefore our estimates based on $\Delta^{13}C$ appear reasonable. We conclude that (a) *E. camaldulensis* and *C. opaca* are accessing groundwater (because annual water use greatly exceeded annual rainfall); (b) $\Delta^{13}C$ can be used as an indicator of groundwater use by vegetation; and (c) $\Delta^{13}C$ can provide estimates of upper and lower bounds for the rate of groundwater use by vegetation.

4.1 Patterns of carbon isotope discrimination and intrinsic water-use efficiency along Allungra Creek transects

For the two deep-rooted species (*E. camaldulensis* and *C. opaca*), a significant decline in $\Delta^{13}C$ (and hence an increase in WUE_i) was observed with increasing distance from the creek (Fig. 5a, b). In contrast to the results of O'Grady et al. (2006c) for a steeply rising topography in the Daly River, this is unlikely to be attributable to increased elevation since the change in elevation was minimal across each entire transect (< 3 m), and even smaller near Allungra Creek where most of the change in $\Delta^{13}C$ was recorded. Therefore the cause of the change in $\Delta^{13}C$ with distance from the creek was most likely to be a function of the frequency with which trees receive flood water and hence the amount of recharge into the soil profile (Ehleringer and Cooper, 1988; Thorburn et al., 1994; Villeneuve et al., 2015; Singer et al., 2014). We therefore further conclude that foliar $\Delta^{13}C$ can be used as an indicator of access to additional water to that of rainfall, regardless of the source of that additional water (e.g. groundwater, flood recharged soil water storage, irrigation).

4.2 Leaf-vein density across the depth-to-groundwater gradient

In our study, LVD was independent of DTGW for *A. aneura*, but a breakpoint (9.4 m) was apparent across the combined data of the two deep-rooted species. Increased DTGW reflects a declining availability of water resources (Zolfaghar et al., 2015), especially in arid zones. Uhl and Mosbrugger (1999) concluded that water availability is the most important factor determining LVD. Sack and Scoffoni (2013) also showed LVD to be negatively correlated with mean annual precipitation in a 796-species meta-analysis. Therefore increasing LVD with increasing DTGW is consistent with an increased LVD with a declining water supply, despite similar amounts of rainfall being received along the DTGW gradient.

Both *A. aneura* and *C. opaca* in the present study showed LVDs close to the higher end of the global spectrum (Sack and Scoffoni, 2013), consistent with larger LVDs observed in "semi-desert" species. Higher LVDs allow for a more even spatial distribution of water across the phyllode or lamina during water stress, which contributes to a greater consistency of mesophyll hydration in species of arid and semi-arid regions (Sommerville et al., 2012). In turn, this allows continued photosynthetic carbon assimilation during water stress (Sommerville et al., 2010). Presumably, a large LVD also decreases the resistance to water flow from minor veins to mesophyll cells, which is likely to be beneficial for leaf hydration as water availability declines while also facilitating rapid rehydration following rain in these arid-zone species. Large LVDs for *A. aneura* of semi-arid regions in Australia have been associated with rapid up-regulation of phyllode function with the return of precipitation following drought (Sommerville et al., 2010), and such rapid up-regulation is crucial for vegetation in regions with unpredictable and pulsed rainfall like Ti Tree (Byrne et al., 2008; Grigg et al., 2010).

LVD was negatively correlated with bulk-leaf $\Delta^{13}C$ (and thus positively correlated with WUE_i) in *C. opaca* and the data for *E. camaldulensis* appeared to conform to the regression for *C. opaca* (Fig. 7b). What mechanism can explain the significant relationship observed between a structural leaf trait (LVD) and a functional trait (WUE_i)? The stomatal optimization model (Medlyn et al., 2011) is based on the fact that transpiration (E) and CO_2 assimilation (A) are linked via stomatal function. In order to gain carbon most economically while minimizing water loss (i.e. optimization of the ratio A / E), stomata should function such that the marginal water cost of carbon assimilation $\left(\frac{\partial A}{\partial E}\right)$ remains constant (Cowan and Farquhar, 1977; Farquhar and Sharkey, 1982). This aspect of stomatal control couples the structural traits involved with water flow with traits associated with primary production (Brodribb and Holbrook, 2007) and explains observed correlations between K_{leaf} and A_{max} in a number of studies (Brodribb et al., 2007, 2010, 2005; Brodribb and Jordan, 2008; Sack and Holbrook, 2006; Sack and Scoffoni, 2013). The length of the hydraulic pathway is directly proportional to K_{leaf} (Brodribb et al., 2007) and the A/g_s ratio determines foliar $\Delta^{13}C$ (and WUE_i) signatures in leaves. Thus, the constraint on K_{leaf} by LVD affects the coordination between the processes of A and E and thereby might explain significant relationships between structural (LVD) and functional (WUE_i) traits. For the two deep-rooted species, having access to groundwater resulted in convergence to a common solution for optimizing water supply through veins with respect to E (Brodribb and Holbrook, 2007).

Several robust and testable predictions arise from the conclusion that *E. camaldulensis* and *C. opaca* are functioning at a semi-arid site as though they have access to ca. 1700 or 900 mm rainfall, respectively. Species growing in high-rainfall zones possess a suite of traits, including low-density sapwood, large-diameter xylem vessels, small resistance to vessel implosion, large SLA, and a large maximum stomatal conductance, compared to species growing in arid regions (O'Grady et al., 2006b, 2009; Wright et al., 2004). Therefore, these attributes should be present in *E. camaldulensis* and *C. opaca* if they are functioning as though they are growing in a mesic environment. We have previously established (Eamus et al., 2016; Santini et al., 2016) that these predictions are confirmed by field data and therefore conclude that analyses of foliar $\Delta^{13}C$ have global application to the preservation and understanding of GDEs.

5 Conclusions

We posed five questions regarding depth-to-groundwater (DTGW), foliar discrimination against ^{13}C ($\Delta^{13}C$), and leaf-vein density (LVD) as underpinning the rationale for this study. We confirmed that access to shallow groundwater by *E. camaldulensis* and *C. opaca* (DTGW ca. 0–11 m) resulted in smaller WUE_i than *A. aneura*. We also demonstrated that LVD correlated with DTGW for the shallower depths (< 10 m) in *E. camaldulensis* and *C. opaca*, but not in *A. aneura*. We further demonstrated that there was correlation between LVD and $\Delta^{13}C$ (and hence WUE_i) for *E. camaldulensis* and *C. opaca*, but not in *A. aneura*. Similarly, as distance increased from a creek near a flood-out associated with aquifer recharge, foliar $\Delta^{13}C$ decreased and WUE_i increased for *E. camaldulensis* and *C. opaca*, but not for *A. aneura*. Finally, we conclude that foliar $\Delta^{13}C$ can be used as an indicator of utilization of groundwater or stored soil water by vegetation in arid regions, providing an inexpensive and rapid alternative to the stable isotopes of water that have been used in many previous studies. The observation that the $\Delta^{13}C$ of the two groundwater-using species was distant from the continental regression of $\Delta^{13}C$ against rainfall (Fig. 8) is strong evidence of the value of $\Delta^{13}C$ as an indicator of utilization of water that is additional to rainfall, and this supplemental water can be derived from either groundwater or soil recharge arising in flood-out zones of creeks.

Author contributions. All the authors contributed to field data sampling. RR undertook all the isotope analyses and statistical analyses and wrote the thesis that formed the basis of this paper. DE oversaw the design and implementation of the entire project. All the authors contributed to writing this paper and interpreting the data.

Competing interests. The authors declare that they have no conflict of interest.

Acknowledgements. The authors would like to acknowledge the financial support of the Australian Research Council for a Discovery grant awarded to Derek Eamus.

Edited by: Theresa Blume

References

Beer, C., Ciais, P., Reichstein, M., Baldocchi, D., Law, B. E., Papale, D., Soussana, J. F., Ammann, C., Buchmann, N., Frank, D., Gianelle, D., Janssens, I. A., Knohl, A., Kostner, B., Moors, E., Roupsard, O., Verbeeck, H., Vesala, T., Williams, C. A., and Wohlfahrt, G.: Temporal and among-site variability of inherent water use efficiency at the ecosystem level, Global Biogeochem. Cy., 23, GB2018, https://doi.org/10.1029/2008GB003233, 2009.

Benyon, R. G., Theiveyanathan, S., and Doody, T. M.: Impacts of tree plantations on groundwater in south-eastern Australia, Aust. J. Bot., 54, 181–192, https://doi.org/10.1071/bt05046, 2006.

Box, J. B., Duguid, A., Read, R. E., Kimber, R. G., Knapton, A., Davis, J., and Bowland, A. E.: Central Australian waterbodies: The importance of permanence in a desert landscape, J. Arid Environ., 72, 1395–1413, https://doi.org/10.1016/j.jaridenv.2008.02.022, 2008.

Brodribb, T. J., Feild, T. S. and Jordan, G. J.: Leaf maximum photosynthetic rate and venation are linked by hydraulics, Plant Phys., 144, 1890–1898, 2007.

Brodribb, T. J., Feild, T. S., and Sack, L.: Viewing leaf structure and evolution from a hydraulic perspective, Funct. Plant Biol., 37, 488-498, 2010.

Brodribb, T. J. and Holbrook, N. M.: Changes in leaf hydraulic conductance during leaf shedding in seasonally dry tropical forest, New Phytol., 158, 295–303, 2003.

Brodribb, T. J. and Holbrook, N. M.: Forced depression of leaf hydraulic conductance in situ: effects on the leaf gas exchange of forest trees, Func. Ecol., 21, 705–712, 2007.

Brodribb, T. J., Holbrook, N. M., Zwieniecki, M. A., and Palma, B.: Leaf hydraulic capacity in ferns, conifers and angiosperms: impacts on photosynthetic maxima, New Phytol., 165, 839–846, 2005.

Brodribb, T. J. and Jordan, G. J.: Internal coordination between hydraulics and stomatal control in leaves, Plant Cell Environ., 31, 1557–1564, 2008.

Bucci, S., Goldstein, G., Meinzer, F., Scholz, F., Franco, A., and Bustamante, M.: Functional convergence in hydraulic architecture and water relations of tropical savanna trees: from leaf to whole plant, Tree Physiol., 24, 891–899, 2004.

Budyko, M. I.: Climate and Life, Academic Press, New York, 508 pp., 1974.

Byrne, M., Yeates, D. K., Joseph, L., Kearney, M., Bowler, J., Williams, M. A. J., Cooper, S., Donnellan, S. C., Keogh, J. S., and Leys, R.: Birth of a biome: insights into the assembly and maintenance of the Australian arid zone biota, Molec. Ecol., 17, 4398–4417, 2008.

Calf, G. E., McDonald, P. S., and Jacobson, G.: Recharge mechanism and groundwater age in the Ti-Tree basin, Northern Territory, Aust. J. Earth Sci., 38, 299–306, https://doi.org/10.1080/08120099108727974, 1991.

Cernusak, L. A., Hutley, L. B., Beringer, J., Holtum, J. A., and Turner, B. L.: Photosynthetic physiology of eucalypts along a sub-continental rainfall gradient in northern Australia, Agr. Forest Meteorol., 151, 1462–1470, 2011.

Clarke, R.: Water: the international crisis, Earthscan, London, UK, 1991.

Cleverly, J.: Alice Springs Mulga OzFlux site, TERN OzFlux: Australian and New Zealand Flux Research and Monitoring Network, hdl:102.100.100/14217, 2011.

Cleverly, J.: Ti Tree East OzFlux Site, TERN OzFlux: Australian and New Zealand Flux Research and Monitoring Network, hdl:102.100.100/11135, 2013.

Cleverly, J., Boulain, N., Villalobos-Vega, R., Grant, N., Faux, R., Wood, C., Cook, P. G., Yu, Q., Leigh, A., and Eamus, D.: Dynamics of component carbon fluxes in a semi-arid *Acacia* woodland, central Australia, J. Geophys. Res.-Biogeosci., 118, 1168–1185, 2013.

Cleverly, J., Eamus, D., Van Gorsel, E., Chen, C., Rumman, R., Luo, Q., Coupe, N. R., Li, L., Kljun, N., and Faux, R.: Productivity and evapotranspiration of two contrasting semiarid ecosystems following the 2011 global carbon land sink anomaly, Agr. Forest Meteorol., 220, 151–159, 2016a.

Cleverly, J., Eamus, D., Restrepo Coupe, N., Chen, C., Maes, W., Li, L., Faux, R., Santini, N. S., Rumman, R., Yu, Q., and Huete, A.: Soil moisture controls on phenology and productivity in a semi-arid critical zone, Sci. Total Environ., 568, 1227–1237, https://doi.org/10.1016/j.scitotenv.2016.05.142, 2016b.

Cook, P. G., O'Grady, A. P., Wischusen, J. D. H., Duguid, A., Fass, T., Eamus, D., and Palmerston, N. T.: Ecohydrology of sandplain woodlands in central Australia, Department of Natural Resources, Environment and The Arts, Northern Territory Government, Darwin, 2008a.

Cook, P. G., Hatton, T. J., Pidsley, D., Herczeg, A. L., Held, A., O'Grady, A., and Eamus, D.: Water balance of a tropical woodland ecosystem, northern Australia: a combination of micro-meteorological, soil physical and groundwater chemical approaches, J. Hydrol. 210, 161–177, https://doi.org/10.1016/S0022-1694(98)00181-4, 2008b.

Cordry, K.: HydraSleeve: A new no-purge groundwater sampler for all contaminants, Interstate Technology and Regulatory Council Fall Conference, Monterey, California, USA, 2003.

Cowan, I. R. and Farquhar, G. D.: Stomatal function in relation to leaf metabolism and environment: Stomatal function in the regulation of gas exchange, Symposia of the Society for Experimental Biology, 31, 471–505, 1997.

Craig, H.: Isotopic variations in meteoric waters, Science, 133, 1702–1703, 1961.

Cullen, L. E. and Grierson, P. F.: A stable oxygen, but not carbon, isotope chronology of *Callitris columellaris* reflects recent climate change in north-western Australia, Clim. Change, 85, 213–229, https://doi.org/10.1007/s10584-006-9206-3, 2007.

Donohue, R. J., McVicar, T., and Roderick, M. L.: Climate-related trends in Australian vegetation cover as inferred from satellite observations, 1981–2006, Glob. Change Biol., 15, 1025–1039, 2009.

Eamus, D., O'Grady, A. P., and Hutley, L.: Dry season conditions determine wet season water use in the wet–tropical savannas of northern Australia, Tree Physiol., 20, 1219–1226, 2000.

Eamus, D., Hatton, T., Cook, P., and Colvin, C.: Ecohydrology: vegetation function, water and resource management, CSIRO Publishing, Melbourne, 348 pp., 2006.

Eamus, D., Cleverly, J., Boulain, N., Grant, N., Faux, R., and Villalobos-Vega, R.: Carbon and water fluxes in an arid-zone Acacia savanna woodland: An analyses of seasonal patterns and responses to rainfall events, Agr. Forest Meteorol., 182, 225–238, 2013.

Eamus, D., Zolfaghar, S., Villalobos-Vega, R., Cleverly, J., and Huete, A.: Groundwater-dependent ecosystems: recent insights from satellite and field-based studies, Hydrol. Earth Syst. Sci., 19, 4229–4256, https://doi.org/10.5194/hess-19-4229-2015, 2015.

Eamus, D., Huete, A., Cleverly, J., Nolan, R. H., Ma, X., Tarin, T., and Santini, N. S.: Mulga, a major tropical dry open forest of Australia: recent insights to carbon and water fluxes, Environ. Res. Lett., 11, 125011, https://doi.org/10.1088/1748-9326/11/12/125011, 2016.

Ehleringer, J. R. and Cooper, T. A.: Correlation between carbon isotope ratio and microhabitat in desert plants, Oecologia, 76, 562–566, 1988.

Farquhar, G. D. and Sharkey, T. D.: Stomatal conductance and photosynthesis, Ann. Rev. Plant Physio., 33, 317–345, 1982.

Gardner, R. O.: An overview of botanical clearing technique, Stain Tech., 50, 99–105, 1975.

Grigg, A. M., Lambers, H., and Veneklaas, E. J.: Changes in water relations for *Acacia ancistrocarpa* on natural and mine-rehabilitation sites in response to an experimental wetting pulse in the Great Sandy Desert, Plant Soil, 326, 75–96, 2010.

Hacke, U. G., Sperry, J. S., Ewers, B. E., Ellsworth, D. S., Schäfer, K. V. R., and Oren, R.: Influence of soil porosity on water use in *Pinus taeda*, Oecologia, 124, 495–505, 2000.

Howe, P., O'Grady, A. P., Cook, P. G., and Fas, T.: Project REM1 – A Framework for Assessing Environmental Water Requirements for Groundwater Dependent Ecosystems Report-2 Field Studies, Land and Water Australia, 2007.

IAEA: Reference Sheet for VSMOW2 and SLAP2 international measurement standards, International Atomic Energy Agency, Vienna, available at: http://nucleus.iaea.org/rpst/document/vsmow2_slap2.pdf (last access: April 2018), 2009.

Ingraham, N. L. and Shadel, C.: A comparison of the toluene distillation and vacuum/heat methods for extracting soil water for stable isotopic analysis, J. Hydrol., 140, 371–387, 1992.

Karan, M., Liddell, M., Prober, S., Arndt, S., Beringer, J., Boer, M., Cleverly, J., Eamus, D., Grace, P., van Gorsel, E., Hero, J.-M., Hutley, L., Macfarlane, C., Metcalfe, D., Meyer, W., Pendall, E., Sebastian, A., and Wardlaw, T.: The Australian SuperSite Network: a continental, long-term terrestrial ecosystem observatory, Sci. Total Environ., 568, 1263–1274, https://doi.org/10.1016/j.scitotenv.2016.05.170, 2016.

Leffler, A. J. and Evans, A. S.: Variation in carbon isotope composition among years in the riparian tree *Populus fremontii*, Oecologia, 119, 311–319, 1999.

Maslin, B. R. and Reid, J. E.: A taxonomic revision of Mulga (*Acacia aneura* and its close relatives: Fabaceae) in Western Australia, Nuytsia, 22, 129–167, 2012.

Medlyn, B. E., Duursma, R. A., Eamus, D., Ellsworth, D. S., Prentice, I. C., Barton, C. V. M., Crous, K. Y., De Angelis, P., Freeman, M., and Wingate, L.: Reconciling the optical and empirical approaches to modelling stomatal conductance, Glob. Change Biol., 17, 2134–2144, 2011.

Meinzer, F. C. and Grantz, D. A.: Stomatal and hydraulic conductance in growing sugarcane: stomatal adjustment to water transport capacity, Plant Cell Environ., 13, 383–388, 1990.

Miller, J. M., Williams, R. J., and Farquhar, G. D.: Carbon isotope discrimination by a sequence of *Eucalyptus* species along a sub-continental rainfall gradient in Australia, Func. Ecol., 15, 222–232, 2001.

Morton, S., Smith, D. S., Dickman, C. R., Dunkerley, D., Friedel, M., McAllister, R., Reid, J., Roshier, D., Smith, M., and Walsh, F.: A fresh framework for the ecology of arid Australia, J. Arid Environ., 75, 313–329, 2011.

Niinemets, Ü., Portsmuth, A., and Tobias, M.: Leaf size modifies support biomass distribution among stems, petioles and mid-ribs in temperate plants, New Phytol., 171, 91–104, 2006.

Niinemets, Ü., Portsmuth, A., Tena, D., Tobias, M., Matesanz, S., and Valladares, F.: Do we underestimate the importance of leaf size in plant economics? Disproportional scaling of support costs within the spectrum of leaf physiognomy, Ann. Bot., 100, 283–303, 2007.

Niklas, K. J., Cobb, E. D., Niinemets, Ü., Reich, P. B., Sellin, A., Shipley, B., and Wright, I. J.: "Diminishing returns" in the scaling of functional leaf traits across and within species groups, P. Natl. Acad. Sci., 104, 8891–8896, 2007.

Nolan, R. H., Fairweather, K. A., Tarin, T., Santini, N. S., Cleverly, J., Faux, R., and Eamus, D.: Divergence in plant water-use strategies in semiarid woody species, Funct. Plant Biol., 44, 1134–1146, https://doi.org/10.1071/FP17079, 2017.

NRETA: The Ti Tree Basin Aquifer. Department of Natural Resources, Environment and the Arts, Water Resources Branch, Land and Water Division, available at: https://supersites.tern.org.au/images/resource/TiTree_Basin_Groundwater.pdf (last access: April 2018), 2007.

NRETAS: Ti Tree Basin Water Resource Report, Department of Natural Resources, Environment, the Arts and Sport, Natural Resource Management Division, Water Management Branch Document No. 04/2009A, ISBN:978-1-921519-21-5, 2009.

O'Grady, A. P. and Holland, K.: Review of Australian groundwater discharge studies of terrestrial systems, CSIRO National Water Commission, Report, 60 pp., https://doi.org/10.4225/08/5852dc73abbe4, 2010.

O'Grady, A. P., Cook, P. G., Howe, P., and Werren, G.: Groundwater use by dominant tree species in tropical remnant vegetation communities, Aust. J. Bot., 54, 155–171, 2006a.

O'Grady, A. P., Eamus, D., Cook, P. G., and Lamontagne, S.: Comparative water use by the riparian trees *Melaleuca argentea* and *Corymbia bella* in the wet–dry tropics of northern Australia, Tree Phys., 26, 219–228, 2006b.

O'Grady, A. P., Eamus, D., Cook, P. G., and Lamontagne, S.: Groundwater use by riparian vegetation in the wet–dry tropics of northern Australia, Aust. J. Bot., 54, 145–154, 2006c.

O'Grady, A. P., Cook, P. G., Eamus, D., Duguid, A., Wischusen, J. D. H., Fass, T., and Worldege, D.: Convergence of tree water use within an arid-zone woodland, Oecologia, 160, 643–655, 2009.

O'Grady, A. P., Carter, J. L., and Bruce, J.: Can we predict groundwater discharge from terrestrial ecosystems using existing eco-hydrological concepts?, Hydrol. Earth Syst. Sci., 15, 3731–3739, https://doi.org/10.5194/hess-15-3731-2011, 2011.

Pressland, A. J.: Productivity and management of mulga in south-western Queensland in relation to tree structure and density, Aust. J. Bot., 23, 965–976, 1975.

Prior, L. D., Eamus, D., and Duff, G. A.: Seasonal and diurnal patterns of carbon assimilation, stomatal conductance and leaf water potential in *Eucalyptus tetrodonta* saplings in a wet–dry savanna in northern Australia, Aust. J. Bot., 45, 241–258, 1997.

Reynolds, J. F., Smith, D. M. S., Lambin, E. F., Turner, B., Mortimore, M., Batterbury, S. P., Downing, T. E., Dowlatabadi, H., Fernández, R. J., and Herrick, J. E.: Global desertification: building a science for dryland development, Science, 316, 847–851, 2007.

Rumman, R., Atkin, O. K., Bloomfield, K. J., and Eamus, D.: Variation in bulk-leaf ^{13}C discrimination, leaf traits and water use efficiency – trait relationships along a continental-scale climate gradient in Australia, Glob. Change Biol., 24, 1186–1200, 2017.

Sack, L. and Frole, K.: Leaf structural diversity is related to hydraulic capacity in tropical rain forest trees, Ecology, 87, 483–491, 2006.

Sack, L. and Holbrook, N. M.: Leaf hydraulics, Ann. Rev. Plant Biol., 57, 361–381, 2006.

Sack, L. and Scoffoni, C.: Leaf venation: structure, function, development, evolution, ecology and applications in the past, present and future, New Phytol., 198, 983–1000, 2013.

Sack, L., Cowan, P., Jaikumar, N., and Holbrook, N.: The "hydrology" of leaves: co-ordination of structure and function in temperate woody species, Plant Cell Environ., 26, 1343–1356, 2003.

Santini, N. S., Cleverly, J., Faux, R., Lestrange, C., Rumman, R., and Eamus, D.: Xylem traits and water-use efficiency of woody species co-occurring in the Ti Tree Basin arid zone, Trees, 30, 295–303, 2016.

Schmidt, S., Lamble, R. E., Fensham, R. J., and Siddique, I.: Effect of woody vegetation clearing on nutrient and carbon relations of semi-arid dystrophic savanna, Plant Soil, 331, 79–90, 2010.

Shanafield, M., Cook, P. G., Gutiérrez-Jurado, H. A., Faux, R., Cleverly, J., and Eamus, D.: Field comparison of methods for estimating groundwater discharge by evaporation and evapotranspiration in an arid-zone playa, J. Hydrol., 527, 1073–1083, https://doi.org/10.1016/j.jhydrol.2015.06.003, 2015.

Singer, M. B., Sargean, C. I., Piegay, H., Riquier, J., Wilson, R. J. S., and Evans, C. N.: Floodplain ecohydrology: Climatic, anthropogenic, and local physical controls on partitioning of water sources to riparian trees, Water Resour. Res., 50, 4490–4513, 2014.

Sommerville, K. E., Gimeno, T. E., and Ball, M. C.: Primary nerve (vein) density influences spatial heterogeneity of photosynthetic response to drought in two *Acacia* species, Funct. Plant Biol., 37, 840–848, 2010.

Sommerville, K. E., Sack, L., and Ball, M. C.: Hydraulic conductance of *Acacia* phyllodes (foliage) is driven by primary nerve (vein) conductance and density, Plant Cell Environ., 35, 158–168, 2012.

Sperry, J. S.: Hydraulic constraints on plant gas exchange, Agr. Forest Meteorol., 104, 13–23, 2000.

Stewart, G. R., Turnbull, M., Schmidt, S., and Erskine, P.: ^{13}C natural abundance in plant communities along a rainfall gradient: a biological integrator of water availability, Funct. Plant Biol., 22, 51–55, 1995.

Taylor, D.: Tree, leaf and branch trait coordination along an aridity gradient, PhD Thesis, University of Technology, Sydney, 2008.

Thomas, D. S. and Eamus, D.: The influence of predawn leaf water potential on stomatal responses to atmospheric water content at constant C_i and on stem hydraulic conductance and foliar ABA concentrations, J. Exp. Bot., 50, 243–251, 1999.

Thorburn, P. J., Mensforth, L. J., and Walker, G. R.: Reliance of creek-side river red gums on creek water, Mar. Freshwater Res., 45, 1439–1443, 1994.

Uhl, D. and Mosbrugger, V.: Leaf venation density as a climate and environmental proxy: a critical review and new data, Palaeogeogr. Palaeocl., 149, 15–26, 1999.

UTS library: Application of stable isotope analyses to examine patterns of water uptake, water use strategies and water use efficiency on contrasting ecosystems in Australia, available at: http://hdl.handle.net/10453/102763, last access: August 2018

Villeneuve, S., Cook, P. G., Shanafield, M., Wood, C., and White, N.: Groundwater recharge via infiltration through an ephemeral riverbed, central Australia, J. Arid Environ., 117, 47–58, https://doi.org/10.1016/j.jaridenv.2015.02.009, 2015.

Warren, C. R., Tausz, M., and Adams, M. A.: Does rainfall explain variation in leaf morphology and physiology among populations of red ironbark (*Eucalyptus sideroxylon* subsp. *tricarpa*) grown in a common garden?, Tree Physiol., 25, 1369–1378, 2005.

Werner, C., Schnyder, H., Cuntz, M., Keitel, C., Zeeman, M. J., Dawson, T. E., Badeck, F.-W., Brugnoli, E., Ghashghaie, J., Grams, T. E. E., Kayler, Z. E., Lakatos, M., Lee, X., Máguas, C., Ogée, J., Rascher, K. G., Siegwolf, R. T. W., Unger, S., Welker, J., Wingate, L., and Gessler, A.: Progress and challenges in using stable isotopes to trace plant carbon and water relations across scales, Biogeosciences, 9, 3083–3111, https://doi.org/10.5194/bg-9-3083-2012, 2012.

West, A. G., Patrickson, S. J., and Ehleringer, J. R.: Water extraction times for plant and soil materials used in stable isotope analysis, Rapid Comm. Mass Spec., 20, 1317–1321, 2006.

Wright, I. J., Groom, P. K., Lamont, B. B., Poot, P., Prior, L. D., Reich, P. B., Schulze, E. D., Veneklaas, E. J., and Westoby, M.: Leaf trait relationships in Australian plant species, Funct. Plant Biol., 31, 551–558, https://doi.org/10.1071/fp03212, 2004.

Wright, B. R., Latz, P. K., and Zuur, A. F.: Fire severity mediates seedling recruitment patterns in slender mulga (*Acacia aptaneura*), a fire-sensitive Australian desert shrub with heat-stimulated germination, Plant Ecol., 217, 789–800, https://doi.org/10.1007/s11258-015-0550-0, 2016.

Zolfaghar, S., Villalobos-Vega, R., Cleverly, J., Zeppel, M., Rumman, R., and Eamus, D.: The influence of depth-to-groundwater on structure and productivity of *Eucalyptus* woodlands, Aust. J. Bot., 62, 428–437, 2014.

Zolfaghar, S., Villalobos-Vega, R., Cleverly, J., and Eamus, D.: Co-ordination among leaf water relations and xylem vulnerability to embolism of Eucalyptus trees growing along a depth-to-groundwater gradient, Tree Physiol., 35, 732–743, https://doi.org/10.1093/treephys/tpv039, 2015.

Zolfaghar, S., Villalobos-Vega, R., Zeppel, M., Cleverly, J., Rumman, R., Hingee, M., Boulain, N., Li, Z., Eamus, D., and Tognetti, R.: Transpiration of *Eucalyptus* woodlands across a natural gradient of depth-to-groundwater, Tree Physiol., 37, 961–975, https://doi.org/10.1093/treephys/tpx024, 2017.

6

Effects of microarrangement of solid particles on PCE migration and its remediation in porous media

Ming Wu[1,2], Jianfeng Wu[1], Jichun Wu[1], and Bill X. Hu[2]

[1]Key Laboratory of Surficial Geochemistry, Ministry of Education, Department of Hydrosciences, School of Earth Sciences and Engineering, Nanjing University, Nanjing 210023, China
[2]Institute of Groundwater and Earth Sciences, Jinan University, Guangzhou 510632, China

Correspondence: Jianfeng Wu (jfwu@nju.edu.cn) and Jichun Wu (jcwu@nju.edu.cn)

Abstract. Groundwater can be stored abundantly in granula-composed aquifers with high permeability. The microstructure of granular materials has important effect on the permeability of aquifers and the contaminant migration and remediation in aquifers is also influenced by the characteristics of porous media. In this study, two different microscale arrangements of sand particles are compared to reveal the effects of microstructure on the contaminant migration and remediation. With the help of fractal theory, the mathematical expressions of permeability and entry pressure are conducted to delineate granular materials with regular triangle arrangement (RTA) and square pitch arrangement (SPA) at microscale. Using a sequential Gaussian simulation (SGS) method, a synthetic heterogeneous site contaminated by perchloroethylene (PCE) is then used to investigate the migration and remediation affected by the two different microscale arrangements. PCE is released from an underground storage tank into the aquifer and the surfactant is used to clean up the subsurface contamination. Results suggest that RTA can not only cause more groundwater contamination, but also make remediation become more difficult. The PCE remediation efficiency of 60.01–99.78 % with a mean of 92.52 and 65.53–99.74 % with a mean of 95.83 % is achieved for 200 individual heterogeneous realizations based on the RTA and SPA, respectively, indicating that the cleanup of PCE in aquifer with SPA is significantly easier. This study leads to a new understanding of the microstructures of porous media and demonstrates how microscale arrangements control contaminant migration in aquifers, which is helpful to design successful remediation scheme for underground storage tank spill.

1 Introduction

Groundwater is an essential natural resource for water supply to domestic, agricultural, and industrial activities as well as ecosystem health (Boswinkel, 2000; Valipour, 2012, 2015; Yannopoulos et al., 2015; Valipour and Singh, 2016). Unfortunately, with the rapid development of economic activities such as mining, agriculture, landfills and industrial activities (Bakshevskaia and Pozdniakov, 2016; Cui et al., 2016; H. Liu et al., 2016; An et al., 2016; Shen et al., 2017), more and more contaminants released from human activities are contaminating the precious groundwater resource and subsurface environment (Dawson and Roberts, 1997; Liu, 2005; Hadley and Newell, 2014; Carroll et al., 2015; Essaid et al., 2015; Huang et al., 2015; Y. Liu et al., 2016; Schaefer et al., 2016; Weathers et al., 2016). Among the contaminants detected in groundwater, dense nonaqueous phase liquids (DNAPLs) such as perchloroethylene (PCE) and other polycyclic aromatic hydrocarbons (PAHs), are highly toxic and carcinogenic (Dawson and Roberts, 1997; Hadley and Newell, 2014). When DNAPLs are released into aquifer from underground storage tanks, they will infiltrate through the entire aquifer and form residual ganglia and pools of DNAPLs due to their large densities, high interfacial tension, and low solubility. The residual ganglia and pools of DNAPLs can serve as long-term sources of groundwater contamination which are harmful to the subsurface environment and human beings (Bob et al., 2008; Liang and Lai, 2008; Liang and Hsieh, 2015). Consequently, it is very important to explore DNAPL migration in aquifers and associated remediation of groundwater contamination.

When DNAPLs migrate in aquifers on a macroscopic scale, the transport properties such as permeability, diffusivity and dispersivity are closely related to the aquifer's microstructures and can affect DNAPL behavior (Yu and Li, 2004; Yu, 2005; Yun et al., 2005; Feng and Yu, 2007; Yu et al., 2009). Therefore, characterizing the effect of microstructures on macroscopic properties is a key point of research on heterogeneity of porous media (Mishra et al., 2016). In the classical Kozeny–Carman equation, the permeability K is related to porosity n, surface area S and the Kozeny constant c, where c is affected by the porosity, solid particles and microgeometric structures (Bear, 1972; Yu et al., 2009). According to fractal theory, natural porous media can be treated as fractal objects (Pfeifer and Avnir, 1983; Katz and Thompson, 1985; Krohn, 1988). For example, the tortuosity of flow paths in porous media is deeply studied by various proposed fractal models (Yu and Cheng, 2002; Yu et al., 2009; Cai et al., 2010), indicating the effectiveness of fractal methods. Based on fractal concepts, mathematic models are proposed to depict the permeability and invasion of fluids in some special porous media (Yu and Cheng, 2002; Yu et al., 2009; Cai et al., 2010). Furthermore, fractal method is also used to explore the effect of microstructure of biological media on associated thermal conductivity while this kind of material has a complex randomly distributed vascular-tree structure at microscale (Li and Yu, 2013).

In this study, we focus on the effect of microarrangement of sand particles on macroscopic DNAPL migration and associated remediation for underground storage tank spills. With the help of fractal theory, the microstructures of two different microscale arrangements of sand particles are explored. Afterwards, the mathematical relationships between porosity, permeability, and entry pressure are derived for regular triangle arrangement (RTA) and square pitch microscale arrangement (SPA). An idealized heterogeneous contaminated site is generated using a sequential Gaussian simulation (SGS) method. An underground storage tank releases PCE into heterogeneous aquifer composed of granular material. After a long-term migration, PCE contamination is alleviated using a surfactant remediation method. A multicomponent, multiphase model simulator, UTCHEM, is then used to simulate the entire process of DNAPL migration and remediation. Effects of arrangements of sand particles on migration and remediation of DNAPLs are comparatively analyzed based on the simulations to reveal how the microstructure of porous media controls the contaminant migration and remediation on a macroscopic scale.

2 Methodology

2.1 Fractal models of two different microscale arrangements of sand particles

The porous media can be treated as the bundle of tortuous capillary tubes, and the relationship between the diameter and the length of the capillary tube is as follows (Yu and Cheng, 2002):

$$L_t(\lambda) = \lambda^{1-D_t} L_s^{D_t}, \tag{1}$$

where L_s is the straight length between the tortuous flow path's end point, λ is the diameter of the capillary tube, and D_t is the fractal dimension of tortuosity for porous media, $1 < D_t < 2$ (Yu and Cheng, 2002).

If an infinitesimal element consisting of a bundle of tortuous capillary tubes from porous media is selected, the total number of capillary tubes in infinitesimal element can be calculated by the power–law relation:

$$N(L \geq \lambda) = \left(\frac{\lambda_{max}}{\lambda}\right)^{D_f}, \tag{2}$$

where D_f is the fractal dimension for pore areas in porous media, $1 < D_f < 2$ (Yu and Cheng, 2002); λ_{max} is the maximum diameter of capillary tubes.

Afterward, the derivative of Eq. (2) can be achieved:

$$-dN(L \geq \lambda) = D_f \lambda_{max}^{D_f} \lambda^{-(D_f+1)} d\lambda. \tag{3}$$

The total number of capillary tubes in an infinitesimal element can be derived from Eq. (3):

$$N_t(L \geq \lambda_{min}) = \left(\frac{\lambda_{max}}{\lambda_{min}}\right)^{D_f}, \tag{4}$$

where λ_{min} is the minimum diameter of capillary tubes.

Dividing Eq. (3) by Eq. (4) can achieve the following:

$$-\frac{d_{N(L \geq \lambda)}}{N_t} = D_f \lambda_{min}^{D_f} \lambda^{-(D_f+1)} d\lambda = f(\lambda) d\lambda, \tag{5}$$

where $f(\lambda)$ is the probability density function, $f(\lambda) = D_f \lambda_{min}^{D_f} \lambda^{-(D_f+1)}$.

The probability density function satisfies the following relationship:

$$\int_{-\infty}^{+\infty} f(\lambda) d\lambda = 1 - \left(\frac{\lambda_{min}}{\lambda_{max}}\right)^{D_f}. \tag{6}$$

Considering $\left(\frac{\lambda_{min}}{\lambda_{max}}\right)^{D_f} = 0$, the above Eq. (6) becomes the following:

$$\int_{-\infty}^{+\infty} f(\lambda) d\lambda = \int_{\lambda_{min}}^{\lambda_{max}} f(\lambda) d\lambda = 1 - \left(\frac{\lambda_{min}}{\lambda_{max}}\right)^{D_f} = 1. \tag{7}$$

With regard to fluid flow in capillary tubes, the flow rate Q can be calculated by the Hagen–Poiseulle equation:

$$Q = \frac{\pi r^4 \Delta P}{8\mu L_s} = \frac{\pi \left(\frac{\lambda}{2}\right)^4 \Delta P}{8\mu L_s} = \frac{\pi \lambda^4 \Delta P}{128\mu L_s}, \tag{8}$$

where μ is fluid's viscosity, and ΔP is the pressure gradient across the capillary tube.

The differentiation of flow rate of capillary tubes is as follows (Yu and Cheng, 2002):

$$\begin{aligned}
d_q &= [-d_{N(L \geq \lambda)}] \frac{\pi \lambda^4 \Delta P}{128\mu L_t(\lambda)} \\
&= D_f \lambda_{max}^{D_f} \lambda^{-(D_f+1)} d_\lambda \cdot \frac{\pi \lambda^4 \Delta P}{128\mu L_t(\lambda)} \\
&= \frac{\pi}{128} \frac{\Delta P}{\mu} \frac{D_f \lambda_{max}^{D_f}}{L_t(\lambda)} \lambda^{3-D_f} d_\lambda \\
&= \frac{\pi}{128} \frac{\Delta P}{\mu} \frac{D_f \lambda_{max}^{D_f}}{\lambda^{1-D_t} L_s^{D_t}} \lambda^{3-D_f} d_\lambda \\
&= \frac{\pi}{128} \frac{\Delta P}{\mu} \frac{D_f \lambda_{max}^{D_f}}{L_s^{D_t}} \lambda^{2+D_t-D_f} d_\lambda.
\end{aligned} \tag{9}$$

Integrating the individual flow rate from λ_{min} to λ_{max} can achieve the total flow rate (Yu and Cheng, 2002):

$$\begin{aligned}
Q &= \int d_q = \int_{\lambda_{min}}^{\lambda_{max}} \frac{\pi}{128} \frac{\Delta P}{\mu} \frac{D_f \lambda_{max}^{D_f}}{L_s^{D_t}} \lambda^{2+D_t-D_f} d_\lambda \\
&= \frac{\pi}{128} \frac{\Delta P}{\mu} \frac{D_f}{3-D_f+D_t} \frac{1}{L_s^{D_t}} \lambda_{max}^{D_f} \\
&\quad \cdot \left(\lambda_{max}^{3-D_f+D_t} - \lambda_{min}^{3-D_f+D_t}\right) \\
&= \frac{\pi}{128} \frac{\Delta P}{\mu} \frac{D_f}{3-D_f+D_t} \frac{1}{L_s^{D_t}} \lambda_{max}^{3+D_t} \\
&\quad \cdot \left[1 - \left(\frac{\lambda_{min}}{\lambda_{max}}\right)^{D_f} \left(\frac{\lambda_{min}}{\lambda_{max}}\right)^{3+D_t-2D_f}\right].
\end{aligned} \tag{10}$$

Due to $1 < D_t < 2$ and $1 < D_f < 2$, $3+D_t-2D_f > 0$. Simultaneously, $\left(\frac{\lambda_{min}}{\lambda_{max}}\right)^{D_f} \cong 0$, $0 < \left(\frac{\lambda_{min}}{\lambda_{max}}\right)^{3+D_t-D_f} < 1$. Therefore, Eq. (10) can be simplified as follows:

$$Q = \int d_q = \frac{\pi}{128} \frac{\Delta P}{\mu} \frac{D_f}{3-D_f+D_t} \frac{1}{L_0^{D_t}} \lambda_{max}^{3+D_t}. \tag{11}$$

Substituting Darcy's law, $Q = \frac{kA\Delta P}{\mu L_0}$, in Eq. (11) will obtain the permeability of porous media:

$$k = \frac{\pi}{128} \frac{D_f}{3+D_t-D_f} \frac{L_0^{1-D_t}}{A} \lambda_{max}^{3+D_t}. \tag{12}$$

To obtain the fractal dimension of tortuosity D_t, the expression of tortuosity (τ) can be obtained from Eq. (1):

$$\tau = \frac{L_t(\lambda)}{L_s} = \frac{\lambda^{1-D_t} L_s^{D_t}}{L_s} = \left(\frac{L_s}{\lambda}\right)^{D_t-1}. \tag{13}$$

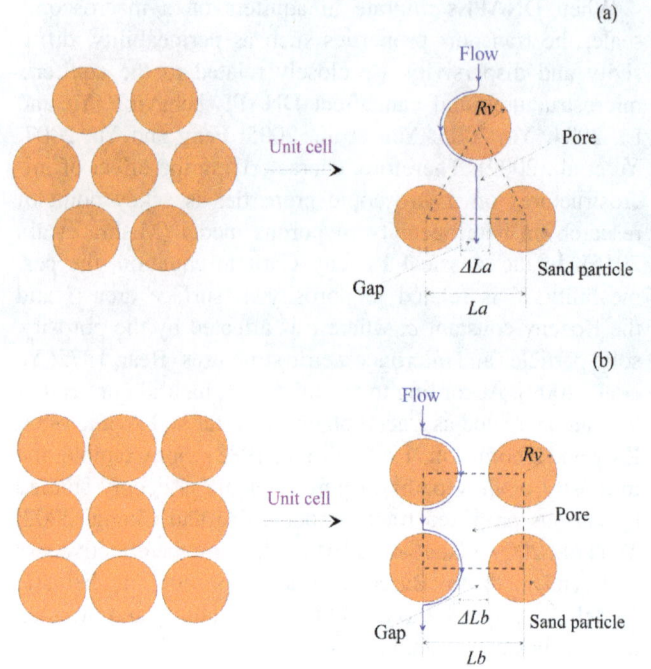

Figure 1. Two different microscale arrangements of solid particles: (a) RTA and (b) SPA.

Then D_t is given by the following (Yu and Li, 2004):

$$D_t = 1 + \frac{\ln \tau}{\ln \left(\frac{L_s}{\lambda}\right)}. \tag{14}$$

RTA and SPA are shown in Fig. 1. An equilateral triangle and a square are selected from the two microstructures as unit cells (Fig. 1a and b). The unit cell of the equilateral triangle is composed of three solid particles with the pore among them, while the unit cell of square is composed of four solid particles. For the unit cell of RTA in Fig. 1a, corresponding porosity is given by the following:

$$n = \frac{A_a - \pi R_v^2/2}{A_a}, \tag{15}$$

where n is porosity, A_a is the total area of equilateral triangle, and R_v is the average radius of solid particles. The total area of equilateral triangle can be achieved:

$$A_a = \frac{\pi R_v^2}{2(1-n)}. \tag{16}$$

The side length of the equilateral triangle in Fig. 1a can be calculated as follows:

$$L_a = R_v \sqrt{\frac{2\pi}{\sqrt{3}(1-n)}}, \tag{17}$$

where L_a is the side length.

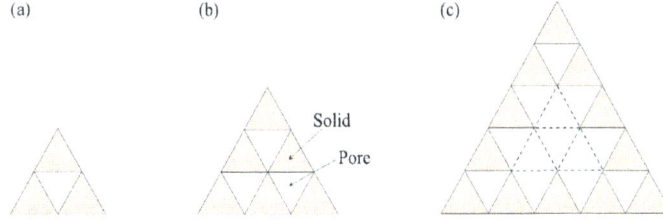

Figure 2. Three kinds of Sierpinski gasket [30]: **(a)** $L_a = 2$; **(b)** $L_a = 3$; and **(c)** $L_a = 5$.

The area of irregular pore among solid particles is given by the following:

$$A_{ap} = A_a - \frac{\pi R_v^2}{2} = \frac{\pi R_v^2 n}{2(1-n)}, \tag{18}$$

where A_{ap} is the area of pore in the unit cell.

Approximate the pore in the equilateral triangle as a circle, then the maximum diameter of pore can be obtained:

$$\lambda_{\max,a} = R_v \sqrt{\frac{2n}{1-n}}, \tag{19}$$

where $\lambda_{\max,a}$ is the diameter of the capillary tube in equilateral triangle. The fluid does not only pass the central pore of the unit cell, but also flow through the gap between adjacent particles. The gap length and the average diameter of the capillary tube perpendicular to the plane of equilateral triangle are calculated as follows:

$$\Delta L_a = L_a - 2R_v = R_v \left(\sqrt{\frac{2\pi}{\sqrt{3}(1-n)}} - 2 \right), \tag{20}$$

$$\lambda_a = \frac{\lambda_{\max,a} + \Delta L_a}{2} = \frac{R_v}{2} \left(\sqrt{\frac{2n}{1-n}} + \sqrt{\frac{2\pi}{\sqrt{3}(1-n)}} - 2 \right), \tag{21}$$

where ΔL_a is the gap length between solid particles; λ_a is the average diameter of capillary tubes in the equilateral triangle.

Generally, the tortuosity of flow path in porous media is the ratio of the length of tortuous flow path to the straight length of flow path along the flow direction (Taiwo et al., 2016):

$$\tau = \frac{L_t}{L_s}, \tag{22}$$

where L_t is the length of tortuous flow path, and L_s is the straight length of flow path along the flow direction.

For the flow path shown in Fig. 1a, L_t and L_s are respectively as follows:

$$L_t = (h_o - R_v) + \frac{\pi R_v}{2} = R_v \left(\sqrt{\frac{\sqrt{3}\pi}{2(1-n)}} + \frac{\pi}{2} - 1 \right) \tag{23}$$

$$L_s = h_o = R_v \sqrt{\frac{\sqrt{3}\pi}{2(1-n)}}, \tag{24}$$

where h_o is the altitude of the equilateral triangle, $h_o = \frac{\sqrt{3}}{2} L_a = R_v \sqrt{\frac{\sqrt{3}\pi}{2(1-n)}}$.

Consequently, the tortuosity of RTA is yielded:

$$\tau = \frac{L_t}{L_s} = 1 + \frac{\frac{\pi}{2} - 1}{\sqrt{\frac{\sqrt{3}\pi}{2(1-n)}}}. \tag{25}$$

The D_f is determined using Sierpinski gasket (Fig. 2) in fractal theory (Yu and Cheng, 2002). The shaded area represents solids of porous media and the white area represents pore. The pore area fractal dimensions in Fig. 2a–c are 0.000, 1.000 and 1.594, respectively ($1 = L_a^{D_f} = 2^{D_f}$, $3 = L_a^{D_f} = 3^{D_f}$, $13 = L_a^{D_f} = 5^{D_f}$). Based on the Sierpinski gasket, the dimensionless pore area in RTA (Fig. 1a) is approximated as follows:

$$A_{apd} = (L_a^+)^{D_f}, \tag{26}$$

where A_{apd} is the dimensionless pore area of RTA; $L_a^+ = L_a/\lambda_{\min}$. Equation (26) can be solved to achieve D_f:

$$D_f = \frac{\ln A_{apd}}{\ln L_a^+}. \tag{27}$$

The porosity equals to the ratio of the dimensionless pore area of RTA (A_{apd}) to the dimensionless total area of RTA (A_a^+):

$$n = \frac{A_{apd}}{A_a^+}, \tag{28}$$

where $A_a^+ = \frac{A_a}{\pi \lambda_{\min}^2/4} = \frac{\frac{\pi R_v^2}{2(1-n)}}{\pi \frac{\lambda_{\min}^2}{4}} = \frac{2R_v^2}{\lambda_{\min}^2} \frac{1}{1-n} = \frac{(d^+)^2}{2} \frac{1}{1-n}$; $d^+ = \frac{2R_v}{\lambda_{\min}}$, $L_a^+ = \sqrt{A_a^+}$.

From Eq. (28), the dimensionless pore area of RTA (A_{apd}) is given by the following:

$$A_{\mathrm{apd}} = n \cdot A_{\mathrm{a}}^{+}. \tag{29}$$

The dimensionless total area of RTA (A_{a}^{+}) can be written as follows:

$$A_{\mathrm{a}}^{+} = (L_{\mathrm{a}}^{+})^{2}. \tag{30}$$

Afterward, L_{a}^{+} is calculated as follows:

$$L_{\mathrm{a}}^{+} = \sqrt{A_{\mathrm{a}}^{+}} = \sqrt{\frac{(d^{+})^{2}}{2}\frac{1}{1-n}} = d^{+}\sqrt{\frac{1}{2(1-n)}}. \tag{31}$$

Substituting Eqs. (29) and (31) into Eq. (27) will derive D_{f} of RTA:

$$
\begin{aligned}
D_{\mathrm{f}} &= \frac{\ln A_{\mathrm{apd}}}{\ln L_{\mathrm{a}}^{+}} = \frac{\ln(n \cdot A_{\mathrm{a}}^{+})}{\ln\left(\sqrt{A_{\mathrm{a}}^{+}}\right)} = 2 + \frac{\ln(n)}{\ln\left(\sqrt{A_{\mathrm{a}}^{+}}\right)} \\
&= 2 + \frac{\ln(n)}{\ln\left(d^{+}\sqrt{\frac{1}{2(1-n)}}\right)}.
\end{aligned}
\tag{32}
$$

For the unit cell of square shown in Fig. 1b, the porosity is as follows:

$$n = \frac{A_{\mathrm{b}} - \pi R_{\mathrm{v}}^{2}}{A_{\mathrm{b}}}, \tag{33}$$

where A_{b} is the total area of the square. Equation (33) can also be expressed as the area of unit cell:

$$A_{\mathrm{b}} = \frac{\pi R_{\mathrm{v}}^{2}}{1-n}. \tag{34}$$

Again, the side length of the square is as follows:

$$L_{\mathrm{b}} = \sqrt{A_{\mathrm{b}}} = R_{\mathrm{v}}\sqrt{\frac{\pi}{1-n}}. \tag{35}$$

Consequently, the area of an irregular pore in the square is given by the following:

$$A_{\mathrm{bp}} = A_{\mathrm{b}} - \pi R_{\mathrm{v}}^{2} = \frac{n\pi R_{\mathrm{v}}^{2}}{1-n}, \tag{36}$$

where A_{bp} is the area of pore in the square.

Using the following equation, the pore as a circle and obtain corresponding maximum diameter can be approximated:

$$\lambda_{\max,b} = 2R_{\mathrm{v}}\sqrt{\frac{n}{1-n}}, \tag{37}$$

where $\lambda_{\max,b}$ is the maximum diameter of the capillary tube perpendicular to the plane of the square. Similarly, fluid flows through the central pore in the square and the gap between adjacent particles. As a result, the gap and average diameter of the capillary tube are expressed as follows:

$$\Delta L_{\mathrm{b}} = L_{\mathrm{b}} - 2R_{\mathrm{v}} = R_{\mathrm{v}}\left(\sqrt{\frac{\pi}{1-n}} - 2\right), \tag{38}$$

$$\lambda_{\mathrm{b}} = \frac{\lambda_{\max,b} + \Delta L_{\mathrm{b}}}{2} = \frac{R_{\mathrm{v}}}{2}\left(2\sqrt{\frac{n}{1-n}} + \sqrt{\frac{\pi}{1-n}} - 2\right), \tag{39}$$

where ΔL_{b} is the gap length between the adjacent two solid particles, and λ_{b} is the average diameter of the capillary tube.

For the tortuous flow path in Fig. 1b, L_{t} and L_{s} are respectively given by the following:

$$L_{\mathrm{t}} = \Delta L_{\mathrm{b}} + \pi R_{\mathrm{v}} = R_{\mathrm{v}}\left(\sqrt{\frac{\pi}{1-n}} - 2 + \pi\right), \tag{40}$$

$$L_{\mathrm{s}} = L_{\mathrm{b}} = R_{\mathrm{v}}\sqrt{\frac{\pi}{1-n}}. \tag{41}$$

Afterward, the tortuosity of SPA yields the following:

$$\tau = \frac{L_{\mathrm{t}}}{L_{\mathrm{s}}} = 1 + \frac{\pi - 2}{\sqrt{\frac{\pi}{1-n}}}, \tag{42}$$

The procedure of deriving D_{f} of SPA is similar to the procedure of calculating D_{f} of RTA. Similarly, the D_{f} and porosity of SPA (Fig. 1b) are given by the following:

$$D_{\mathrm{f}} = \frac{\ln A_{\mathrm{bpd}}}{\ln L_{\mathrm{b}}^{+}}, \tag{43}$$

$$n = \frac{A_{\mathrm{bpd}}}{A_{\mathrm{b}}^{+}}, \tag{44}$$

where A_{bpd} is the dimensionless pore area of SPA, $L_{\mathrm{b}}^{+} = L_{\mathrm{b}}/\lambda_{\min}$, A_{b}^{+} is the dimensionless total area of SPA, and

$$A_{\mathrm{b}}^{+} = \frac{A_{\mathrm{b}}}{\pi \lambda_{\min}^{2}/4} = \frac{\frac{\pi R_{\mathrm{v}}^{2}}{1-n}}{\pi \frac{\lambda_{\min}^{2}}{4}} = \frac{4R_{\mathrm{v}}^{2}}{\lambda_{\min}^{2}}\frac{1}{1-n} = (d^{+})^{2}\frac{1}{1-n}.$$

The dimensionless pore area of SPA (A_{bpd}) can be yielded from Eq. (44):

$$A_{\mathrm{bpd}} = n \cdot A_{\mathrm{b}}^{+}. \tag{45}$$

L_{b}^{+} can be calculated as follows:

$$L_{\mathrm{b}}^{+} = \sqrt{A_{\mathrm{b}}^{+}} = \sqrt{(d^{+})^{2}\frac{1}{1-n}} = d^{+}\sqrt{\frac{1}{1-n}}. \tag{46}$$

Substituting Eqs. (45) and (46) into Eq. (43), D_{f} of SPA can be derived:

$$
\begin{aligned}
D_{\mathrm{f}} &= \frac{\ln A_{\mathrm{bpd}}}{\ln L_{\mathrm{b}}^{+}} = \frac{\ln\left(n \cdot A_{\mathrm{b}}^{+}\right)}{\ln\left(\sqrt{A_{\mathrm{b}}^{+}}\right)} \\
&= 2 + \frac{\ln(n)}{\ln\left(\sqrt{A_{\mathrm{b}}^{+}}\right)} = 2 + \frac{\ln(n)}{\ln\left(d^{+}\sqrt{\frac{1}{1-n}}\right)}.
\end{aligned}
\tag{47}
$$

The entry pressure of a tortuous capillary tube (P_c) is defined by Young–Laplace equation as follows (Ahn and Seferis, 1991):

$$P_c = \frac{\omega}{\lambda} \frac{1-n}{n}, \tag{48}$$

where P_c is the entry pressure, λ is the diameter of the capillary tube, ω equals to $F\sigma\cos\theta$ in which θ is the contact angle between fluid and solid, σ is the surface tension of the wetting fluid, and F is the form factor depending on the capillary tube alignment and the flow direction.

2.2 Dealing with the heterogeneity of porous media

In this study, SGS is used to generate random realization of a heterogeneous porosity field. SGS is a stochastic simulation method combining sequential principles and a Gaussian method. It assumes variable fit to a Gaussian random field. The Gaussian distribution function is constructed at each simulated spatial location based on the characteristics of variation function and afterward randomly selects a value as the variable at the location. In the SGS method, observation data are transformed to Gaussian distributions or normal distributions. Based on current sample data, the conditional probability distribution of points to be simulated is calculated by the SGS method and then simulation is performed based on a semivariogram model. Each simulated value, together with measured data and previous simulation data, becomes the conditional data set for the next step. As simulation proceeds, the conditional data set increases. Previous research suggests 50–400 realizations are required to obtain a statistically stable mean realization (Eggleston et al., 1996; Hu et al., 2007).

2.3 Modeling PCE migration and its remediation

The DNAPL migration and remediation are modeled using a multicomponent, multiphase, and multicomposition simulator named UTCHEM (University of Texas Chemical Compositional Simulator) (Delshad et al., 1996). As an extension to Delshad's work, UTCHEM was developed by the University of Texas as a comprehensive and practical tool. In numerous applications, UTCHEM has proved to be particularly useful in simulation of contaminant migrations and has been a popular multiphase-flow, multiconstituent, reactive transport model used widely in groundwater simulations. UTCHEM accounts for chemical, physical and biological reactions; complex non-equilibrium sorption; decay and geochemical reactions; and surfactant-enhanced solubilization and mobilization of DNAPLs. Moreover, heterogeneous properties of porous media are also considered. As a result, UTCHEM has been adapted for a variety of environmental applications such as surfactant-enhanced aquifer remediation (SEAR). In this study, DNAPL migration and remediation for cleaning up DNAPL contamination in idealized heterogeneous sites are simulated by UTCHEM.

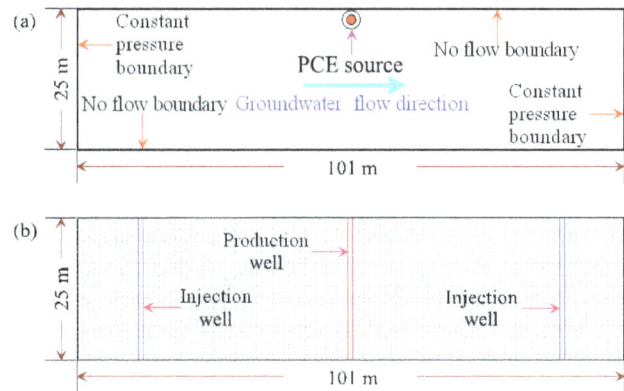

Figure 3. (a) Two-dimensional view of contaminated domain and **(b)** locations of injection and extraction wells.

3 Application to a synthetic heterogeneous PCE contaminated site

3.1 Site description

The idealized domain of synthetic application is a two-dimensional confined aquifer (Fig. 3). The length, width and depth of aquifer are 101, 25 and 25 m, respectively. Idealized aquifer is discretized into 101 grids horizontally and 25 layers vertically (Fig. 3b). The spacing of each grid is uniformly 1 m along x and z axes, and the longitudinal and transverse dispersivities are set as to 1.0 and 0.1 m, respectively. Horizontal and vertical correlation length values are each 5 m. The top and bottom borders of aquifer are defined as no-flow boundaries, while the left and right borders are defined as constant potential boundaries to create a groundwater flow from left to right under a low hydraulic gradient of 0.005 m m^{-1} (Liu et al., 2003; Liu, 2005; Qin et al., 2007). The porous media of idealized aquifer is assumed to be heterogeneous and mixed by different grades of sands.

The porosity of aquifer is assumed to be spatially and uniformly distributed with an average value of 0.220 and SD (standard deviation) of 0.060. In this study, porosity follows normal distribution and its SD represents the enhanced geological heterogeneity. A total of 200 realizations of the porosity field are generated using SGS. One of the 200 realizations of heterogeneous field is shown in Fig. 4a. Simultaneously, statistical assessment is undertaken on the individual realization of the porosity field, and corresponding histograms are shown in Fig. 4b. We find that the frequency of the individual realization of the porosity field is close to normal distribution, which conforms to the situation that most characteristics of natural aquifer can be expressed as a normal distribution (Montgomery et al., 1987). Based on the heterogeneous porosity field, the fractal dimension of tortuosity D_t, the fractal dimension for pore areas D_f and the diameter of capillary tubes in porous media, permeability is obtained by Eq. (12). Figure 4c shows the individual heteroge-

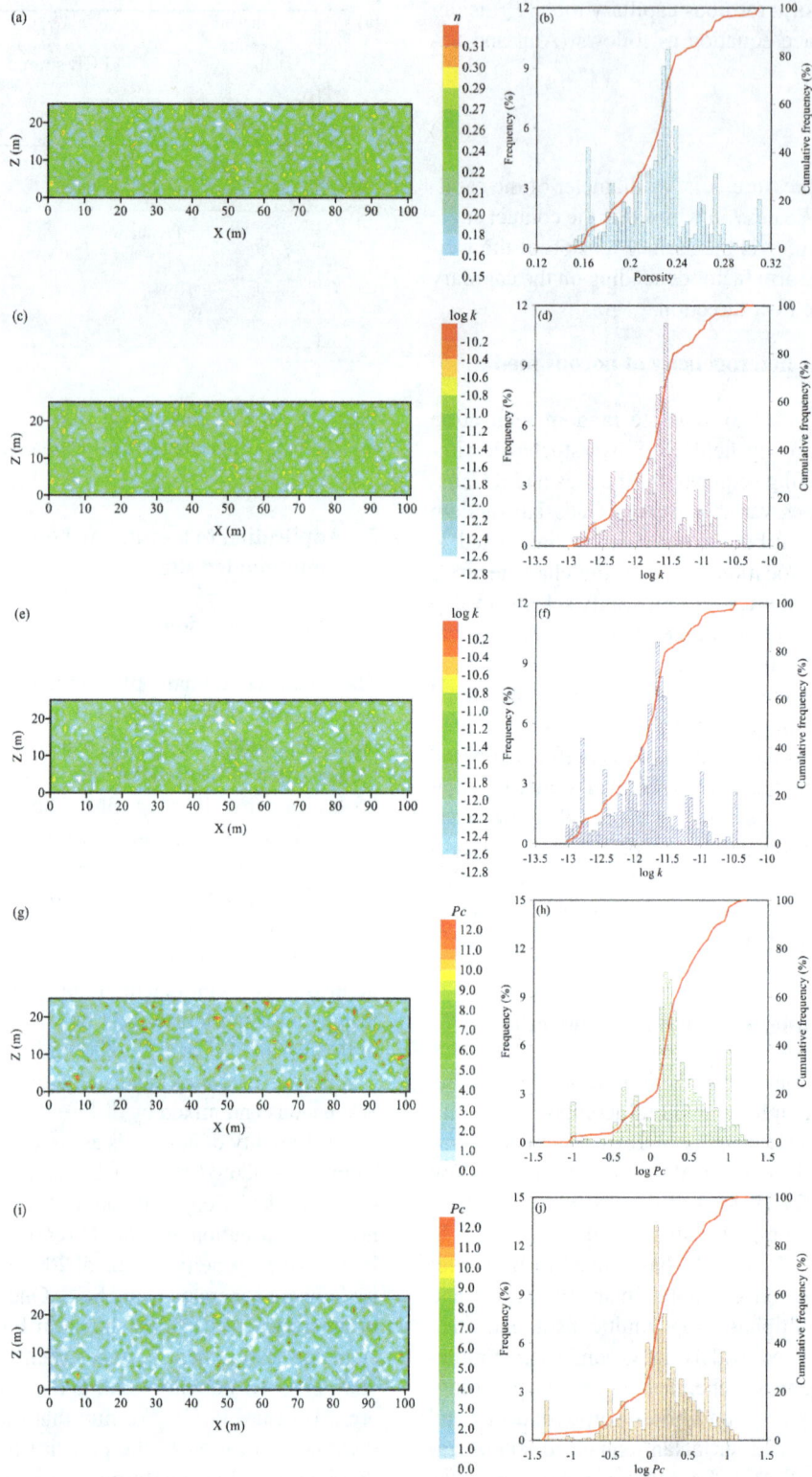

Figure 4. (a) The individual porosity field generated by the sequential Gaussian simulation (SGS) method; (b) the frequency of an individual porosity field; (c) the individual permeability field of RTA obtained from an individual porosity field; (d) the frequency of individual permeability field for RTA; (e) the individual permeability field of SPA obtained from an individual porosity field; (f) the frequency of individual permeability field for SPA; (g) the obtained individual entry-pressure field of RTA; (h) the frequency of individual entry-pressure field of RTA; (i) the obtained individual entry-pressure field of SPA; and (j) the frequency of individual entry pressure of SPA.

neous permeability field selected from the 200 realizations of RTA, and the result of the associated frequency analysis is shown in Fig. 4d. The permeability field fits the lognormal distribution well, which has been presented by many studies which show that the parameter of aquifer penetrability follows lognormal distribution (Montgomery et al., 1987; Veneziano and Tabaei, 2004). Compared to the histogram of the porosity field in Fig. 4b, the shape of permeability is similar. The individual heterogeneous permeability field of SPA is shown in Fig. 4e. Corresponding frequency analysis of SPA reveals that the permeability field follows lognormal distribution, while some difference appears compared with RTA (Fig. 4f). The average permeability of individual realization of RTA is $2.012 \times 10^{-12}\,\mathrm{m}^2$, and the average permeability of individual realization of SPA is $1.618 \times 10^{-12}\,\mathrm{m}^2$. For 200 realizations, the average permeability of RTA and SPA are 2.120×10^{-12} and $1.706 \times 10^{-12}\,\mathrm{m}^2$, indicating the permeability of RTA is slightly bigger than SPA.

The average pore diameters of two different microscale arrangements of particles are derived using corresponding fractal models. In detail, the average diameter of RTA is calculated by Eq. (21) and the average diameter of SPA is calculated by Eq. (39). Consequently, the entry pressure of the two kinds of microscale arrangements can be obtained by Eq. (48). The individual entry-pressure fields of two microscale arrangements and associated frequency analysis are shown in Fig. 4g–j. From the frequency of entry pressure in Fig. 4h and j, the entry pressures of both RTA and SPA are the lognormal distributions. However, the average entry pressure of individual realization of RTA is 1.980 kPa, while the average entry pressure of SPA is 1.481 kPa. For 200 realizations of the entry-pressure field, the average entry pressure of RTA is 1.922 kPa and the average entry pressure of SPA is 1.442 kPa. The differences of average entry pressure between RTA and SPA imply the microstructure of aquifer has an effect on the macroscopic characteristics.

The purpose of this study is to explore the effects of microstructure of aquifer on DNAPL migration and remediation. A PCE spill event (the leaking of underground storage tank) occurs on the top of the aquifer and a surfactant remediation is designed to clean up the contaminated aquifer. The total duration of 300 days is divided into four stages: (1) $300\,\mathrm{m}^3$ PCE is released from underground storage tank into aquifer at the top layer of spill position shown in Fig. 3a during 0–30 days, (2) PCE migrates in aquifer freely during 30–100 days, (3) surfactant is injected into aquifer during 100–150 days, and (4) water is flushed during 150–300 days. In the first stage, PCE is released as a point pollution source in the center grid block at the top layer of the aquifer, in which spill is at a constant rate of $10\,\mathrm{m}^3\,\mathrm{day}^{-1}$. After PCE coming into the heterogeneous aquifer, PCE is migrating freely under the effects of gravity and the natural hydraulic gradient condition. The PCE not only migrates downward through the aquifer, but can also be trapped by capillary forces as residual ganglia and globules. During

Table 1. Parameters used in the simulation.

Parameter	Value
Average value of porosity	0.22
SD of porosity	0.06
Longitudinal dispersivity	1.0 m
Transverse dispersivity	0.1 m
Hydraulic gradient	$0.005\,\mathrm{m}\,\mathrm{m}^{-1}$
Water density	$1.00\,\mathrm{g}\,\mathrm{cm}^{-3}$
PCE density	$1.63\,\mathrm{g}\,\mathrm{cm}^{-3}$
Surfactant density	$1.15\,\mathrm{g}\,\mathrm{cm}^{-3}$
Water viscosity	1.00 cp
PCE viscosity	0.89 cp
PCE–water interfacial tension	$45\,\mathrm{dyn}\,\mathrm{cm}^{-1}$
PCE solubility in water	$240\,\mathrm{mg}\,\mathrm{L}^{-1}$
Residual water saturation	0.24
Residual PCE saturation	0.17
Endpoint of water (BC model)	0.486
Endpoint of PCE (BC model)	0.65
Exponent of water (BC model)	2.85
Exponent of PCE (BC model)	2.7
Exponent of capillary pressure	−0.52

the long-term PCE migration period, PCE is contaminating groundwater and expanding plume. To clean up the contaminated aquifer, 4 % surfactant solution is injected into aquifer through the two injection wells (Fig. 3b) at a constant rate of $80\,\mathrm{m}^3\,\mathrm{day}^{-1}$, and, simultaneously, contaminated groundwater is extracted through production well at a constant rate of $160\,\mathrm{m}^3\,\mathrm{day}^{-1}$. Surfactant can reduce the interfacial tension between the DNAPL and aqueous phase to promote solubilization and mobilization of DNAPL. After surfactant injection, the contaminated aquifer is flushed by water over a long period of 150 days. Based on the distributions of porosity, permeability and entry pressure of two microscale arrangements, the entire PCE migration and remediation process is simulated by a multicomponent, multiphase model simulator, UTCHEM (Delshad et al., 1996). The parameters used in the simulation are listed in Table 1. Simulation results of two different microscale arrangements are compared to reveal the effect of microstructure on the DNAPL migration and remediation.

3.2 Results and discussion

3.2.1 PCE migration and its remediation based on single realizations

The simulation results of PCE migration for individual realization of the porosity field for RTA are shown in Fig. 5a–f. When PCE is released into an aquifer at the top layer of spill position, PCE almost infiltrates vertically under the effect of gravity (Fig. 5a). Due to the heterogeneity of the aquifer, some preferential flow appears and the PCE plume becomes irregular (Fig. 5b). After 30 days, PCE plume almost touches

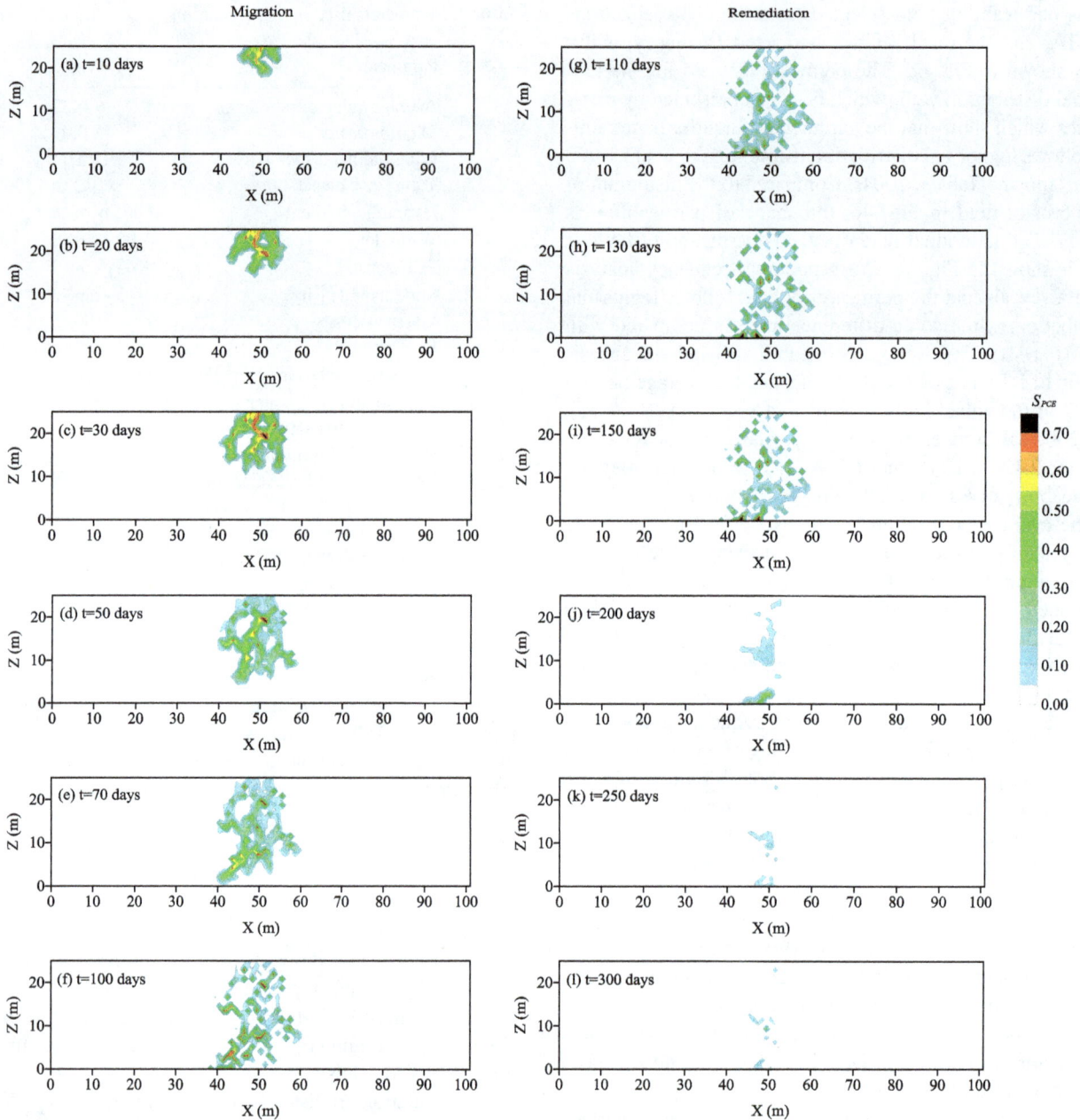

Figure 5. Simulated PCE saturation for individual realization of RTA over the entire migration and remediation periods (0 ∼ 300 day).

the bottom of aquifer (Fig. 5c). When the PCE leakage is stopped, PCE migrates continuously in aquifer for 70 days (Fig. 5d–f). The released PCE is migrating downward and entrapped by capillary forces as residual ganglia and globules. The heterogeneity of the aquifer makes PCE migrate along a preferential pathway. When the PCE plume touches the zones of low permeability and high entry pressure, it will bypass these zones and migrate continuously, which leads to an increasing variability in PCE distribution. After the PCE plume reaches the bottom of aquifer, PCE begin accumulate and form a contaminant pool at the bottom. At $t = 100$ days,

a PCE pool is formed at the bottom of aquifer, moving toward the right boundary.

Figure 6a–f show the simulated PCE saturation for individual realization of porous media for SPA during migration period. Under the effects of gravity and the natural hydraulic gradient, PCE is migrating and the contaminant plume becomes larger and larger. The heterogeneity of the aquifer significantly changes the migration paths and leads to irregular morphology of the PCE plume (Fig. 6a–c). However, due to the different microarrangements of the aquifer, the entry-pressure distribution is also different, which leads

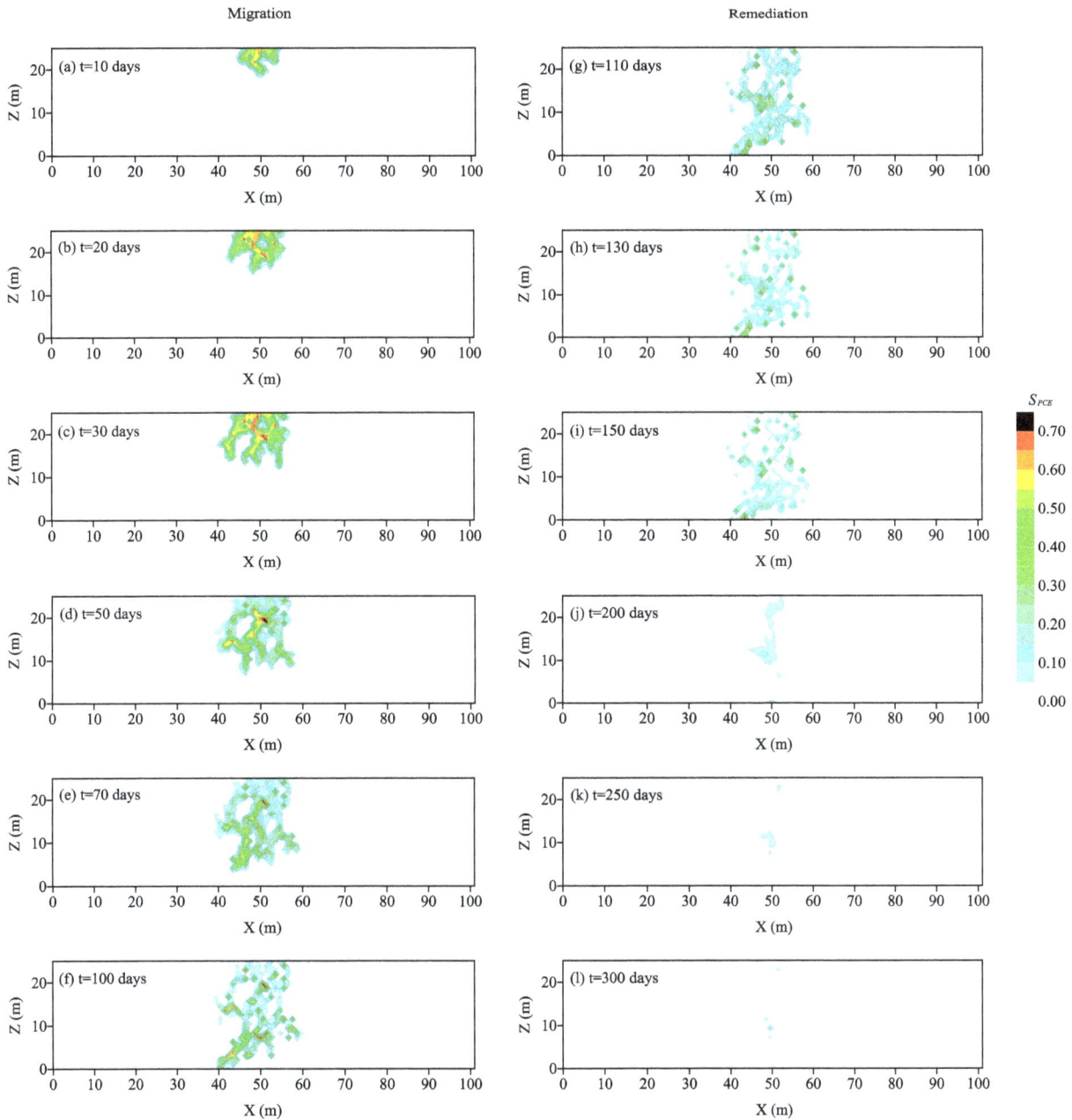

Figure 6. Simulated PCE saturation for individual realization of SPA over the entire migration and remediation periods (0 ~ 300 day).

to some differences. After the PCE injection, the simulated PCE saturation in Fig. 6d–f indicates that further trapping and spreading of the PCE occurs during this period. Compared with the simulation results of RTA in Fig. 5, the PCE plume slightly seems similar in Fig. 6. Moreover, PCE infiltrates more quickly in porous media of RTA in Fig. 5. After 70 days, the PCE plume has touched the bottom for RTA (Fig. 5e), while the PCE plume based on SPA still keeps a significant distance from the bottom (Fig. 6e).

To clean up the DNAPL, 4 % surfactant solution is injected

through two injection wells at a constant rate of $80\,m^3\,day^{-1}$ over 50 days. Afterwards, water flushing is applied during 150–300 days. The locations of the injection and production wells are presented in Fig. 3b. The production well is rightly installed at the location of the PCE spill position and two injection wells are located 39 m to the left and right of the production well. Figure 5g–l show the PCE remediation results of individual realization for RTA. During the early remediation period, the effect of cleaning up DNAPL is not yet apparent (Fig. 5g–i). When the water flushing begins, the surfactant solution circulates throughout the contaminated aquifer (Fig. 5j–l). At $t = 200$ days, $237.01\,\mathrm{m}^3$ PCE is removed from contaminated aquifer, occupying 79.00 % of the total released PCE (Fig. 5j). As time goes on, $268.30\,\mathrm{m}^3$ PCE is removed from the aquifer and remediation efficiency reaches 89.43 %.

The same surfactant remediation is also conducted for individual realization of SPA. Compared with the remediation for RTA, the remediation effect is more apparent for SPA (Fig. 6g–l). As the remediation progresses, more DNAPL is removed and less DNAPL remains at the bottom of aquifer. At $t = 200$ days, $267.68\,\mathrm{m}^3$ PCE is removed from the contaminated aquifer and the corresponding remediation efficiency rises to 89.23 %. At $t = 300$ days, $285.32\,\mathrm{m}^3$ PCE is cleaned up and remediation efficiency reaches 95.11 %. From the results of remediation, it is obvious that microstructure has an effect on remediation for macroscopic-scale aquifer. Results suggest contaminated aquifer of RTA is hard to clean up by surfactant remediation while SPA can improve DNAPL remediation efficiency.

3.2.2 PCE migration and SGS realizations

PCE migration and remediation processes are simulated for 200 realizations of the porosity field for porous media of RTA and SPA. The variations of contaminant mass, the ganglia-to-pool (GTP) ratio and moments of PCE plume vs. time are presented in Fig. 7a–h. During 0–30 days, the PCE in aquifer increases linearly at a constant rate of $10\,\mathrm{m}^3\,\mathrm{day}^{-1}$ (Fig. 7a), which corresponds to the contaminant spill stage. Afterward, PCE volume keeps constant during the second stage (30–100 days), while PCE volume in aquifer is reduced when surfactant is injected into aquifer. After surfactant insertion and water flushing of the contaminated aquifer, most DNAPL is cleaned up. The residual DNAPL mass remained in aquifer of 0.67–$119.89\,\mathrm{m}^3$ with a mean of 22.42 and 0.79–$103.33\,\mathrm{m}^3$ with a mean of $12.51\,\mathrm{m}^3$ are achieved for 200 heterogeneous realizations based on the RTA and SPA, respectively. The average remediation efficiency of SPA is undoubtedly higher than RTA, indicating the aquifer of SPA is easier to clean up. PCE plume architectures are quantified by measuring the GTP ratio in Fig. 7b. Over entire periods, curves of GTP value show obvious oscillations. Surfactant has the ability to promote solubilization, and mobilization of DNAPL can reduce GTP value. As a result, when surfactant is injected at $t = 100$ day, the GTP value reduces quickly. When surfactant injection is ended and water flushing begins, the GTP value increases with steep flank slope. Finally, GTP values reach 0.10–0.41 with a mean of 0.21 and 0.15–0.42 with a mean of 0.28 for 200 heterogeneous realizations based on the RTA and SPA, respectively.

Figure 7c shows cumulative PCE removal from the contaminated aquifer vs. flushing time for RTA and SPA. During the surfactant injection period, 100–150 days, the DNAPL removal is not apparent. However, DNAPL is removed effec-

tively and quickly during the water-flushing period. Through long-term remediation, the removal of PCE from the contaminated aquifer reached 179.89–$298.98\,\mathrm{m}^3$ with a mean of 277.29 and 196.45–$298.87\,\mathrm{m}^3$ with a mean of $287.21\,\mathrm{m}^3$ for 200 realizations based on RTA and SPA, respectively. Average remediation efficiency of SPA (95.83 %) is noticeably higher than average remediation efficiency of RTA (92.52 %).

Figure 7d shows the GTP value as a function of cumulative PCE removal for the contaminated aquifer. The GTP remains at a relatively low level before 30 % of the DNAPL is removed from the aquifer. When 40 % of the total $300\,\mathrm{m}^3$ PCE is removed, GTP values are increasing and corresponding curves appear as a wave crest because the high-saturation zones of the PCE plume are dissolved and turned into ganglia. After the wave crest, the GTP values decline quickly with steep flank slope due to PCE ganglia removal through water flushing. Finally, GTP values increase at the end of the remediation process for 200 realizations, indicating that most of PCE is removed and most of residual PCE turns into ganglia.

For the center of the PCE plume on the horizontal axis, associated variations vs. time are similar for 200 realizations based on RTA and SPA (Fig. 7e). Significantly, the PCE plume vertical-infiltration rate in aquifer of RTA is slightly faster than PCE infiltration in the aquifer of SPA for 200 realizations (Fig. 7f). Simultaneously, the second PCE plume moments in the horizontal direction of RTA are different from the second PCE plume moments in the horizontal direction of SPA (Fig. 7g). After PCE migration under natural conditions at $t = 100$ days, the second PCE plume moments in the horizontal direction are 10.61–$40.50\,\mathrm{m}^2$ with a mean of 21.51 and 10.99–$36.38\,\mathrm{m}^2$ with a mean of $20.75\,\mathrm{m}^2$ for 200 realizations based on RTA and SPA, respectively. At $t = 300$ day, the second PCE plume moments in the horizontal direction change to 0.81–$34.88\,\mathrm{m}^2$ with a mean of 5.79 and 1.03–$24.57\,\mathrm{m}^2$ with a mean of $4.64\,\mathrm{m}^2$ for RTA and SPA, respectively. The horizontal second moment of RTA is always larger than the horizontal second moment of SPA, indicating that the PCE plume in the aquifer of RTA is wider than the PCE plume in the aquifer of SPA, and RTA can cause more groundwater contamination. Similarly, the second moments in vertical direction for RTA are larger than the second moments in vertical direction for SPA.

This study takes an important step toward exploring how microscale arrangements control contaminant migration on a small-aquifer scale. Results are essential to the macroscopic aquifer composed of porous media without large heterogeneity, such as sandy aquifers containing rich groundwater resources. However, there are many problems associated with the upscaling of aquifers in real-world conditions (Dagan et al., 2013; Pacheco, 2013; Pacheco et al., 2015). Due to the large heterogeneity of natural aquifers, research results may be very different and can not be extrapolated to complex regional aquifer on a large scale. However, the finding

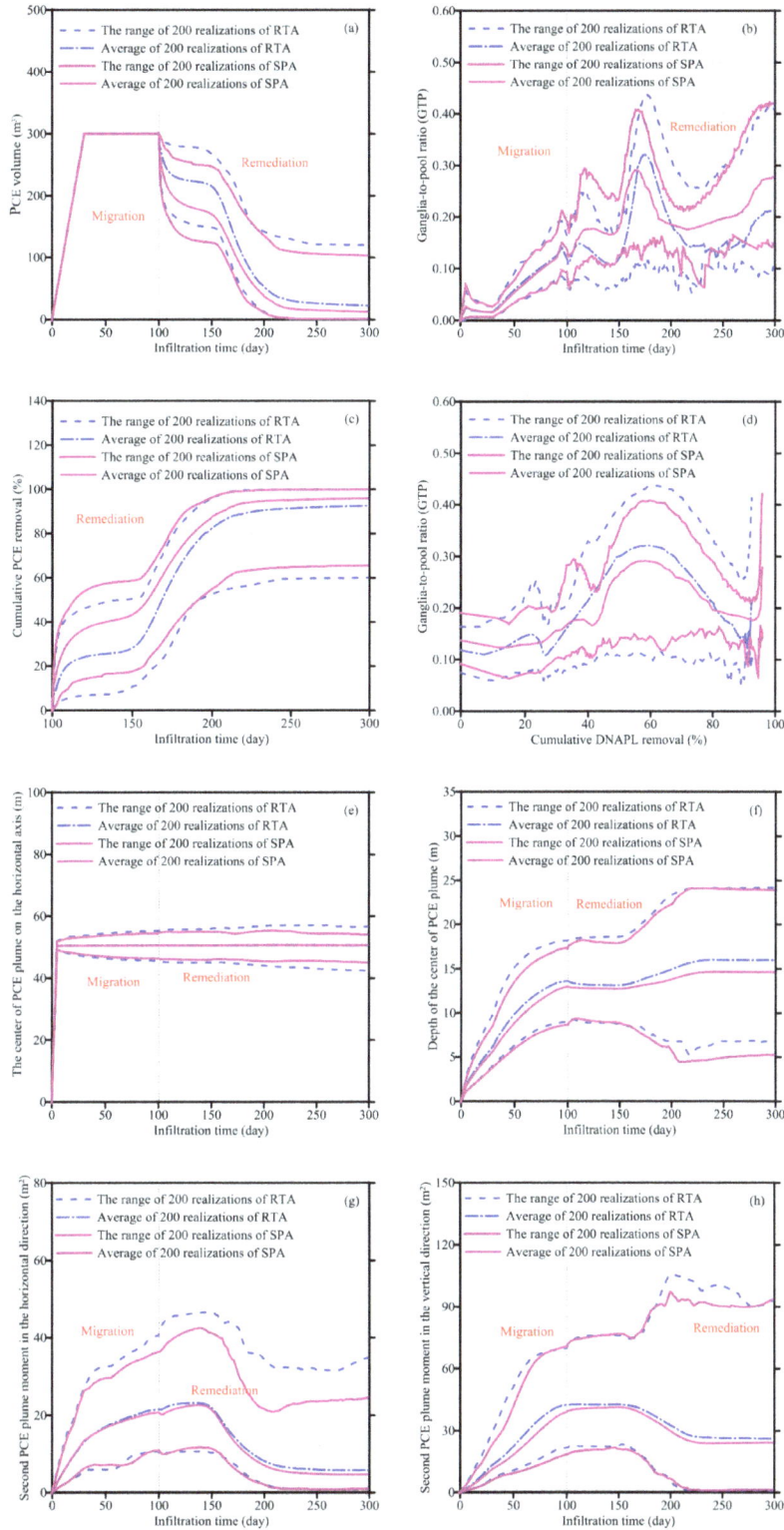

Figure 7. (a) PCE volume in aquifer vs. time, RTA represents RTA and SPA represents SPA; **(b)** changes in GTP as a function of time; **(c)** cumulative DNAPL removal as a function of time; **(d)** variation of GTP value as a function of cumulative DNAPL removal percent; **(e)** the change of the center of PCE plume during the entire periods of migration and remediation; **(f)** the change of the depth of PCE plume center during the entire periods; **(g)** variation of second PCE plume moment on the horizontal axis; and **(h)** variation of second PCE plume moment in vertical axis.

in this study is absolutely applicable for natural aquifers with similar heterogeneities. If the heterogeneity and anisotropy of natural aquifers are very different, the effect of the microscale arrangements on the macroscopic contaminant migration and remediation will be different. Although real-world conditions are complex, the new findings achieved from this research are very significant for understanding the effect of microscale arrangement on contaminant behaviors on an aquifer scale. The upscaling problem of the results obtained on the simulation scale ($100 \, \text{m} \times 25 \, \text{m} \times 25 \, \text{m}$) is the basis and the upscaling problem with more complex heterogeneity conditions is needed to be further investigated. Various research on the upscaling problem is done through experiment and simulation (Wu et al., 2017a–d). Based on this research, the microstructure of porous media is developed and the contaminates migration in porous media is explored using fractal methods in this study, implying the experimental results are very significant for real-world problems on an aquifer scale. Our next procedure involves applying these models in a real-world aquifer with complex heterogeneity conditions and modifying our models and method according to realistic conditions.

4 Conclusions

The microstructure of aquifers has an important effect on macroscopic-scale characteristics of contaminant migration and remediation. In this study, we focus on the DNAPL migration and remediation in heterogeneous aquifers composed of granular porous media with RTA and SPA. The microscale models of RTA and SPA are developed to obtain the mathematical expressions of permeability and entry pressure using fractal methods. A total of 200 realizations of the porosity field are generated using the SGS method, and PCE is released from an underground storage tank into a heterogeneous aquifer. To clean up contamination caused by the underground storage tank spill, a surfactant remediation technique is used to remove contaminants in the aquifer. The entire process of DNAPL migration and remediation is simulated by a multicomponent, multiphase model simulator, UTCHEM. Results suggest RTA not only cause more groundwater contamination than RTA, but also the contaminated aquifer of RTA is harder to clean up compared with SPA. The second PCE plume moments in the horizontal direction are 10.61–$40.50 \, \text{m}^2$ with a mean of 21.51 and 10.98–$36.38 \, \text{m}^2$ with a mean of $20.75 \, \text{m}^2$ for 200 realizations based on RTA and SPA, respectively, after long-term migration at $t = 100$ days. Furthermore, the second PCE plume moments in the horizontal direction at $t = 300$ day are 0.807–$34.88 \, \text{m}^2$ with a mean of 5.79 and 1.025–$24.57 \, \text{m}^2$ with a mean of $4.64 \, \text{m}^2$ for RTA and SPA, respectively, after long-term remediation. Simultaneously, the residual DNAPL mass remaining in the aquifer is 0.67–$119.89 \, \text{m}^3$ with a mean of 22.42 for RTA and 0.79–$103.33 \, \text{m}^3$ with a mean of $12.51 \, \text{m}^3$

for SPA, indicating that the remediation efficiency of SPA (65.53–99.74% with a mean of 95.83%) is mostly higher than the remediation efficiency of RTA (60.01–99.78% with a mean of 92.52%). This study reveals that the microstructure of an aquifer has an important effect on contaminant movement and associated remediation efficiency on a macroscopic scale, which is very essential and significant for dealing with accidental underground storage tank spills and for identifying subsurface contaminant sources in the future.

Competing interests. The authors declare that they have no conflict of interest.

Acknowledgements. This research was financially supported by the National Key Research and Development Plan of China (2016YFC0402800), the National Natural Science Foundation of China (41772254 and 41372235), and the National Natural Science Foundation of China-Xianjiang (project U1503282). The authors are also profoundly grateful to Pacheco Fla and an anonymous reviewer whose precious suggestions and constructive comments helped to improve the paper significantly.

Edited by: Sabine Attinger

References

Ahn, K. J. and Seferis, J. C.: Simultaneous measurements of permeability and capillary pressure of thermosetting matrices in woven fabric reinforcements, Polym. Composite., 12, 146–152, 1991.

An, C. J., McBean, E., Huang, G. H., Yao, Y., Zhang, P., Chen, X. J., and Li, Y. P.: Multi-soil-layering systems for wastewater treatment in small and remote communities, J. Environ. Inform., 27, 131–144, 2016.

Bakshevskaia, V. A. and Pozdniakov, S. P.: Simulation of hydraulic heterogeneity and upscaling permeability and dispersivity in sandy-clay foormations, Math. Geosci., 48, 45–64, 2016.

Bear, J.: Dynamics of Fluids in Porous Media, Dover, New York, 1972.

Bob, M. M., Brooks, M. C., Mravik, S. C., and Wood, A. L.: A modified light transmission visualization method for DNAPL saturation measurements in 2-D models, Adv. Water Resour., 31, 727–742, 2008.

Boswinkel, J. A.: Information Note, International Groundwater Resources Assessment Centre (IGRAC), Netherland Institute of Applied Geoscience, Netherlands, in: UNEP (2002), Vital Water Graphics – An Overview of the State of the World's Fresh and Marine Waters, UNEP, Nairobi, Kenya, 2000.

Carroll, K. C., McDonald, K., Marble, J., Russo, A. E., and Brusseau, M. L.: The impact of transitions between two-fluid and three-fluid phases on fluid configuration and fluid-fluid interfacial area in porous media, Water Resour. Res., 51, 7189–7201, 2015.

Cai, J. C., Yu, B. M., Zou, M. Q., and Mei, M. F.: Fractal analysis of invasion depth of extraneous fluids in porous media, Chem. Eng. Sci., 65, 5178–5186, 2010.

Cui, Q. L., Wu, H. N., Shen, S. L., Yin, Z. Y., and Horpibulsuk, S.: Protection of neighbour buildings due to construction of shield tunnel in mixed ground with sand over weathered granite, Environ. Earth Sci., 75, 458, https://doi.org/10.1007/s12665-016-5300-7, 2016.

Dagan, G., Fiori, A., and Jankovic, I.: Upscaling of flow in heterogeneous porous formations: critical examination and issues of principle, Adv. Water Resour., 51, 67–85, 2013.

Dawson, H. E. and Roberts, P. V.: Influence of viscous, gravitational, and capillary forces on DNAPL saturation, Groundwater, 35, 261–269, 1997.

Delshad, M., Pope, G. A., and Sepehrnoori, K.: A compositional simulator for modeling surfactant enhanced aquifer remediation, 1 Formation, J. Contam. Hydrol., 23, 303–327, 1996.

Eggleston, J. R., Rojstaczer, S. A., and Peirce, J. J.: Identification of hydraulic conductivity structure in sand and gravel aquifers: Cape Cod data set, Water Resour. Res., 32, 1209–1222, 1996.

Essaid, H. I., Bekins, B. A., and Cozzarelli, I. M.: Organic contaminant transport and fate in the subsurface: evolution of knowledge and understanding, Water Resour. Res., 51, 4861–4902, 2015.

Feng, Y. J. and Yu, B. M.: Fractal dimension for tortuous streamtubes in porous media, Fractals, 15, 385–390, 2007.

Hadley, P. W. and Newell, C.: The new potential for understanding groundwater contaminant transport, Groundwater, 52, 174–186, 2014.

Hu, K., White, R., Chen, D., Li, B., and Li, W.: Stochastic simulation of water drainage at the field scale and its application to irrigation management, Agr. Water Manage., 89, 123–130, 2007.

Huang, J. Q., Christ, J. A., Goltz, M. N., and Demond, A. H.: Modeling NAPL dissolution from pendular rings in idealized porous media, Water Resour. Res., 51, 8182–8197, 2015.

Katz, A. J. and Thompson, A. H.: Fractal sandstone: implications for conductivity and pore formation, Phys. Rev. Lett., 54, 325–332, 1985.

Krohn, C. E.: Sandstone fractal and Euclidean pore volume distributions, J. Geophys. Res., 93, 3286–3296, 1988.

Li, L. and Yu, B. M.: Fractal analysis of the effective thermal conductivity of biological media embedded with randomly distributed vascular trees, Int. J. Heat Mass Tran., 67, 74–80, 2013.

Liang, C. and Hsieh, C. L.: Evaluation of surfactant flushing for remediating EDC-tar contamination, J. Contam. Hydrol., 177–178, 158–166, 2015.

Liang, C. and Lai, M. C.: Trichloroethylene degradation by zero valent iron activated persulfate oxidation, Envrion. Eng. Sci., 25, 1071–1077, 2008.

Liu, H., Li, Y. X., He, X., Sissou, Z., Tong, L., Yarnes, C., and Huang, X.: Compound-specific carbon isotopic fractionation during transport of phthalate esters in sandy aquifer, Chemosphere, 144, 1831–1836, 2016.

Liu, L.: Modeling for surfactant-enhanced groundwater remediation processes at DNAPLs-contaminated sites, J. Environ. Inform., 5, 42–52, 2005.

Liu, L., Hao, R. X., and Cheng, S. Y.: A possibilistic analysis approach for assessing environmental risks from drinking groundwater at petroleum-contaminated sites, J. Environ. Inform., 2, 31–37, 2003.

Liu, Y., Wang, S., McDonough, C. A., Khairy, M., Muir, D. C. G., Helm, P. A., and Lohmann, R.: Gaseous and freely-dissolved PCBs in the lower great lake based on passive sampling: spatial trends and air-water exchange, Environ. Sci. Technol., 50, 4932–4939, 2016.

Mishra, A. K., Kumar, B., and Dutta, J.: Prediction of hydraulic conductivity of soil bentonite mixture using Hybrid-ANN approach, J. Environ. Inform., 27, 98–105, 2016.

Montgomery, R.H ., Loftis, J. C., and Harris, J.: Statistical characteristics of ground-water quality variables, Ground Water, 25, 176–184, 1987.

Pacheco, F. A. L.: Hydraulic diffusivity and macrodispersivity calculations embedded in a geographic information system, Hydrolog. Sci. J., 58, 930–943, 2013.

Pacheco, F. A. L., Landim, P. M. B., and Szocs, T.: Bridging hydraulic diffusivity from aquifer to particle-size scale: a study on loess sediments from southwest Hungary, Hydrolog. Sci. J., 60, 269–284, 2015.

Pfeifer, P. and Avnir, D.: Chemistry in nonintegral dimensions between two and three. I. Fractal theory of heterogeneous surface, J. Chem. Phys., 79, 3558–3565, 1983.

Qin, X. S., Huang, G. H., Chakma, A., Chen, B., and Zeng, G. M.: Simulation-based process optimization for surfactant-enhanced aquifer remediation at heterogeneous DNAPL-contaminated sites, Sci. Total Environ., 381, 17–37, 2007.

Schaefer, C. E., White, E. B., Lavorgna, G. M., and Annable, M. D.: Dense nonaqueous-phase liquid architecture in fractured bedrock: implications for treatment and plume longevity, Environ. Sci. Technol., 50, 207–213, 2016.

Shen, J., Huang, G., An, C. J., Zhao, S., and Rosendahl, S.: Immobilization of Tetrabromobisphenol A by pinecone-derived biochars at solid–liquid interface_Synchrotron-assisted analysis and role of inorganic fertilizer ions, Chem. Eng. J., 321, 346–357, 2017.

Taiwo, O. O., Finegan, D. P., Eastwood, D. S., Fife, J. L., Brown, L. D., Darr, J. A., Lee, P. D., Brett, D. J. L., and Shearing, P. R.: Comparison of three-dimensional analysis and stereological techniques for quantifying lithium-ion battery electrode microstructures, J. Microsc.-Oxford, 263, 280–292, 2016.

Valipour, M.: Comparison of surface irrigation simulation models: full hydrodynamic, zero inertia, kinematic wave, J. Agr. Sci., 4, 68–74, 2012.

Valipour, M.: Future of agricultural water management in Africa, Arch. Agron. Soil Sci., 61, 907–927, 2015.

Valipour, M. and Singh, V. P.: Global Experiences on Wastewater Irrigation: Challenges and Prospects, in: Balanced Urban Development: Options and Strategies for Liveable Cities, edited by: Maheshwari, B., Singh, V., and Thoradeniya, B., Water Science and Technology Library, Springer, Cham, 72, 289–327, 2016.

Veneziano, D. and Tabaei, A.: Nonlinear spectral analysis of flow through porous media with isotropic lognormal hydraulic conductivity, J. Hydrol., 294, 4–17, 2004.

Weathers, T. S., Harding-Marjanovic, K., Higgins, C. P., Alvarez-Cohen, L., and Sharp, J. O.: Perfluoroalkyl acids inhibit reductive dechlorination of trichloroethene by repressing dehalococcoides, Environ. Sci. Technol., 50, 240–248, 2016.

Wu, M., Cheng, Z., Wu, J. F., and Wu, J. C.: Quantifying representative elementary volume of connectivity for translucent granular materials by light transmission micro-tomography, J. Hydrol., 545, 12–27, 2017a.

Wu, M., Cheng, Z., Wu, J. F., and Wu, J. C.: Estimation of representative elementary volume for DNAPL saturation and DNAPL-water interfacial areas in 2-D heterogeneous porous media, J. Hydrol., 549, 12–26, 2017b.

Wu, M., Wu, J. F., and Wu, J. C.: Simulation of DNAPL migration in heterogeneous translucent porous media based on estimation of representative elementary volume, J. Hydrol., 553, 276–288, 2017c.

Wu, M., Cheng, Z., Wu, J. F., and Wu, J. C.: Precise simulation of long-term DNAPL migration in heterogeneous porous media based on light transmission micro-tomography, Journal of Environmental Chemical Engineering, 5, 725–734, https://doi.org/10.1016/j.jece.2016.12.039, 2017d.

Yannopoulos, S. I., Lyberatos, G., Theodossiou, N., Li, W., Valipour, M., Tamburrino, A., and Angelakis, A. N.: Evolution of water lifting devices (pumps) over the centuries worldwide, Water, 7, 5031–5060, 2015.

Yu, B. M.: Fractal character for tortuous streamtubes in porous media, Chinese Phys. Lett., 22, 158–160, 2005.

Yu, B. M. and Cheng, P.: Fractal models for the effective thermal conductivity of bidispersed porous media, J. Thermophys. Heat Tr., 16, 22–29, 2002.

Yu, B. M. and Li, J. H.: A geometry model for tortuosity of flow path in porous media, Chinese Phys. Lett., 21, 1569–1571, 2004.

Yu, B. M., Cai, J. C., and Zou, M. Q.: On the physical properties of apparent two-phase fractal porous media, Vadose Zone J., 8, 177–186, 2009.

Yun, M. J., Yu, B. M., Zhang, B., and Huang, M. T.: A geometry model for tortuosity of streamtubes in porous media with spherical particles, Chinese Phys. Lett., 22, 1464–1467, 2005.

7

Estimating long-term groundwater storage and its controlling factors in Alberta, Canada

Soumendra N. Bhanja[1], Xiaokun Zhang[2], and Junye Wang[1]

[1]Athabasca River Basin Research Institute (ARBRI), Athabasca University, 1 University Drive, Athabasca, Alberta T9S 3A3, Canada
[2]School of Computing & Information System, Athabasca University, 1 University Drive, Athabasca, Alberta T9S 3A3, Canada

Correspondence: Junye Wang (junyew@athabascau.ca)

Abstract. Groundwater is one of the most important natural resources for economic development and environmental sustainability. In this study, we estimated groundwater storage in 11 major river basins across Alberta, Canada, using a combination of remote sensing (Gravity Recovery and Climate Experiment, GRACE), in situ surface water data, and land surface modeling estimates ($GWSA_{sat}$). We applied separate calculations for unconfined and confined aquifers, for the first time, to represent their hydrogeological differences. Storage coefficients for the individual wells were incorporated to compute the monthly in situ groundwater storage ($GWSA_{obs}$). The $GWSA_{sat}$ values from the two satellite-based products were compared with $GWSA_{obs}$ estimates. The estimates of $GWSA_{sat}$ were in good agreement with the $GWSA_{obs}$ in terms of pattern and magnitude (e.g., RMSE ranged from 2 to 14 cm). While comparing $GWSA_{sat}$ with $GWSA_{obs}$, most of the statistical analyses provide mixed responses; however the Hodrick–Prescott trend analysis clearly showed a better performance of the GRACE-mascon estimate. The results showed trends of $GWSA_{obs}$ depletion in 5 of the 11 basins. Our results indicate that precipitation played an important role in influencing the $GWSA_{obs}$ variation in 4 of the 11 basins studied. A combination of rainfall and snowmelt positively influences the $GWSA_{obs}$ in six basins. Water budget analysis showed an availability of comparatively lower terrestrial water in 9 of the 11 basins in the study period. Historical groundwater recharge estimates indicate a reduction of groundwater recharge in eight basins during 1960–2009. The output of this study could be used to develop sustainable water withdrawal strategies in Alberta, Canada.

1 Introduction

Fresh water is an important resource for economic development and social sustainability around the world. Approximately 1.2 billion people live in water-scarce areas across the globe (UN-Water/FAO, 2007). More than a billion people lack access to safe drinking water and this number is increasing due to an increasing population (Connor, 2015). However, the effects of climate change on glaciers and snowpack and the effects of human activities, such as overuse and overextraction of resources, can result in lowering water tables and groundwater depletion (Scanlon et al., 2016; Bhanja et al., 2017b). In situ monitoring of wells is the traditional approach for estimating groundwater storage. However, well monitoring is spatially not continuous and has a high cost for a large region. There are only scant observation stations in some areas, especially in semiarid and arid environments, or cold climate regions covered by glacier and snowpack, due to difficulties of access and monitoring. As a result, proper groundwater management and decision-making are hampered considerably by the scarcity of data.

Remote-sensing data from the Gravity Recovery and Climate Experiment (GRACE) satellite mission could be used to estimate groundwater storage at a continuous and large scale across the globe and offer a new opportunity for groundwater storage assessment (Rodell et al., 2007). Although the GRACE satellite mission currently provides global-scale data for the detection of temporal gravity changes (Tapley et al., 2004), these temporal gravity changes are not a direct measurement of groundwater storage. A relationship would have to be established between temporal gravity changes and groundwater storage variations through the continuously evolving algorithms (Watkins et al., 2015). Estimates of groundwater storage using the remote sensing have been performed around the globe (Swenson et al., 2006; Rodell et al., 2007, 2009; Strassberg et al., 2007; Tiwari et al., 2009; Scanlon et al., 2012; Shamsudduha et al., 2012; Voss et al., 2013; Bhanja et al., 2014, 2016, 2017b, 2018; Richey et al., 2015; Panda and Wahr, 2016; Chen et al., 2016; Long et al., 2016). Huang et al. (2016) used remote-sensing data for computing the groundwater storage anomalies (GWSAs) in order to estimate groundwater storage in Alberta. They used groundwater levels (GWLs) at 36 wells, mostly confined to the southern Alberta region, and were correlated with both the GRACE terrestrial water storage (TWS) and groundwater storage variations. Then they compared the TWS with GWLs instead of the groundwater storage and without considering surface water data due to the lack of available high-resolution data.

Recent studies (e.g., Huang et al., 2015; Nanteza et al., 2016) have considered both confined and unconfined aquifers for in situ GWSA computation but they have not separated the data from the two types. The two types of aquifers have different recharge and storage patterns. Confined aquifers are overlain by relatively impermeable rock or clay, which limits vertical water infiltration, while in unconfined aquifers, vertical water infiltration can occur from precipitation, snowmelt, surface water, etc. The two types of aquifers also respond differently to effects from pumping (Alley et al., 1999). Therefore, these should be studied separately for estimating groundwater storage in a region. Further, Rodell et al. (2007) indicated the importance of surface water factors in the GWSA estimation and sought for inclusion of surface water storage variations in GWSA disaggregation. They also pointed out the importance of separating contributions by temporal mass variability using auxiliary observations and numerical models when estimating groundwater storage changes in large-scale regions. In cold climate regions, such as in Alberta, the surface water could make a significant contribution to groundwater storage variations due to the effects of climate change on snowpack, glaciers, permafrost, and wetlands. Therefore, more efforts are required to properly evaluate groundwater storage for aquifer storage coefficients in transforming GWL information to groundwater storage in cold climate regions (Feng et al., 2013). The main objectives of this study are

1. to investigate the long-term groundwater storage conditions in cold climate regions, such as the 11 river basins in Alberta, Canada, by combining all of the processing steps, such as the surface water storage estimates.

2. to validate the remote-sensing estimates from two different remote-sensing products using the maximum available in situ observation well data; the in situ groundwater storage has been estimated by combining the storage coefficients and aquifer thickness (for confined aquifers) with the water table fluctuation.

3. to find the role of natural hydrological components (e.g., precipitation, snowmelt, evapotranspiration) in influencing groundwater storage variations; we have also studied long-term groundwater recharge trends from a global-scale hydrological model for inferring long-term variabilities in groundwater recharge rates.

2 Materials and methods

2.1 Study area

This study has been conducted in the major river basins (the map has been made following Lemay and Guha, 2009; AEP, 2011; AEP, 2017) within the province of Alberta (Fig. 1a, b). The Peace River basin is the largest basin in the province, followed by the Athabasca River basin and Hay River basin (Table 1). Most parts of the study region have been characterized as cold climate regions (Peel et al., 2007). Basin-scale annual average precipitation varies within 330 to 570 mm year^{-1} (Table 1). We used Global Land Cover Facility (GLCF) native-resolution data (resolution: $\sim 460\,\mathrm{m} \times 460\,\mathrm{m}$; http://www.landcover.org/, last access: 2 October 2017) for characterizing land cover (Channan et al., 2014). Most of the land areas in Alberta are covered by natural vegetation (i.e., forest, shrubland, mixture of shrub and grassland, and grassland; Fig. 1c, Table S1 in the Supplement). The second most prevalent land-cover type is cropland (Fig. 1c, Table S1). Surface water bodies (water and permanent wetland) cover less than 6 % of the area of all the river basins (Fig. 1c, Table S1).

We used monthly mean precipitation data from the archives of the Climatic Research Unit (CRU), University of East Anglia. The quality-controlled, gridded $0.25° \times 0.25°$, monthly mean TS4.0 total precipitation products are used here (Harris et al., 2014). The precipitation-gauge-based data were collected through the World Meteorological Organization (WMO), National Oceanographic and Atmospheric Administration (NOAA), and other international and national agencies across the globe for preparing this dataset (Harris et al., 2014). The precipitation data are spatially averaged in order to provide basin-scale data. CRU data have been found to have the best match of other available products while comparing with in situ precipitation measurements in China (Zhao and Fu, 2006). Precipitation data exhibit temporal as well as spatial variations in the study period with values of 150 to >1000 mm year^{-1} (Fig. 2). In general, the lowest precipitation was observed in 2004 and the highest in 2010 (Fig. 2). Spatially averaged basin-scale precipitation values

Table 1. Details of the river basins used (within Alberta only), number of wells used, precipitation, and $GWSA_{obs}$ trends (statistically significant (p value < 0.01) trend estimates are shown in bold).

Basin ID	Basin name	Ocean	Basin area (m^2)	Number of wells	Precipitation ($mm\ year^{-1}$)	$GWSA_{obs}$ trends ($cm\ year^{-1}$)	$P - ET$ trends (km^3 in 2003–2015)
1	Hay River basin	Arctic Ocean	66 196 942 347	3	401	**1.22**	−0.17
2	Peace River basin	Arctic Ocean	213 025 952 509	15	429	−0.19	−0.41
3	Athabasca River basin	Arctic Ocean	144 499 671 762	8	508	**−0.19**	−0.17
4	North Saskatchewan River	Hudson Bay	57 046 775 461	21	573	**0.21**	−0.25
5	Battle River basin	Hudson Bay	36 561 280 700	28	424	**0.43**	−0.06
6	Red Deer River basin	Hudson Bay	50 024 664 775	21	425	**0.12**	−0.11
7	Bow River basin	Hudson Bay	25 639 800 168	15	546	−0.04	−0.08
8	Oldman River basin	Hudson Bay	27 023 265 616	10	486	−0.08	−0.03
9	South Saskatchewan River basin	Hudson Bay	13 504 374 212	6	334	**−0.10**	−0.01
10	Milk River basin	Hudson Bay	11 833 516 877	9	353	**0.52**	0.00
11	Beaver River basin	Hudson Bay	16 904 014 071	21	469	**0.24**	0.01

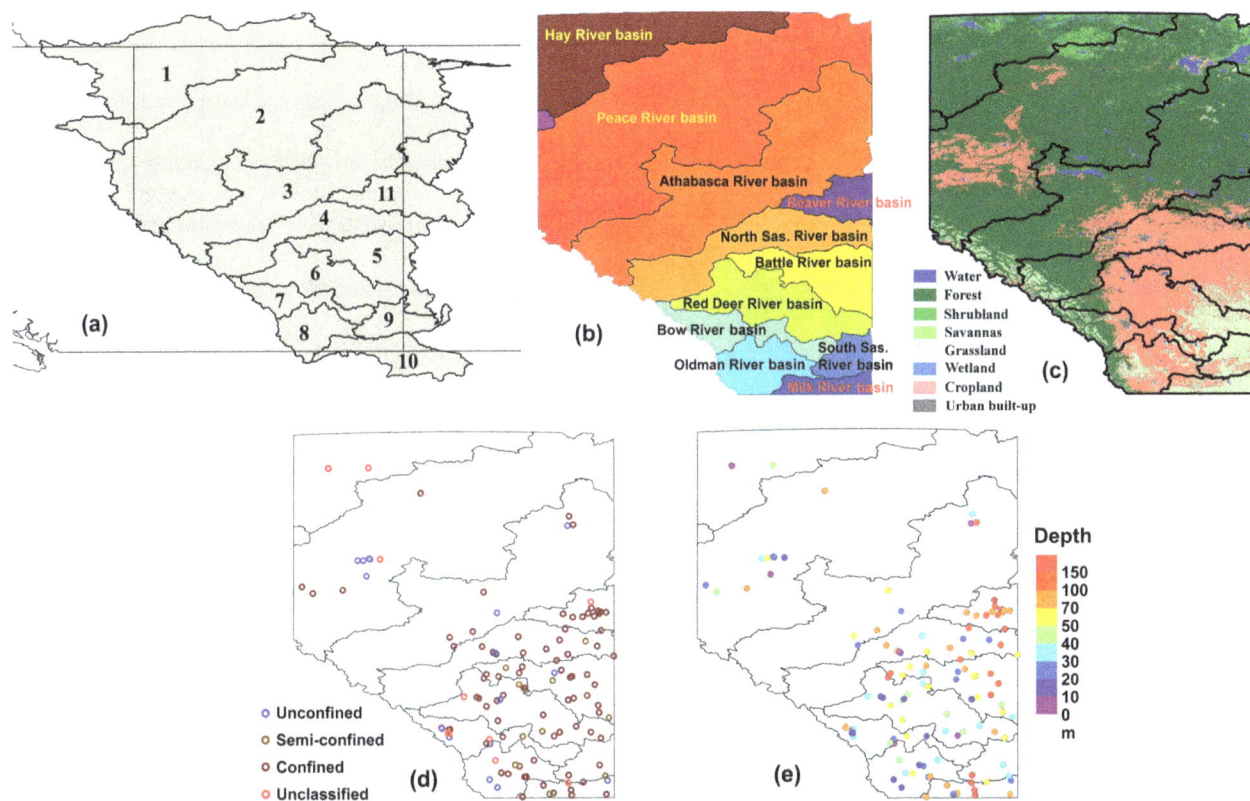

Figure 1. Major river basins in Alberta. (**a**) Full basin extent; (**b**) Alberta only; (**c**) dominant land cover types; (**d**) aquifer types represented through the studied wells; (**e**) depth of wells screened in Alberta, overlaid by basin boundaries.

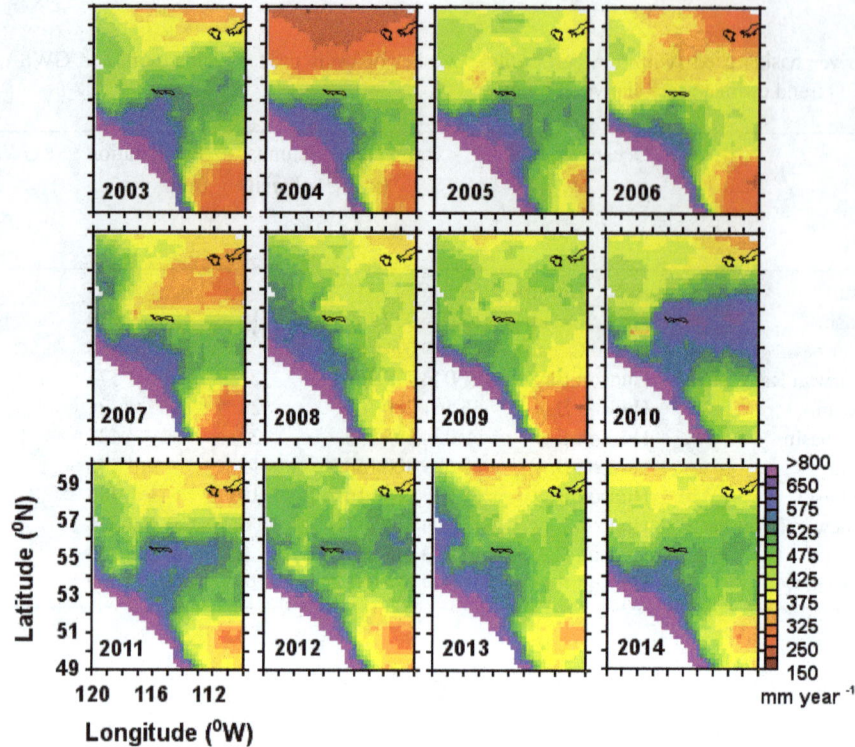

Figure 2. Annual precipitation rates (mm year^{-1}) in Alberta between 2003 and 2014.

indicate the highest precipitation rates prevail in the North Saskatchewan River basin (basin 4, 573 mm year^{-1}; Table 1) and the lowest rates prevail in the South Saskatchewan River basin (basin 9, 334 mm year^{-1}; Table 1). Precipitation rates are highly seasonal in Alberta (Fig. 3).

2.2 In situ measurements of groundwater storage

GWL depth data are obtained from the Alberta Environment and Parks, Government of Alberta (http://environment. alberta.ca/apps/GOWN/#, last access: 14 September 2017). Daily GWL depth data are obtained for 470 monitoring wells distributed across the province of Alberta. The data are screened for data continuity (at least 80 % of the data are present in each location) within the study period of 2003–2015, resulting in the use of GWL data from 157 measurement locations. Daily GWL information is converted to monthly GWL at individual locations. Because of the differences in groundwater storage variations within different types of aquifers, these wells need to be classified as unconfined, semi-confined, and confined. Out of the 157 measurement locations used in the study, 24 are located in unconfined aquifers, 17 are located within semi-confined aquifers, 100 are located within confined aquifers, and 16 are unclassified (Fig. 1d). Basin-wide details of the distribution of wells are provided in Table S2. The screen depth of the wells varies from 6 to 220 m (Fig. 1e). The wells located in unclassified

or semi-confined aquifers are characterized as either confined or unconfined based on their location hydrogeology and screen depth. For example, a well screened at a semi-confined aquifer with shallower depth and underlain by permeable materials can be classified as an unconfined aquifer for storage calculations and vice versa.

We have studied the subsurface hydrogeology in detail using well-specific lithology information from the Alberta Environment and Parks, Government of Alberta (http://environment.alberta.ca/apps/GOWN/#, last access: 14 September 2017). In order to compute groundwater storage anomalies (GWSA$_{obs}$) in an unconfined aquifer, the GWSA needs to be accurately represented using the storage coefficients of the aquifer (Scanlon et al., 2012). We have followed the equation (Todd and Mays, 2005; Bhanja et al., 2017a)

$$GWSA_{obs} = (h_m \times S_y - h_i \times S_y), \qquad (1)$$

where h_m and h_i represent the mean GWL depth and GWL depths at different time periods at a location; S_y represents the specific yield of the aquifer. S_y is assigned to the individual data based on the specific yield of the geologic material in its screen position. Specific yield data corresponding to a specific geologic material are presented in Table S3.

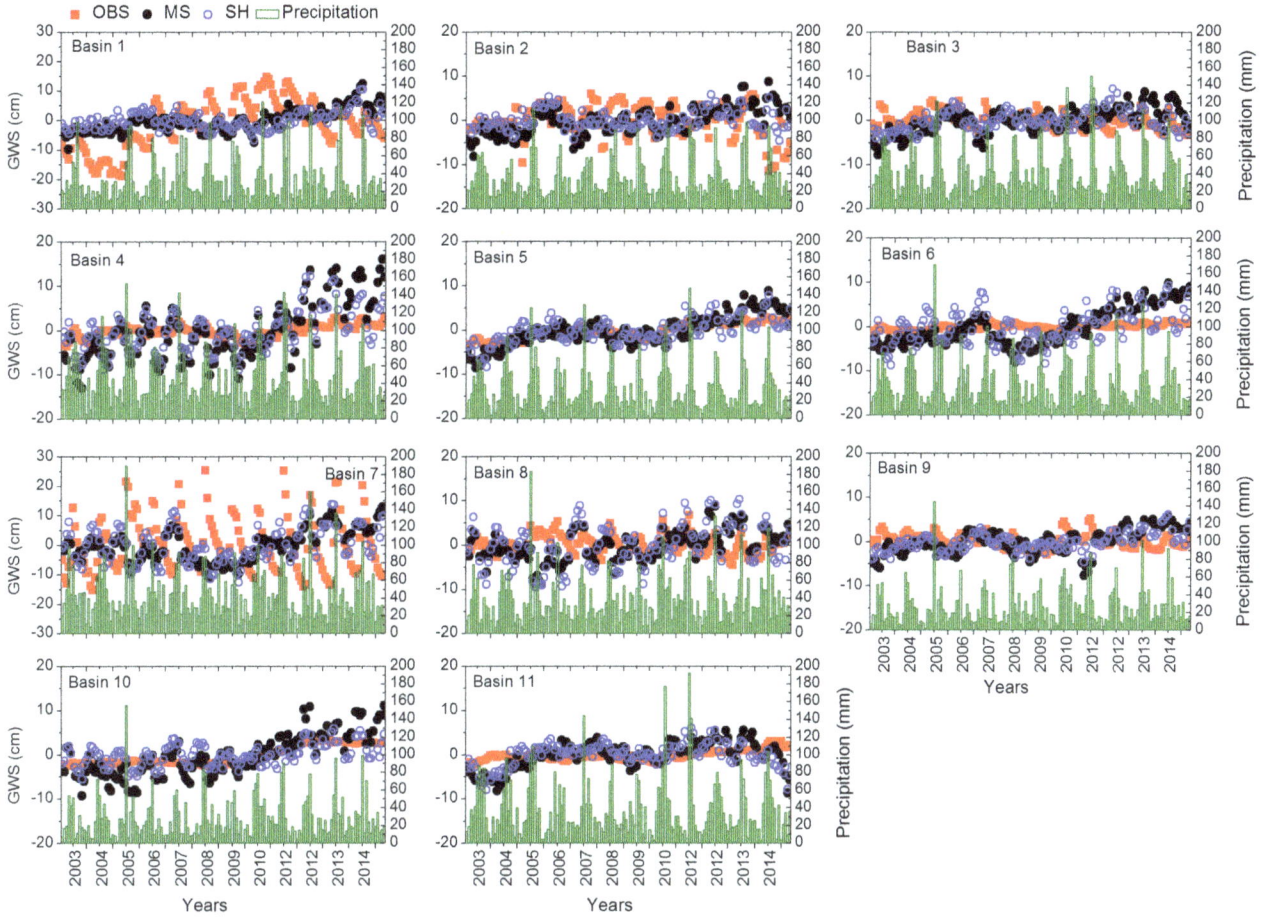

Figure 3. Basin-wide, monthly time series of in situ GWSA (OBS, red squares) and GWSA obtained using the GRACE mascon product (MS, black filled circles) and GRACE SH products (SH, blue open circles). Monthly spatially averaged precipitation data are shown using green columns.

GWSA$_{obs}$ values in a confined aquifer have been estimated following the equation (Todd and Mays, 2005)

$$GWSA_{obs} = (h_m \times S_s \times b - h_i \times S_s \times b), \quad (2)$$

where S_s is the specific storage and b is the thickness of the aquifer. S_s of a material varies over a wide range; the details of material-specific S_s are provided in Table S4. Thickness of the aquifer for the individual aquifer units is obtained from the Alberta Environment and Parks, Government of Alberta. The data are assigned to individual wells based on their screening zone and thickness of the particular aquifer unit.

2.3 Surface water storage processing

Surface water level daily time series are obtained ($n = 393$) from the Water Office, Government of Canada (https://wateroffice.ec.gc.ca/, last access: 25 August 2017) for the study region. After rearranging the data based on near-continuous data availability, we used 65 locations with >80 % of the data availability range. The data are tempo-

rally averaged at each location for estimating monthly mean values. The number of locations that fall within each of the river basins are spatially averaged to obtain the month-scale spatially averaged surface water anomaly. The surface water coverage fraction varies over the study region (Fig. 1c and Table S1). In order to obtain realistic surface water storage variations, surface water area fractions have been multiplied with the spatially averaged surface water anomaly in each river basin.

2.4 Gravity Recovery and Climate Experiment (GRACE)

We have obtained the monthly mean liquid water equivalent thickness $1° \times 1°$ gridded data from the archives of the National Aeronautics and Space Administration (NASA) Jet Propulsion Laboratory (JPL). JPL mascon solutions, version RL05, was used for 137 months (the data were not available for some of the months; details can be found in Watkins et al., 2015) between January 2003 and April 2015.

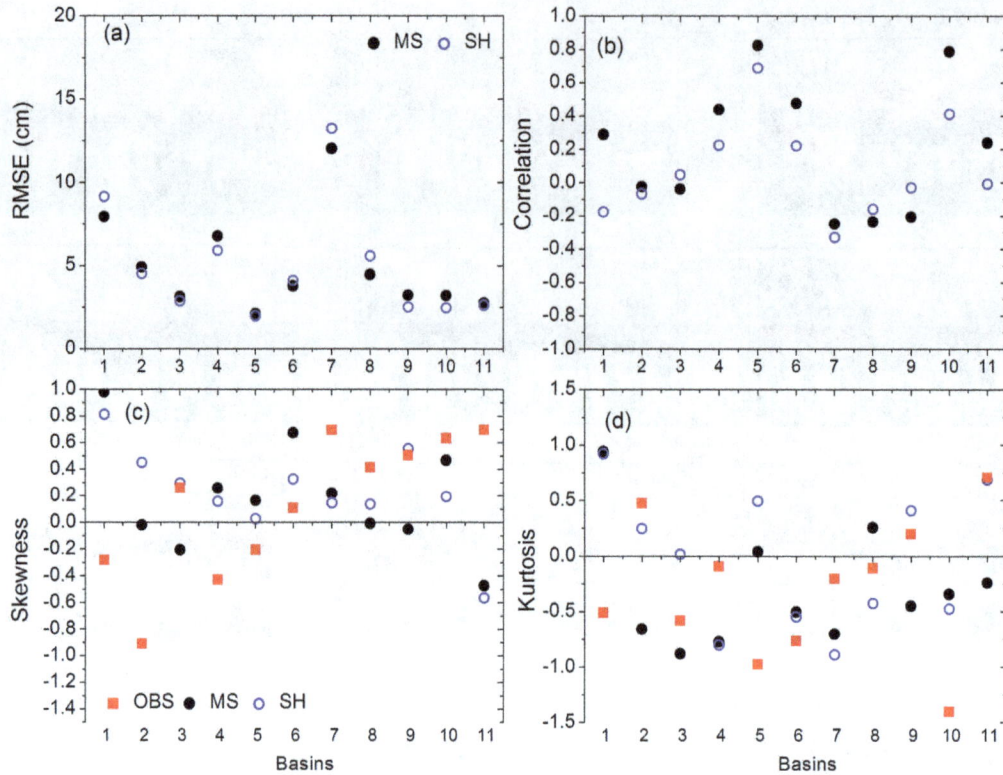

Figure 4. Basin-wide estimates of **(a)** RMSE, **(b)** correlation, **(c)** skewness, and **(d)** kurtosis.

The GRACE mission observes changes in gravity in the Earth's subsurface and provides the data on a continuous basis. The gravity change information has been processed further in order to obtain the TWS change data (details can be found here: http://grace.jpl.nasa.gov/data/get-data/jpl_global_mascons/, last access: 14 November 2017). Satellite laser ranging (SLR) has been incorporated for estimating degree 2 and order 0 coefficients (Cheng and Tapley, 2004). Processes to improve the geocenter correction have been reported by Swenson et al. (2008). The post-glacial rebound signals related to glacial isostatic adjustment (GIA) are removed using the process by Geruo et al. (2013). In the mascon approach, the entire globe is characterized as equally spaced 3° spherical mass concentration blocks (Watkins et al., 2015). In order to improve the TWS estimates, scale factors provided with the data are multiplied (Bhanja et al., 2016). Scale factors are estimated in order to improve the performance of the TWS estimates.

The TWS information related to spherical harmonics (SH) has been obtained for 137 months (between January 2003 and April 2015) from the NASA JPL archive. We used 1° × 1° gridded RL05 datasets of SH solutions (Landerer and Swenson, 2012). Three independent solutions from the Center for Space Research at the University of Texas at Austin, the NASA JPL, and the German Space Agency (GFZ) were retrieved and combined to use

in this study. Like the mascon approach, several similar techniques are applied to obtain the TWS change in the SH approach (source: http://grace.jpl.nasa.gov/data/get-data/monthly-mass-grids-land/, last access: 14 November 2017). Errors associated with N–S stripes in the TWS data are removed using a de-striping filter. A Gaussian filter of 300 km in width is also applied to the data. In order to improve the TWS estimates, the scale factors provided with the data are multiplied (Bhanja et al., 2016).

One advantage of the mascon approach is the introduction of a priori information that leads to the removal of correlated noise (stripes) in the data. As a result, post-processing filters are not required to be applied (Watkins et al., 2015). TWS data obtained from the mascon approach are less dependent on scale factors for estimating basin-scale mass change estimates (Watkins et al., 2015).

2.5 Estimating groundwater storage from remote sensing and global models

Satellite-based groundwater storage anomalies (GWSA$_{sat}$) are estimated using a mass balance approach after removing other components of the hydrological cycle from the TWSA. These components include soil moisture anomaly (SMA), anomalies in snow water equivalents (SNAs), and anomalies in surface water variations (SWAs). Anomalies

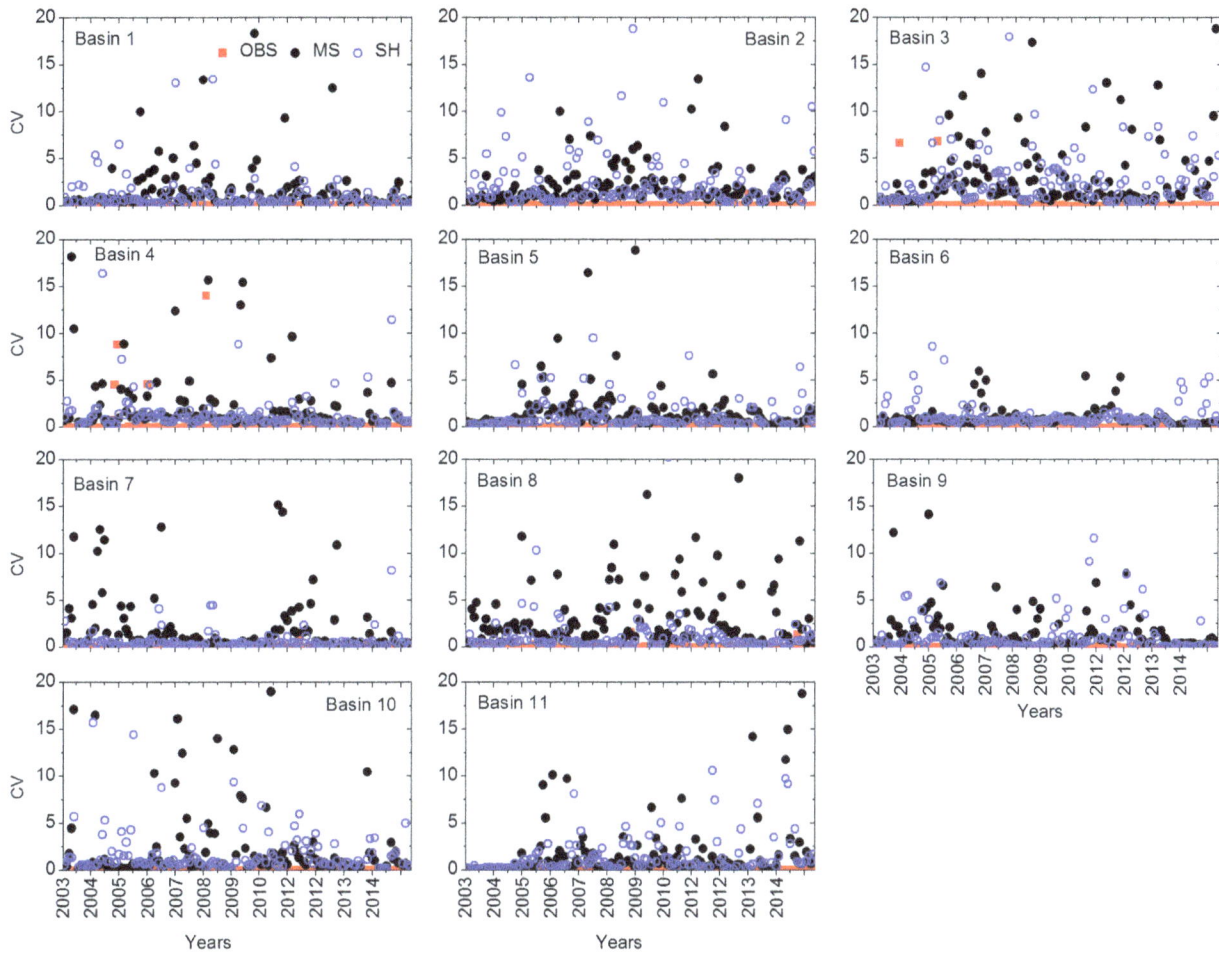

Figure 5. Basin-wide coefficient of variation (CV) analysis for in situ GWSA (OBS, red squares) and GWSA obtained using the GRACE mascon product (MS, black filled circles) and GRACE SH products (SH, blue open circles).

are estimated after removing the all-time mean value from the individual monthly values for all of the components. Soil moisture and snow water equivalent data were retrieved from NASA's Global Land Data Assimilation System (GLDAS) (Rodell et al., 2004) for 148 months in the study period. The GLDAS includes observation data from satellite sensors and ground-based measurements in order to improve the simulation output (Rodell et al., 2015). Bhanja et al. (2016) reported better GWSA estimates while using a combination of data from simulations of three different land surface models (LSMs), comparing the use of any single model's output. We also used a combined estimate from the outputs of the Community Land Model (CLM), Variable Infiltration Capacity (VIC), and Noah (Rodell et al., 2004). Surface water variation plays an important role in estimating GWSA. We have computed the surface water variations using in situ data, described in Sect. 2.3. GWSA can be estimated using the following equation:

$$GWSA_{sat} = TWSA - SMA - SNA - SWA. \qquad (3)$$

2.6 Groundwater recharge from global-scale hydrological model

In order to find the historical groundwater recharge pattern, we used a global-scale hydrological model, WaterGAP (version 2.2) (Doll et al., 2014) to estimate long-term groundwater recharge data (1960–2009). The WaterGAP simulates global-scale water storage and transport including human water use and groundwater recharge from surface water bodies at $0.5° \times 0.5°$ resolution (Doll et al., 2014). Water withdrawal from both groundwater and surface water has also been considered. We used a combination of diffuse groundwater recharge and recharge from the surface water bodies, which we termed "total groundwater recharge". As the WaterGAP simulation considers a simple water balance approach for groundwater recharge estimation, uncertainties may arise as a function of groundwater table gradient (Doll et al., 2014). Furthermore, increasing groundwater recharge from surface water bodies as a function of groundwater with-

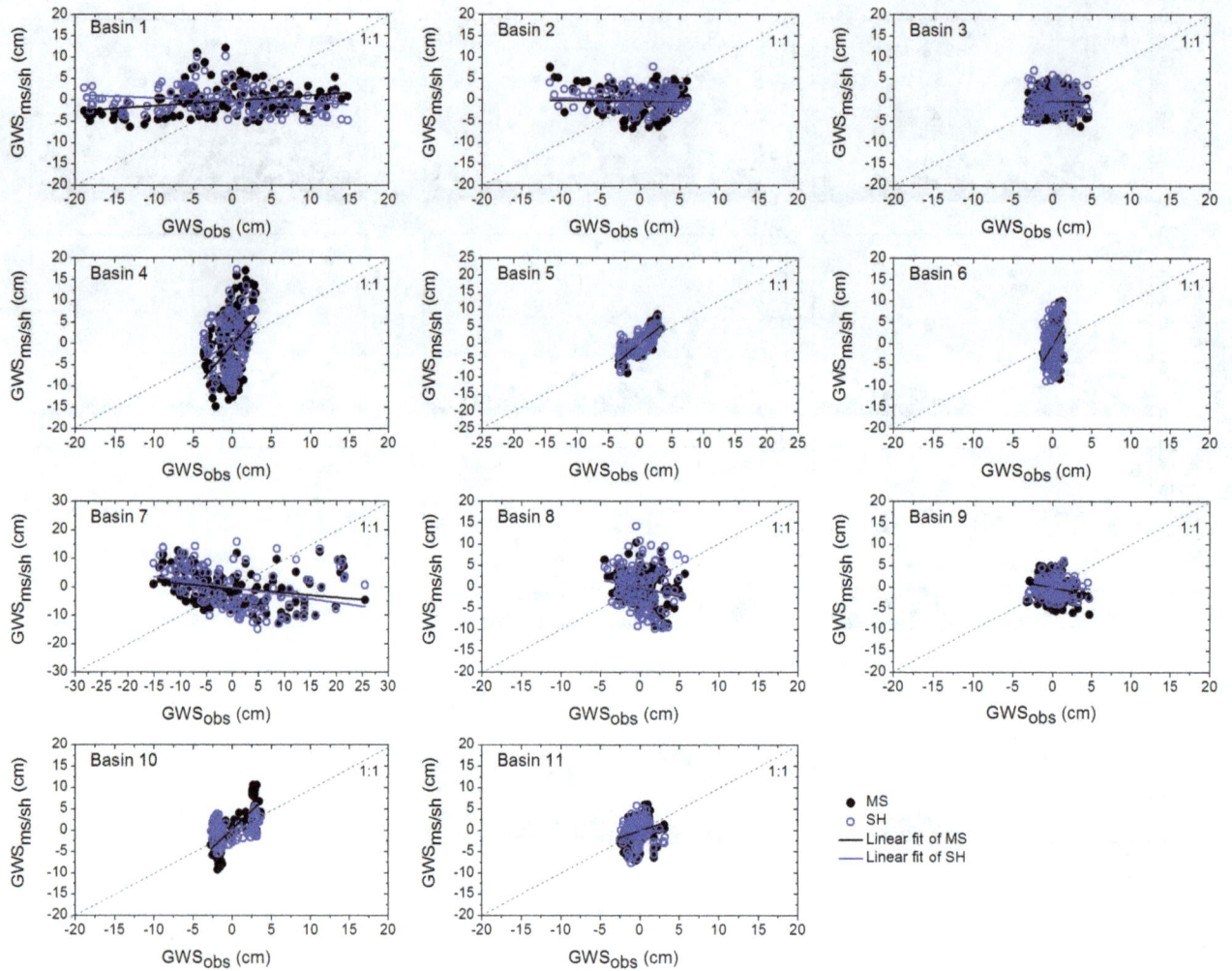

Figure 6. Basin-wide scatter analysis of in situ GWSA with the GWSA obtained using the GRACE mascon product (MS, black filled circles) and GRACE SH products (SH, blue open circles).

drawal has not been considered here (Doll et al., 2014). More information on model processes, data used, and other details can be found in Doll et al. (2014).

data (y_t).

$$y_t = T_t + c_t \tag{4}$$

2.7   Statistical approaches

In order to compare the datasets using statistically robust techniques, we have used the root-mean-square error (RMSE), Pearson's correlation, skewness, kurtosis, and the coefficient of variation. RMSE has been used to show the departures from the true (in situ estimates here) value (Helsel and Hirsch, 2002). The trend analyses are based on the linear regression analysis. In order to represent the nonlinearity present within the data, we used the Hodrick–Prescott (HP) filter (Hodrick and Prescott, 1997), a nonparametric, nonlinear trend analysis. The HP filter employs a specific approach for separating trend (T_t) and cycle (c_t) components in the

In order to estimate the trend and cycle separately, the HP filter solves the following equation (Hodrick and Prescott, 1997):

$$\text{Min}(T)\sum_{t=1}^{T}((y_t - T_t)^2 + ((T_{t+1} - T_t) - (T_t - T_{t-1}))^2, \tag{5}$$

where T_{t+1} and T_{t-1} represent the trend component with time steps of $t+1$ and $t-1$, respectively. The long-term average of the cyclical components is close to zero (Hodrick and Prescott, 1997). The smoothing parameter (λ) is a positive number that reduces the variability within cyclical components (Hodrick and Prescott, 1997). The value of λ was chosen to be 14 400 for monthly data (Hodrick and Prescott, 1997; Ravn and Uhlig, 2002).

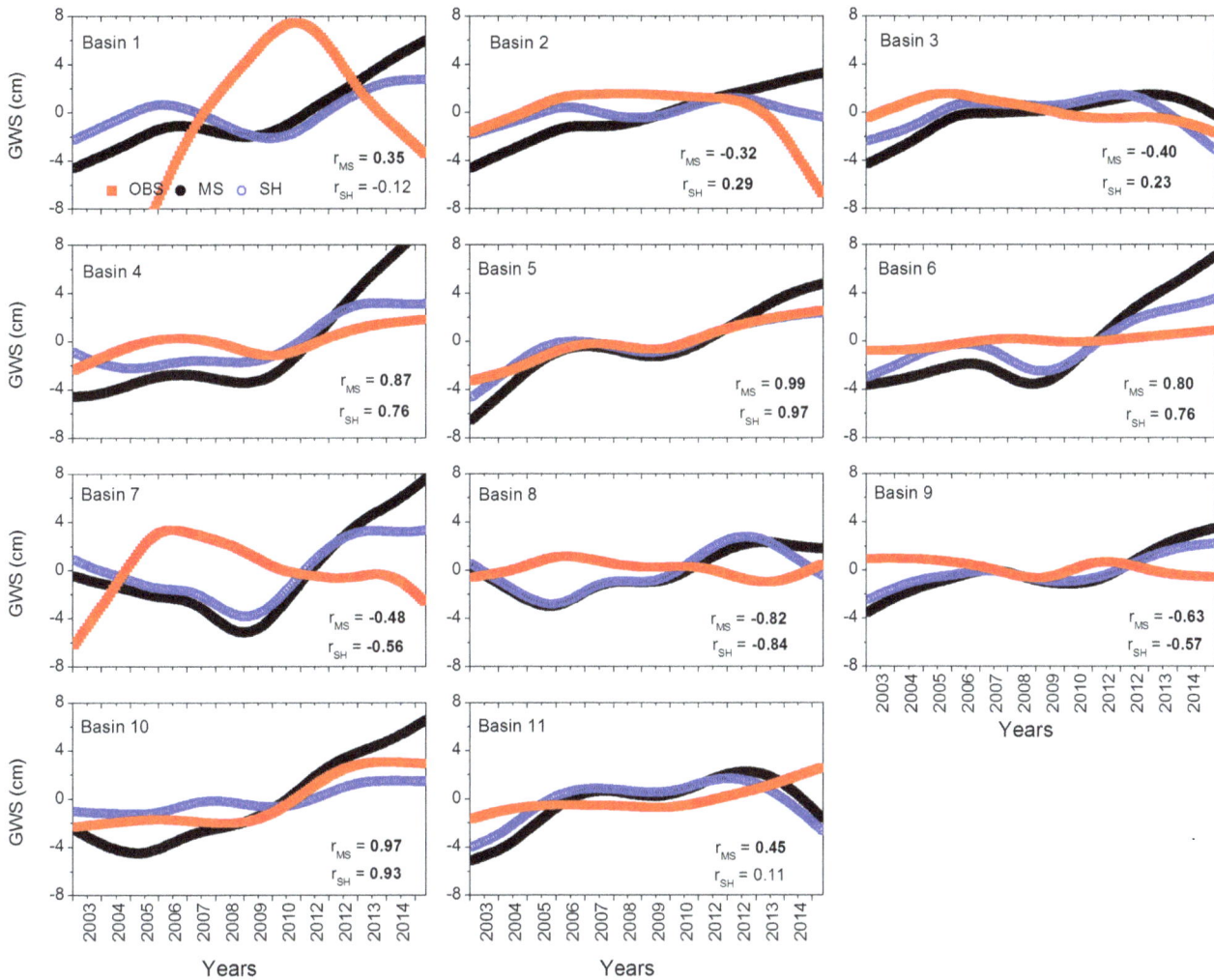

Figure 7. Basin-wide time series of HP filter data for in situ GWSA (OBS, red squares) and GWSA obtained using the GRACE mascon product (MS, black filled circles) and GRACE SH products (SH, blue open circles). Pearson's correlation coefficient values are provided in the inset and statistically significant (p value < 0.01) values are shown in bold.

2.8 Assumptions and limitations

On the basis of data availability, we have not included the entire extent of the river basins (Fig. 1a) in the current analysis. The river basins are selected based only on their geopolitical extent in the province of Alberta (Fig. 1b). For in situ estimates, GWSA information is spatially averaged for providing the basin's GWSA estimates and also to compare them with the satellite-based estimates following Bhanja et al. (2017a) and Scanlon et al. (2018). The time period of the study is restricted by the availability of data. Separation of GWSA signals from TWSA by removing all other components is a challenging task due to the lack of in situ measurements of other components and the large uncertainties associated with LSM-simulated products (Scanlon et al., 2015). We have shown the satellite-based estimates for all of the basins; however, users should be cautious to use GRACE

data in the smallest basins. This is because GRACE's native resolution could not allow users to directly use the data for smaller basins. Other processes such as the use of GRACE and integrated land surface model's operation could make the data available to use for smaller basins (Landerer and Swenson, 2012; Watkins et al., 2015). Data processing methods proposed by Dutt Vishwakarma et al. (2016) could be used to make the data available for smaller basins with GRACE SH products.

3 Results and discussions

3.1 Groundwater storage anomalies

In situ GWSA ($GWSA_{obs}$) values ranged from -30 to 30 cm in all of the basins, with the highest fluctuations observed in

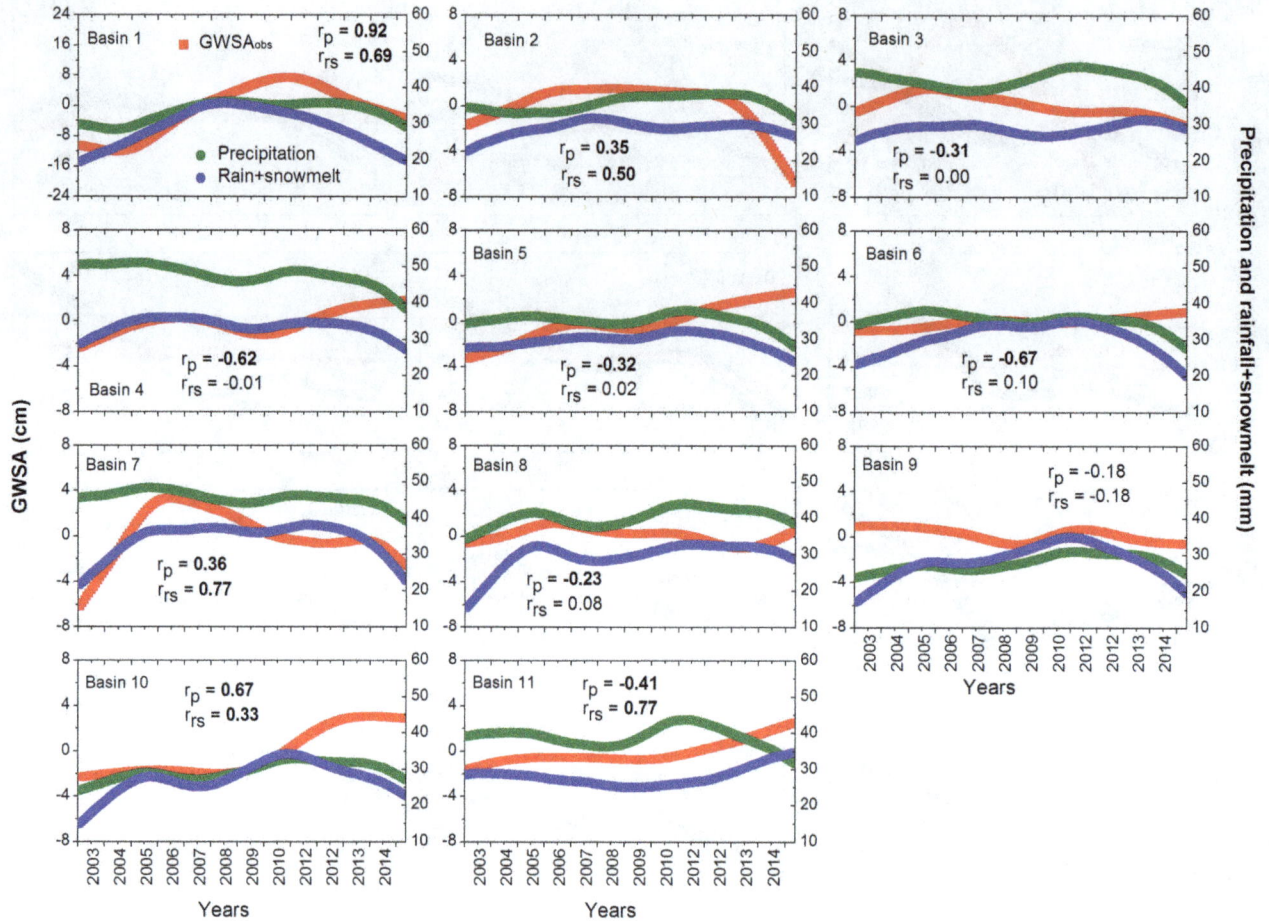

Figure 8. Basin-wide time series of HP filter data for in situ GWSA (OBS, red squares), precipitation data (green circles), and rainfall + snowmelt data (blue circles). Pearson's correlation coefficient (r) values are provided in the inset and statistically significant (p value < 0.01) values are shown in bold. r_p and r_{rs} indicate the correlation between GWSA and precipitation and between GWSA and rainfall + snowmelt, respectively.

basin 7. GWSA_obs exhibits near zero values in basins 5, 6, 10, and 11 (Fig. 3). GWSA_obs magnitudes in different basins can arise as a result of diversity in specific yield values in the underlying material (Table S3). In situ estimates show seasonality, i.e., variations with precipitation rates, in basin 7. Trends in the GWSA_obs show decreasing GWSA between 2003 and 2015 in basins 2, 3, 7, 8, and 9 (Table 1). The results indicate that GWSA_obs depletion is in the range of −0.20 in basins 2 and 3. It is interesting to note that basins 7, 8, and 9 are composed of >25 % cropland (Table S1). Basin 3 has been subjected to the highest amount of licensed groundwater withdrawal allocation in Alberta (basin 3 accounts for 39 % of the total groundwater usage in Alberta). Conversely, an increasing trend has been observed in the remaining basins (Table 1). One probable cause for the groundwater table increase in these basins could be related to precipitation variability. The study region has been subjected to large-scale drought during 1999–2005 (Hanesiak et al., 2011). As a re-

sult, the TWS recovery in 2004–2007 has also been observed by Lambert et al. (2013).

Another important factor influencing groundwater recharge as well as the groundwater storage is the snowmelt processes prevailing in cold regions during the onset of spring–summer. The river basins have been receiving a substantial amount of snowfall during winter months (Fig. 3). This leads to snow accumulation in the region. At the end of the winter season, snowmelt processes majorly account for our observation of increasing GWSA in April onwards (Fig. 3). The observation is in line with the observations from the earlier studies conducted within the study region (Hayashi and Farrow, 2014; Hood and Hayashi, 2015). Comparatively higher rates of precipitation during summer months and the snowmelt during the start of the summer season are the major processes responsible for the observation of higher GWSA during summertime in the entire study region (Fig. 3).

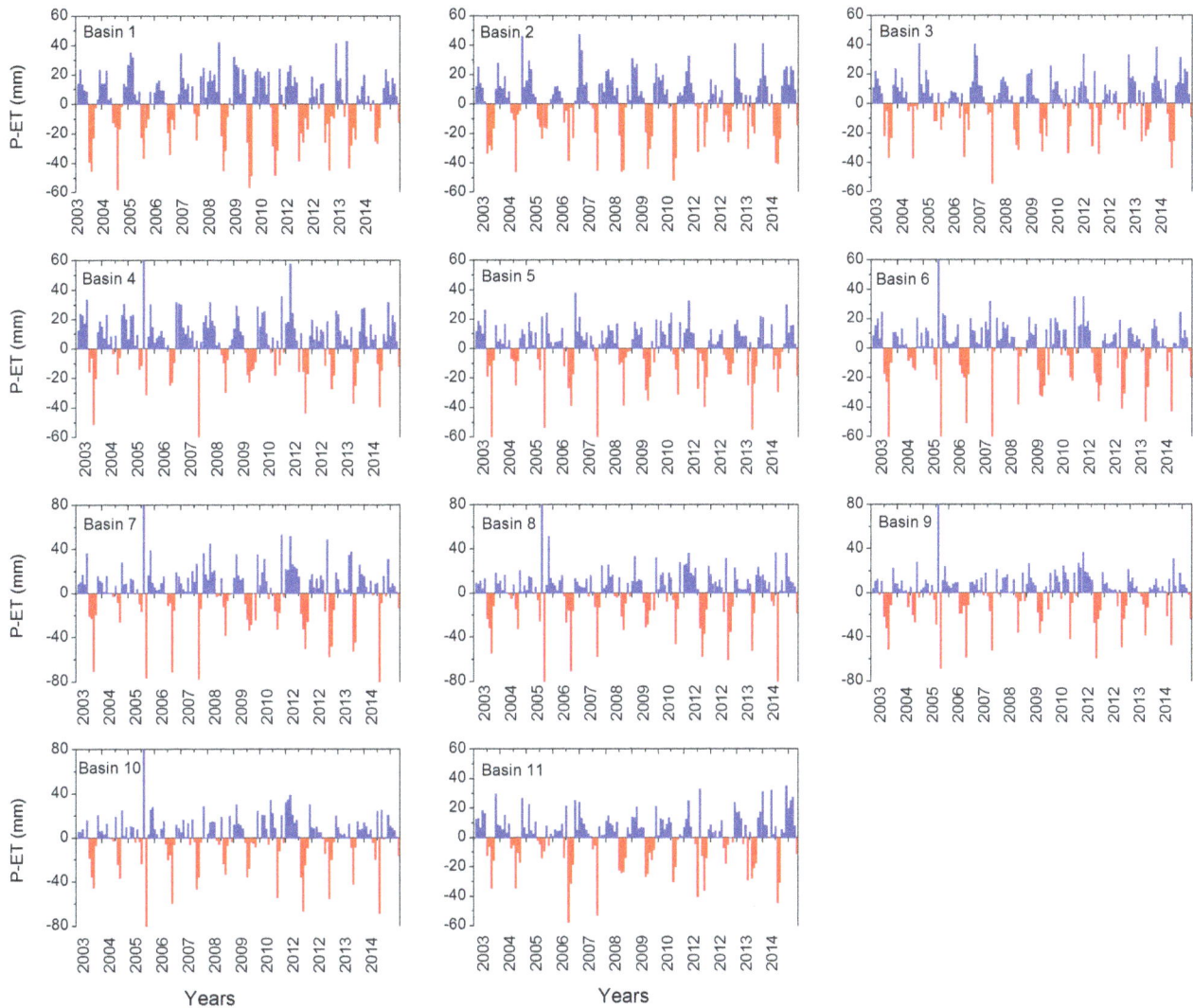

Figure 9. Basin-wide time series of P – ET values. Positive values are shown in blue and negative values are shown in red.

GWSA$_{obs}$ values from the unconfined aquifers reflect a higher magnitude than those in the confined aquifers (Fig. S1). This is because of the intrinsic property of the different types of aquifers. For instance, de-watering from the saturated zone during a pumping event is mainly responsible for the release of water in an unconfined aquifer (Alley et al., 1999). Conversely, a net decrease in groundwater potential and associated reduction in water pressure have occurred during a pumping event in a confined aquifer. The indigenous water expands slightly due to the decrease in water pressure, leading to slight compression in the aquifer material (Alley et al., 1999). This can explain why the groundwater storage change in the confined aquifers is comparatively lower than that in the unconfined aquifers.

Remote-sensing estimates of GWSA (GWSA$_{sat}$) using the two different assessments, GRACE MS and GRACE SH approaches, show temporal variations ranging from −20 to 20 cm. However, the seasonal amplitudes are not similar in different basins (Fig. 3). In general, the magnitude of the GWSA$_{sat}$ compares well with that of the GWSA$_{obs}$ (Fig. 3). GWSA$_{sat}$ exhibits large amplitudes in basins 4, 7, and 8 (Fig. 3). In general, the GWSA$_{sat}$ estimates from the two products match well with the observed estimates (Fig. 3). The estimations are in line with the values reported for the Mackenzie River basin by Scanlon et al. (2018). Overall, the two satellite-based estimates are found to closely match with one another; detailed comparisons are provided in Sect. 3.2.

3.2 Comparison between observed and satellite-based GWSA

Deviations from the observed values are measured by the RMSE, which combines both bias and lack of precision (Helsel and Hirsch, 2002). The RMSE estimates show a good match of satellite-based GWSA estimates in comparing the in situ estimates. RMSE was found to be within 5 cm in most of the basins (Fig. 4a). In general, both of the satellite-based estimates exhibit similar RMSE in basins 2, 3, 5, 6, and 11 (Fig. 4a). The Pearson's correlation coefficient (r) provides information on the linear association between the two variables (Helsel and Hirsch, 2002). Comparing the two products, correlation coefficients are found to be higher for the MS product in most of the basins (Fig. 4b). Skewness has been used to represent the symmetry in the data distribution (Helsel and Hirsch, 2002) and kurtosis has been used to represent the tail length of data distribution. Skewness and kurtosis have been used here in order to compare the GWSA estimates from the two satellite-based estimates with the in situ estimate. Comparing the two estimates, skewness and kurtosis analyses provide mixed results. For example, one product provides better results in some of the basins, the other in the remaining basins (Fig. 4c, d).

Data dispersion can be measured through the coefficients of variation. In general, coefficient of variation data are found to match well for the two satellite-based products and the in situ estimates (Fig. 5). Coefficient of variation data show mixed responses when comparing the two satellite-based estimates. Scatter analysis shows the characteristic of the relationship (i.e., linear, nonlinear) between the two variables (Helsel and Hirsch, 2002). Scatter analysis results do not provide any distinct comparison among the MS, SH, and in situ estimates (Fig. 6). The in situ data contains signatures of individual wells and, as a result, are influenced by local-scale climatic, hydrogeologic, and anthropogenic responses. However, the satellite-based estimates provide responses from a large region and may not be influenced by local-scale fluctuations (Bhanja et al., 2016).

We used a nonparametric filtering (HP filter) approach for computing the trends in GWSA and compared the results with in situ estimates. HP trends in GWSA$_{obs}$ show the recent depleting trends in basins 1, 2, 3, 7, and 9 (Fig. 7). In general, the HP trends in satellite-based estimates are relatively similar to each other. However, a comparatively better match of GWSA for the GRACE MS product and in situ estimates has been observed in basins 4, 5, 6, 10, and 11. Significantly negative (p value <0.01) correlation has been observed for both estimates in basins 7, 8, and 9, which are subjected to irrigation with >25 % land area coverage affected (Fig. 1b and Table S1).

3.3 Precipitation and snowmelt influence on GWSA

In general, precipitation is the major controlling factor for variations in water storage (Scanlon et al., 2012). In this study, we have observed that GWSA values are not directly influenced by the precipitation pattern in some of the basins (Fig. 8). The HP trend analysis shows a good match of GWSA$_{obs}$ with precipitation in basins 1 and 10 only (Fig. 8, Table S5). GWSA$_{obs}$ trends do not follow precipitation patterns in other basins (Fig. 8, Table S5). The cross-correlation analysis among HP trends provides similar inferences (Table S5). In order to investigate the relationship with more detail, the Granger causality analyses (Granger, 1988) were performed with an order of 1 (insignificant results were found when other orders were used). Results show precipitation significantly (p value <0.01) causes GWSA$_{obs}$ in 4 of the 11 studied basins, basin 1, 5, 7, and 11. The results were found to be insignificant or even negatively correlated in other basins (Table S5).

Part of the precipitation, in particular snowfall, has little influence in modulating the groundwater storage, unless it is converted to snow meltwater. Therefore, we have studied the combined influence of rainfall and snow meltwater on GWSA$_{obs}$. Here, the rainfall and the snow meltwater data are retrieved from the three LSMs (CLM, VIC, and Noah) in the GLDAS archive and used in combination. A good match between rainfall and snow meltwater, in combination, and GWSA$_{obs}$ has been obtained in basins 1 and 11. Cross-correlation analyses indicate similar inference (Table S6). Granger causality analyses (order of 1) show the combined effect of rainfall and snow meltwater significantly causes GWSA$_{obs}$ in six basins: 1, 2, 5, 7, 9, and 11. This implies that other factors, such as domestic and industrial water withdrawal, play major roles in influencing the GWSA in other basins.

3.4 GWSA and the natural water budget

Observation of a nonsignificant relationship of precipitation, snow meltwater, and groundwater storage anomalies in most of the basins indicated the influence of other factors controlling groundwater storage. The natural water availability for terrestrial water components (i.e., groundwater, surface water, soil moisture) has been studied by delineating the difference (DIFF) between precipitation (P) and evapotranspiration (ET) in another way, called the net precipitation flux (Syed et al., 2005; Rodell et al., 2015). Long et al. (2014) found outputs from LSMs provide the best ET estimates comparing in situ observations. We retrieved data from the simulation of the Noah land surface model, version 2.1, as a part of the GLDAS simulation (Rodell et al., 2004). The DIFF data exhibit negative values during summer months (Fig. 9). Comparatively lower P and higher ET values are observed in the summer months, making the DIFF negative. The basin-wise DIFF values show reducing estimates in 9 of the 11 basins with the highest estimate observed in the Peace River basin (basin 2), where the DIFF estimate shows a net reduction of 0.41 km^3 of water between 2003 and 2015 (Table 1). Reduction in DIFF values is putting stress on terrestrial water as well as groundwater conditions in the study region. We have studied the long-term (1960–2009) groundwater recharge occurrence from the global-scale model output because of unavailability of direct groundwater recharge measurement in the region. The simulated historical total groundwater recharge was found to be negative in 8 out of 11 basins, suggesting a change in rechargeable water volume. Groundwater storage, being a combination of recharge from precipitation, snow meltwater, and surface water bodies; the inter-aquifer flow; discharge to surface water bodies; and the anthropogenic withdrawal, could be largely impacted by reductions of the first three terms. Increasing human activities linked with groundwater withdrawal could lead to severe groundwater stress if it continues uncontrolled.

4 Conclusions

A network of 157 daily groundwater monitoring wells was used to compute groundwater storage anomalies (GWSAs) in 11 major river basins in Alberta, Canada, between January 2003 and April 2015. Well-specific hydrogeology information and separate treatment of the unconfined and confined aquifers were used for the calculation. Results show that the GWSA trends exhibit depletion in some of the basins that are dominated by anthropogenic groundwater withdrawal, from either irrigational use or domestic and industrial uses. A GWSA depletion rate has been observed to be as high as $-0.20\,\mathrm{cm\,year^{-1}}$ in the Athabasca River basin. The GWSA estimates obtained from remote-sensing probes provided opportunities to evaluate groundwater conditions in remote ungauged regions. We used two recently released satellite products for estimating $GWSA_{sat}$ in the studied basins. A combination of surface water measurements ($n = 393$) and land surface model estimates of soil moisture and snow water equivalents was used. In general, the remote-sensing estimates are in good agreement with those of the observed estimates, implying that remote-sensing estimates could be used in the future to monitor groundwater storage in the region at a near-continuous rate. We have investigated the influence of precipitation and snow meltwater on GWSA variations. Results show that precipitation caused significant GWSA variations in 4 out of 11 studied basins. A combination of rainfall and snow meltwater causes significant GWSA variations in six basins, indicating prevalence of other factors for influencing GWSA in the remaining basins. Water budget analysis of terrestrial water availability shows reductions of available water during the study period in nine basins. Results indicate groundwater recharge rates have been decreasing from 1960 to 2009 in eight of the basins studied. Outputs of this study may be used to frame sustainable water withdrawal strategies in Alberta, keeping in mind the available water for groundwater recharge.

Data availability. This study uses open-source data sets. Groundwater level data were obtained from the Groundwater Observation Well Network (GOWN), Alberta Environment and Parks (http://environment.alberta.ca/apps/GOWN/#, GOWN, 2017). Surface water data were obtained from the Water Office, Government of Canada (https://wateroffice.ec.gc.ca/, Water Office, 2017). The precipitation data from Climatic Research Unit (CRU) were obtained from the CRU archives (http://www.cru.uea.ac.uk/data, CRU, 2017) from the University of East Anglia. GRACE land data were processed by Sean Swenson, supported by the NASA MEaSUREs program, and are available at http://grace.jpl.nasa.gov (GRACE, 2017). The GLDAS data used in this study (https://disc.gsfc.nasa.gov/datasets?keywords=GLDAS, GLDAS, 2017) were acquired as part of the mission of NASA's Earth Science Division and archived and distributed by the Goddard Earth Sciences (GES) Data and Information Services Center (DISC). The WaterGAP model outputs are retrieved from the University of Frankfurt archive: https://www.uni-frankfurt.de/45217892/datensaetze (WaterGAP, 2017). Land cover data were obtained from http://www.landcover.org/ (GLCF, 2017).

Author contributions. SNB and JW conceived the study. SNB retrieved the data and performed the analysis. SNB and JW wrote the paper with input from XZ.

Competing interests. The authors declare that they have no conflict of interest.

Special issue statement. This article is part of the special issue "Assessing impacts and adaptation to global change in water resource systems depending on natural storage from groundwater and/or snowpacks". It is not associated with a conference.

Acknowledgements. The authors would like to thank the Alberta Economic Development and Trade for the Campus Alberta Innovates Program (no. RCP-12-001-BCAIP). We would also like to thank Jim Sellers for looking over a previous version of the paper.

Edited by: Manuel Pulido-Velazquez

References

Alberta Environment and Perk (AEP): Groundwater use, 5 pp., available at: http://aep.alberta.ca/about-us/documents/FocusOn-GroundwaterUse-2014.pdf (last access: 21 November 2017), 2011.

Alberta Environment and Perk (AEP): http://aep.alberta.ca/water/programs-and-services/water-for-life/water-supply/water-allocation-management/water-quantity.aspx, last access 21 November 2017.

Alley, W. M., Reilly, T. E., and Franke, O. L.: Sustainability of ground-water resources, US Department of the Interior, US Geological Survey, 1186, Denver, Colorado, US, 1999.

Bhanja, S., Mukherjee, A., Rodell, M., Velicogna, I., Pangaluru, K., and Famiglietti, J. S.: Regional groundwater storage changes in the Indian Sub-Continent: the role of anthropogenic activities, American Geophysical Union, Fall Meeting, GC21B-0533, 2014.

Bhanja, S. N., Mukherjee, A., Saha, D., Velicogna, I., and Famiglietti, J. S.: Validation of GRACE based groundwater storage anomaly using in situ groundwater level measurements in India, J. Hydrol., 543, 729–738, 2016.

Bhanja, S. N., Rodell, M., Li, B., Saha, D., and Mukherjee, A.: Spatio-temporal variability of groundwater storage in India, J. Hydrol., 544, 428–437, 2017a.

Bhanja, S. N., Mukherjee, A., Rodell, M., Wada, Y., Chattopadhyay, S., Velicogna, I., Pangaluru, K., and Famiglietti, J. S.: Groundwater rejuvenation in parts of India influenced by water-policy change implementation, Sci. Rep., 7, 7453, https://doi.org/10.1038/s41598-017-07058-2, 2017b.

Bhanja, S. N., Mukherjee, A. and Rodell, M.: Groundwater Storage Variations in India, in: Groundwater of South Asia, Springer, Singapore, 49–59, 2018.

Channan, S., Collins, K., and Emanuel, W. R.: Global mosaics of the standard MODIS land cover type data, University of Maryland and the Pacific Northwest National Laboratory, College Park, Maryland, USA, 2014.

Chen, J. L., Wilson, C. R., Tapley, B. D., Scanlon, B., and Güntner, A.: Long-term groundwater storage change in Victoria, Australia from satellite gravity and in situ observations, Global Planet. Change, 139, 56–65, 2016.

Cheng, M. and Tapley, B. D.: Variations in the Earth's oblateness during the past 28 years, J. Geophys. Res., 109, B09402, https://doi.org/10.1029/2004JB003028, 2004.

Climatic Research Unit (CRU): University of East Anglia, available at: http://www.cru.uea.ac.uk/data, last access: 15 September 2017.

Connor, R.: The United Nations world water development report 2015: water for a sustainable world (Vol. 1), UNESCO Publishing, Paris, 2015.

Doll, P., Mueller Schmied, H., Schuh, C., Portmann, F. T., and Eicker, A.: Global-scale assessment of groundwater depletion and related groundwater abstractions: Combining hydrological modeling with information from well observations and GRACE satellites, Water Resour. Res., 50, 5698–5720, 2014.

Dutt Vishwakarma, B., Devaraju, B., and Sneeuw, N.: Minimizing the effects of filtering on catchment scale GRACE solutions, Water Resour. Res., 52, 5868–5890, 2016.

Feng, W., Zhong, M., Lemoine, J. M., Biancale, R., Hsu, H. T., and Xia, J.: Evaluation of groundwater depletion in North China using the Gravity Recovery and Climate Experiment (GRACE) data and ground-based measurements, Water Resour. Res., 49, 2110–2118, 2013.

Geruo, A., Wahr, J., and Zhong, S.; Computations of the viscoelastic response of a 3-D compressible Earth to surface loading: an application to Glacial Isostatic Adjustment in Antarctica and Canada, Geophys. J. Int., 192, 557–572, 2013.

Global Land Cover Facility (GLCF): University of Maryland, available at: http://www.landcover.org/, last access: 14 November 2017.

Global Land Data Assimilation System (GLDAS): Goddard Space Flight Center, NASA, available at: https://disc.gsfc.nasa.gov/datasets?keywords=GLDAS, last access: 14 November 2017.

Granger, C. W.: Some recent development in a concept of causality, Journal of Econometrics, 39, 199–211, 1988.

Gravity Recovery and Climate Experiment (GRACE): Jet Propulsion Laboratory, NASA, available at: http://grace.jpl.nasa.gov, last access: 14 November 2017.

Groundwater Observation Well Network (GOWN): Alberta Environment and Parks, Government of Alberta, available at: http://environment.alberta.ca/apps/GOWN/#, last access: 14 September 2017.

Hanesiak, J. M., Stewart, R. E., Bonsal, B. R., Harder, P., Lawford, R., Aider, R., Amiro, B. D., Atallah, E., Barr, A. G., Black, T. A., and Bullock, P.: Characterization and summary of the 1999–2005 Canadian Prairie drought, Atmos. Ocean, 49, 421–452, 2011.

Harris, I. P. D. J., Jones, P. D., Osborn, T. J., and Lister, D. H.: Updated high-resolution grids of monthly climatic observations – the CRU TS3. 10 Dataset, Int. J. Climatol., 34, 623–642, 2014.

Hayashi, M. and Farrow, C. R.: Watershed-scale response of groundwater recharge to inter-annual and inter-decadal variability in precipitation (Alberta, Canada), Hydrogeol. J., 22, 1825–1839, 2014.

Helsel, D. R. and Hirsch, R. M.: Statistical methods in water resources, Vol. 323, US Department of the Interior, US Geological survey, Reston, VA, 2002.

Hodrick, R. J. and Prescott, E. C.: Postwar US business cycles: an empirical investigation, J. Money Credit Bank., 29, 1–16, 1997.

Hood, J. L. and Hayashi, M.: Characterization of snowmelt flux and groundwater storage in an alpine headwater basin, J. Hydrol., 521, 482–497, 2015.

Huang, Z., Pan, Y., Gong, H., Yeh, P. J. F., Li, X., Zhou, D., and Zhao, W.: Subregional-scale groundwater depletion detected by GRACE for both shallow and deep aquifers in North China Plain, Geophys. Res. Lett., 42, 1791–1799, 2015.

Huang, J., Pavlic, G., Rivera, A., Palombi, D., and Smerdon, B.: Mapping groundwater storage variations with GRACE: a case study in Alberta, Canada, Hydrogeol. J., 24, 1663–1680, 2016.

Lambert, A., Huang, J., Kamp, G., Henton, J., Mazzotti, S., James, T. S., Courtier, N., and Barr, A. G.: Measuring water accumulation rates using GRACE data in areas experiencing glacial isostatic adjustment: The Nelson River basin, Geophys. Res. Lett., 40, 6118–6122, 2013.

Landerer, F. W. and Swenson, S. C.: Accuracy of scaled GRACE terrestrial water storage estimates, Water Resour. Res., 48, W04531, https://doi.org/10.1029/2011WR011453, 2012.

Long, D., Longuevergne, L., and Scanlon, B. R.: Uncertainty in evapotranspiration from land surface modeling, remote sensing, and GRACE satellites, Water Resour. Res., 50, 1131–1151, 2014.

Long, D., Chen, X., Scanlon, B. R., Wada, Y., Hong, Y., Singh, V. P., Chen, Y., Wang, C., Han, Z., and Yang, W.: Have GRACE satellites overestimated groundwater depletion in the Northwest India Aquifer?, Sci. Rep., 6, 24398, https://doi.org/10.1038/srep24398, 2016.

Lemay, T. G. and Guha, S.: Compilation of Alberta groundwater information from existing maps and data sources ERCB/AGS Open File Report 2009-02, available at: http://ags.aer.ca/document/OFR/OFR_2009_02.PDF (last access: 21 November 2017), 2009.

Nanteza, J., de Linage, C. R., Thomas, B. F., and Famiglietti, J. S.: Monitoring groundwater storage changes in complex basement aquifers: An evaluation of the GRACE satellites over East Africa, Water Resour. Res., 52, 9542–9564, 2016.

Panda, D. K. and Wahr, J.: Spatiotemporal evolution of water storage changes in India from the updated GRACE-derived gravity records, Water Resour. Res., 51, 135–149, https://doi.org/10.1002/2015WR017797, 2016.

Peel, M. C., Finlayson, B. L., and McMahon, T. A.: Updated world map of the Köppen-Geiger climate classification, Hydrol. Earth Syst. Sci., 11, 1633–1644, https://doi.org/10.5194/hess-11-1633-2007, 2007.

Ravn, M. O. and Uhlig, H.: On adjusting the Hodrick-Prescott filter for the frequency of observations, Rev. Econ. Stat., 84, 371–376, 2002.

Richey, A. S., Thomas, B. F., Lo, M.-H., Famiglietti, J. S., Reager, J. T., Voss, K., Swenson, S., and Rodell, M.: Quantifying renewable groundwater stress with GRACE, Water Resour. Res., 51, 5217–5238, 2015.

Rodell, M., Houser, P. R., Jambor, U. E. A., Gottschalck, J., Mitchell, K., Meng, C. J., Arsenault, K., Cosgrove, B., Radakovich, J., Bosilovich, M., and Entin, J. K.: The global land data assimilation system, B. Am. Meteorol. Soc., 85, 381–394, https://doi.org/10.1175/BAMS-85-3-381, 2004.

Rodell, M., Chen, J., Kato, H., Famiglietti, J. S., Nigro, J., and Wilson, C.: Estimating groundwater storage changes in the Mississippi River basin (USA) using GRACE, Hydrogeol. J., 15, 159–166, 2007.

Rodell, M., Velicogna, I., and Famiglietti, J. S.: Satellite-based estimates of groundwater depletion in India, Nature, 460, 999–1002, 2009.

Rodell, M., Beaudoing, H. K., L'Ecuyer, T. S., Olson, W. S., Famiglietti, J. S., Houser, P. R., Adler, R., Bosilovich, M. G., Clayson, C. A., Chambers, D., and Clark, E.: The observed state of the water cycle in the early twenty-first century, J. Climate, 28, 8289–8318, https://doi.org/10.1175/JCLI-D-14-00555.1, 2015.

Scanlon, B. R., Longuevergne, L., and Long, D.: Ground referencing GRACE satellite estimates of groundwater storage changes in the California Central Valley, USA, Water Resour. Res., 48, W04520, https://doi.org/10.1029/2011WR011312, 2012.

Scanlon, B. R., Zhang, Z., Reedy, R. C., Pool, D. R., Save, H., Long, D., Chen, J., Wolock, D. M., Conway, B. D., and Winester, D.: Hydrologic implications of GRACE satellite data in the Colorado River Basin, Water Resour. Res., 51, 9891–9903, 2015.

Scanlon, B. R., Zhang, Z., Save, H., Wiese, D. N., Landerer, F. W., Long, D., Longuevergne, L., and Chen, J.: Global evaluation of new GRACE mascon products for hydrologic applications, Water Resour. Res, 52, 9412–9429, 2016.

Scanlon, B. R., Zhang, Z., Save, H., Sun, A. Y., Schmied, H. M., van Beek, L. P., Wiese, D. N., Wada, Y., Long, D., Reedy, R. C., and Longuevergne, L.: Global models underestimate large decadal declining and rising water storage trends relative to GRACE satellite data, P. Natl. Acad. Sci. USA, 115, E1080–E1089, https://doi.org/10.1073/pnas.1704665115, 2018.

Shamsudduha, M., Taylor, R. G., and Longuevergne, L.: Monitoring groundwater storage changes in the Bengal Basin: validation of GRACE measurements, Water Resour. Res., 48, W02508, https://doi.org/10.1029/2011WR010993, 2012.

Strassberg, G., Scanlon, B. R., and Rodell, M.: Comparison of seasonal terrestrial water storage variations from GRACE with groundwater-level measurements from the High Plains Aquifer (USA), Geophys. Res. Lett., 34, L14402, https://doi.org/10.1029/2007GL030139, 2007.

Syed, T. H., Famiglietti, J. S., Chen, J., Rodell, M., Seneviratne, S. I., Viterbo, P., and Wilson, C. R.: Total basin discharge for the Amazon and Mississippi River basins from GRACE and a land-atmosphere water balance, Geophys. Res. Lett., 32, L24404, https://doi.org/10.1029/2005GL024851, 2005.

Swenson, S. P., Yeh, J.-F., Wahr, J., and Famiglietti, J.: A comparison of terrestrial water storage variations from GRACE with in situ measurements from Illinois, Geophys. Res. Lett., 33, L16401, https://doi.org/10.1029/2006GL026962, 2006.

Swenson, S. C., Chambers, D. P., and Wahr, J.: Estimating geocenter variations from a combination of GRACE and ocean model output, J. Geophys. Res.-Sol. Ea., 113, B08410, https://doi.org/10.1029/2007JB005338, 2008.

Tapley, B. D., Bettadpur, S., Ries, J. C., Thompson, P. F., and Watkins, M. M.: GRACE measurements of mass variability in the Earth system, Science, 305, 503–505, 2004.

Tiwari, V. M., Wahr, J., and Swenson, S.: Dwindling groundwater resources in northern India, from satellite gravity observations, Geophys. Res. Lett., 36, L18401, https://doi.org/10.1029/2009GL039401, 2009.

Todd, D. K. and Mays, L. W.: Groundwater hydrology, 3rd edition, John Wiley & Sons, NJ, 636 pp., 2005.

UN-Water/FAO: World Water Day: Coping with Water Scarcity: Challenge of the twenty-first century, available at: http://www.fao.org/nr/water/docs/escarcity.pdf (last access: 21 November 2017), 2007.

Voss, K. A., Famiglietti, J. S., Lo, M., de Linage, C., Rodell, M., and Swenson, S. C.: Groundwater depletion in the Middle East from GRACE with implications for transboundary water management in the Tigris-Euphrates-Western Iran region, Water Resour. Res., 49, 904–914, https://doi.org/10.1002/wrcr.20078, 2013.

WaterGAP model output: University of Frankfurt archive, available at: https://www.uni-frankfurt.de/45217892/datensaetze, last access: 14 November 2017.

Water Office: Government of Canada, available at: https://wateroffice.ec.gc.ca/, last access: 25 August 2017.

Watkins, M. M., Wiese, D. N., Yuan, D.-N., Boening, C., and Landerer, F. W.: Improved methods for observing Earth's time variable mass distribution with GRACE using spherical cap mascons, J. Geophys. Res.-Sol. Ea., 120, 2648–2671, https://doi.org/10.1002/2014JB011547, 2015.

Zhao, T. and Fu, C.: Comparison of products from ERA-40, NCEP-2, and CRU with station data for summer precipitation over China, Adv. Atmos. Sci., 23, 593–604, 2006.

8

Spatial prediction of groundwater spring potential mapping based on an adaptive neuro-fuzzy inference system and metaheuristic optimization

Khabat Khosravi[1], Mahdi Panahi[2], and Dieu Tien Bui[3]

[1]Department of Watershed Management Engineering, Faculty of Natural Resources, Sari Agricultural Science and Natural Resources University, Sari, Iran
[2]Young Researchers and Elites Club, North Tehran Branch, Islamic Azad University, Tehran, Iran
[3]GIS group, Department of Business and IT, University of South-Eastern Norway, Gullbringvegen 36, 3800 Bø i Telemark, Norway

Correspondence: Mahdi Panahi (panahi2012@yahoo.com) and Dieu Tien Bui (dieu.t.bui@usn.no)

Abstract. Groundwater is one of the most valuable natural resources in the world (Jha et al., 2007). However, it is not an unlimited resource; therefore understanding groundwater potential is crucial to ensure its sustainable use. The aim of the current study is to propose and verify new artificial intelligence methods for the spatial prediction of groundwater spring potential mapping at the Koohdasht–Nourabad plain, Lorestan province, Iran. These methods are new hybrids of an adaptive neuro-fuzzy inference system (ANFIS) and five metaheuristic algorithms, namely invasive weed optimization (IWO), differential evolution (DE), firefly algorithm (FA), particle swarm optimization (PSO), and the bees algorithm (BA). A total of 2463 spring locations were identified and collected, and then divided randomly into two subsets: 70 % (1725 locations) were used for training models and the remaining 30 % (738 spring locations) were utilized for evaluating the models. A total of 13 groundwater conditioning factors were prepared for modeling, namely the slope degree, slope aspect, altitude, plan curvature, stream power index (SPI), topographic wetness index (TWI), terrain roughness index (TRI), distance from fault, distance from river, land use/land cover, rainfall, soil order, and lithology. In the next step, the step-wise assessment ratio analysis (SWARA) method was applied to quantify the degree of relevance of these groundwater conditioning factors. The global performance of these derived models was assessed using the area under the curve (AUC). In addition, the Friedman and Wilcoxon signed-rank tests were carried out to check and confirm the best model to use in this study. The result showed that all models have a high prediction performance; however, the ANFIS–DE model has the highest prediction capability (AUC = 0.875), followed by the ANFIS–IWO model, the ANFIS–FA model (0.873), the ANFIS–PSO model (0.865), and the ANFIS–BA model (0.839). The results of this research can be useful for decision makers responsible for the sustainable management of groundwater resources.

1 Introduction

Groundwater is defined as the water in a saturated zone which fills rock and pore spaces (Berhanu et al., 2014; Fitts, 2002), and groundwater potential is the probability of groundwater occurrence in an area (Jha et al., 2010). The occurrence and movement of groundwater in an aquifer are affected by various geo-environmental factors including lithology, topography, geology, fault and fracture and their connectivities, drainage pattern, and land use/land cover (Mukherjee, 1996). Geological strata act like a conduit and reservoir for groundwater, while storage and transmissivity influence the suitability of exploitation of groundwater in a given geological formation. Downhill and depression slopes cause runoff and improve recharge and infiltration, respectively (Waikar and Nilawar, 2014). Globally, groundwater is a major source of drinking water for around 2 billion people (Richey et al., 2015), and in agriculture, about 278.8 million ha of agricultural lands are irrigated by groundwater (Siebert et al., 2013). Due to population and economic growth, the demand for groundwater is anticipated to increase in the future (Ercin and Hoekstra, 2014). For the case of Iran, approximately two-thirds of the land is covered by deserts, and groundwater is still the main water source for drinking and other uses (Nosrati and Van Den Eeckhaut, 2012). According to Rahmati et al. (2016), groundwater in Iran supplies around 65 % of the water use and the remaining 35 % is supplied by surface water. However, groundwater is not an unlimited resource; therefore understanding groundwater potential is crucial to ensure its sustainable use. One of the most efficient methods for the protection and management of groundwater is to identify groundwater potential zoning (Ozdemir, 2011b).

There are a number of methods for groundwater potential zoning and exploitation. Traditional methods, i.e., drilling, geological, geophysical, and hydrogeological methods, are the most widely used (Israil et al., 2006; Jha et al., 2010; Todd and Mays, 1980; Sander et al., 1996; Singh and Prakash, 2002). However, they are time-consuming and costly methods, especially for large areas. In recent years, geographic information systems (GIS) and remote sensing have become effective tools for groundwater potential mapping (Fashae et al., 2014) due to their ability to handle a huge amount of spatial data.

In more recent years, some probabilistic models, such as the frequency ratio (Oh et al., 2011), multi-criteria decision analysis (MCDA; Kaliraj et al., 2014; Rahmati et al., 2015), weights-of-evidence (Pourtaghi and Pourghasemi, 2014), logistic regression (Ozdemir, 2011a; Pourtaghi and Pourghasemi, 2014), evidential belief function (Nampak et al., 2014; Pourghasemi and Beheshtirad, 2015), and Shannon's entropy (Naghibi et al., 2015), have been considered for groundwater potential mapping. Bivariate and multivariate statistical models have disadvantages in measuring the relationship between groundwater occurrence and conditioning factors (Tehrany et al., 2013; Umar et al., 2014), whereas the MCDA technique is a source of bias due to expert opinion. Therefore, machine learning has been considered and has proven efficient due to its ability to handle nonlinear structured data from various sources with different scales. In addition, machine learning requires no statistical assumptions. Among machine learning methods, the artificial neural network (ANN) model is a widely used method for groundwater mapping due to its computational efficiency (Sun et al., 2016; Mohanty et al., 2015; Maiti and Tiwari, 2014). However, the ANN model has a number of weaknesses such as poor prediction and error in the modeling process (Bui et al., 2016); therefore, hybrid models have been proposed. Among hybrid frameworks, an ensemble of fuzzy logic and neural networks, i.e., an adaptive neuro-fuzzy inference system (ANFIS), has proven it is efficient in terms of high accuracy (Lohani et al., 2012; Emamgholizadeh et al., 2014; Zare

and Koch, 2018; Bui et al., 2018; Nourani et al., 2016; Tien Bui et al., 2017; Pham et al., 2018a). It should be noted that although an ANFIS model has a higher accuracy than other models, it is still difficult to find the best internal weight values of ANFIS due to the limited nature of the adaptive algorithm used (Bui et al., 2016). Thus, these weights should be optimized by new metaheuristic optimization algorithms to enhance the prediction accuracy of groundwater models.

The main goal of the current study is to propose and verify integration of new metaheuristic optimization algorithms with ANFIS for groundwater spring potential mapping (GSPM) in the Koohdasht–Nourabad plain, Iran. Accordingly, five new metaheuristic algorithms are investigated, invasive weed optimization (IWO), differential evolution (DE), firefly algorithm (FA), particle swarm optimization (PSO), and the bees algorithm (BA). According to current literature, it is the first time that such a study has been conducted on groundwater potential mapping.

2 Description of the study area

The Koohdasht–Nourabad plain is located in the western part of the Lorestan province (Iran) and covers an area of around $9531.9\,\text{km}^2$. It lies between latitudes $33°3'28''$ and $34°22'55''\,\text{N}$ and between longitudes $46°50'19''$ and $48°21'18''\,\text{E}$ (Fig. 1). The region is located in a semi-arid area, with a mean annual precipitation of about 450 mm (Iran Meteorological Organization, http://www.irimo.ir/, last access: 1 May 2018). The altitude of the study area varies between 531 and 3175 m above sea level. The maximum slope and minimum slope are 64 and $0°$, respectively. Geologically, the study area is located in the Zagros structural zone of Iran and is indicative of a Quaternary and Cretaceous–Paleocene geologic timescale. The dominant land use/land cover of the study area is moderate forest (20 %). Residential areas cover about 3 % of the plain. Rock outcrops/Inceptisols are the dominant soil types in the study area, covering about 51 % of the study area. The population of the plain is 362 000 (according to a 2016 census) and agriculture is the primary occupation. In this plain, groundwater is the main water source for drinking and agricultural activities.

3 Methodology

An overview of the methodological approach is shown in Fig. 2.

3.1 Data preparation

3.1.1 Groundwater spring inventory map

In groundwater modeling, spatial relationships between groundwater springs and conditioning factors should be analyzed and assessed to determine the best subset of these

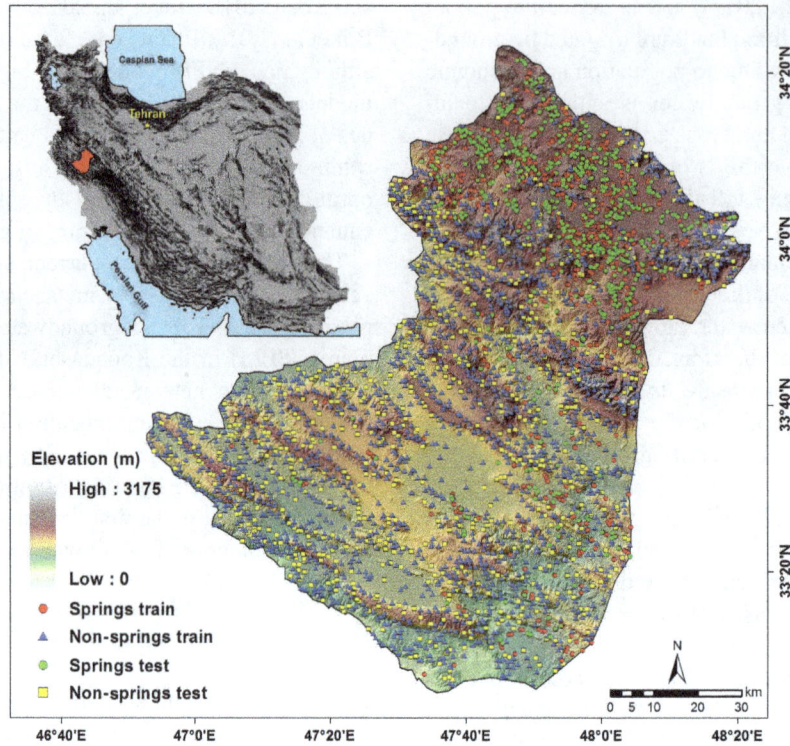

Figure 1. Groundwater well locations with a DEM of the study area.

factors. In the Koohdasht–Nourabad plain, a total of 2463 spring locations were provided by the Iranian Water Resources Management Company. Most of these spring locations were checked during extensive field surveys using a handheld GPS unit.

3.1.2   Construction of the training and testing datasets

Spatial prediction of groundwater potential mapping using a machine learning model is considered a binary classification with two classes, spring and non-spring. Therefore, a total of 2463 non-spring locations were randomly generated using the random point tool in ArcGIS10.2. According to Chung and Fabbri (2003), it is possible to validate the model performance using a cross-validation method that splits the dataset into the two parts of spring and non-spring locations. The first part is used for model building, which is called a training dataset, and the other part is utilized for validation of the model performance, called a testing dataset (Pham et al., 2017a). In this study, a ratio of 70 / 30 was selected randomly for generating the training and testing datasets (Pourghasemi et al., 2012, 2013a, b; Xu et al., 2012). Accordingly, both spring locations and non-spring locations have been randomly divided into two groups for training (1725 locations) and validating (738 locations) purposes (Fig. 1).

Both the training and the testing datasets were converted to a raster format, whereby spring pixels were assigned "1" and non-spring pixels were assigned "0" (Bui et al., 2015), and

then, these pixels were overlaid with 13 groundwater conditioning factors to extract their attribute values.

3.1.3   Groundwater conditioning factor analysis

Selection of the groundwater conditioning factors

After the initial selection of the conditioning factors, these factors should be assessed for multicollinearity problems. Multicollinearity takes place when two or more independent conditioning factors are highly correlated, or, in other words, interdependent (Li et al., 2010). Several methods have been proposed to diagnose multicollinearity, and among them, the variance inflation factor (VIF) and tolerance (TOL) are widely used in environmental modeling (O'brien, 2007; Bui et al., 2016); therefore, they were selected for this research. Factors with VIFs greater than 5 and TOL less than 0.1 indicate that multicollinearity problems existed (O'brien, 2007; Bui et al., 2011). Another method, namely the information gain ratio (IGR) technique, was applied to identify the relative importance of the conditioning factor and also to obtain factors with null effect. These factors must be removed to increase the accuracy of the model (Khosravi et al., 2018).

In the current study, 13 conditioning factors have been selected, namely the slope degree, slope aspect, altitude, plan curvature, stream power index (SPI), topographic wetness index (TWI), terrain roughness index (TRI), distance from fault, distance from river, land use/land cover, rain-

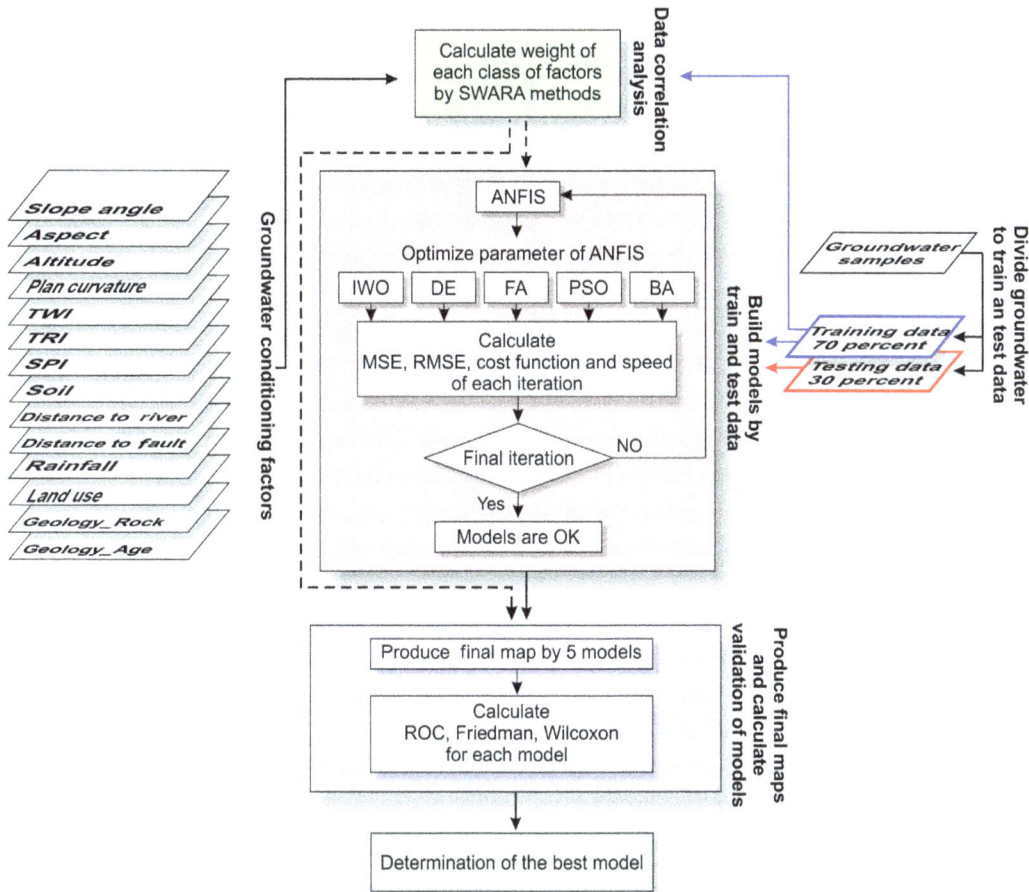

Figure 2. Conceptual model of methodology applied in the current study.

fall, soil order, and lithology units. These factors have been determined based on a literature review, characteristics of the study area, and data availability (Nampak et al., 2014; Mukherjee, 1996; Oh et al., 2011; Ozdemir, 2011b). The process of converting continuous variables into categorical classes was carried out based on our frequency analysis of spring location (Khosravi et al., 2018) in order to define the class intervals.

A digital elevation model (DEM) of the study area was downloaded from ASTER Global DEM (https://asterweb.jpl.nasa.gov/gdem.asp, last access: 20 April 2017) with a grid size of 30×30 m. Based on the DEM, slope degree, slope aspect, altitude, plan curvature, SPI, TWI, and TRI were derived. Slope degree has been divided in five categories using the quantile classification scheme (Tehrany et al., 2013, 2014), namely 0–5.5, 5.5–12.11, 12.11–19.4, 19.4–28.7, and 28.7–64.3° (Fig. 3a). Slope aspect is selected because it controls solar radiation budgets that influence the groundwater potential. Slope aspect has been provided in five different classes, flat, north, west, south, and east (Fig. 3b). Altitude was divided into five classes using the quantile classification scheme, namely 531–1070, 1070–1385, 1385–1703, 1703–2068, and 2068–3175 m (Fig. 3c). Plan curvature was divided

into three classes, namely concave (< -0.05), flat (-0.05 to 0.05), and convex (> 0.05) (Fig. 3d) (Pham et al., 2017a).

SPI is related to the erosive power of surface runoff, whereas TWI relates to the amount of the flow that accumulates at any point in the catchment. In this research, SPI, TWI, and TRI were constructed using the System for Automated Geoscientific Analyses (SAGA) GIS 2.2 software, and finally, were divided into five classes. These classes are 0–48 664, 48 664–227 099, 227 099–583 969, 583 969–1 330 153, and 1 330 153–4 136 452 for SPI (Fig. 3e). For TWI, these classes are 2.1–4.6, 4.6–5.6, 5.6–6.6, 6.6–7.9, and 7.9–11.9 (Fig. 3f), and for TRI, these classes are 0–8.7, 8.7–18.2, 18.2–29.9, 29.9–46.6, and 46.6–185 (Fig. 3g).

Distances from the fault and river for the study area were generated with five classes using the Multiple Ring Buffer tool in ArcGIS 10.2, 0–200, 200–500, 500–1000, 1000–2000, and > 2000 m (Fig. 3h and i). Lithology plays a key role in determining the groundwater potential due to different infiltration rates of formation (Adiat et al., 2012; Nampak et al., 2014). Land use/land cover of the study area was obtained using Landsat 7 Enhanced Thematic Mapper plus (ETM+) images that were downloaded from the US Geological Survey (available at https://earthexplorer.usgs.gov, last

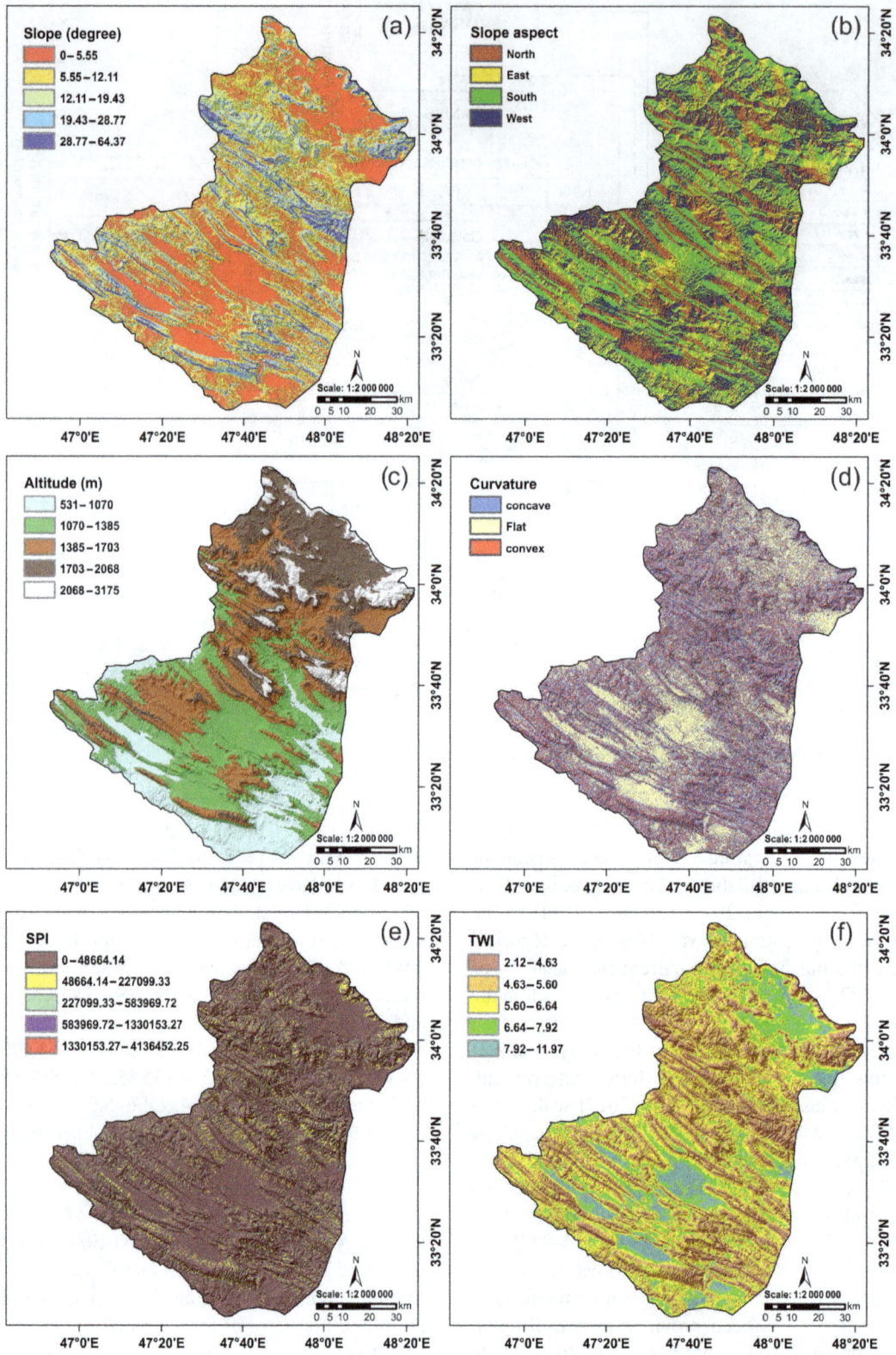

Figure 3.

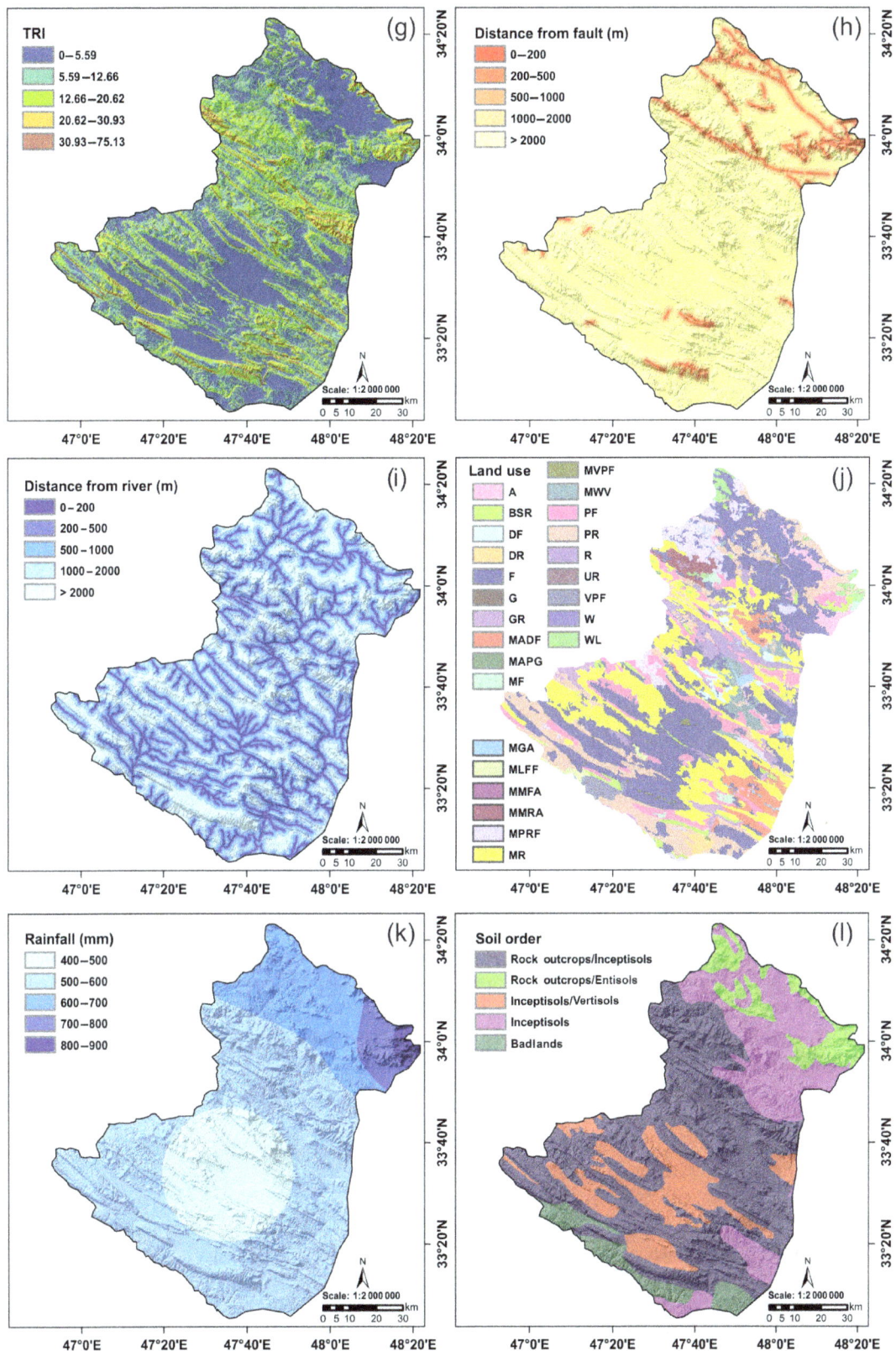

Figure 3.

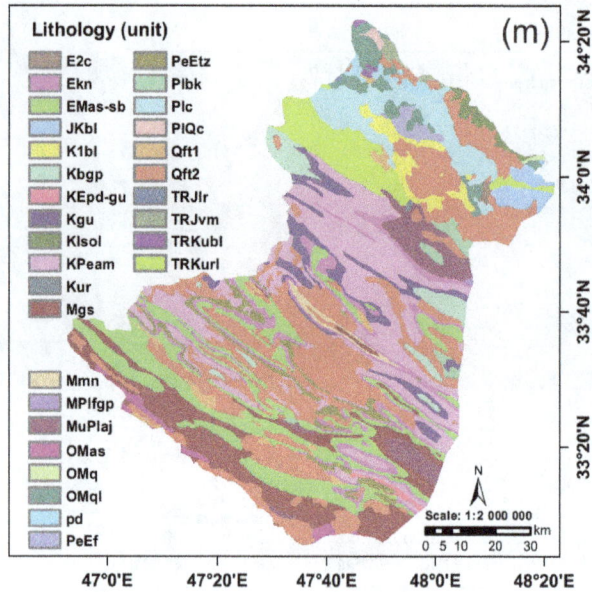

Figure 3. Groundwater conditioning factors for the study area used in this research: **(a)** slope degree, **(b)** slope aspect, **(c)** altitude, **(d)** plan curvature, **(e)** SPI, **(f)** TWI, **(g)** TRI, **(h)** distance from fault, **(i)** distance from river, **(j)** land use/land cover, **(k)** rainfall, **(l)** soil order, and **(m)** lithology units.

access: 1 May 2018). Accordingly, 25 land use/land cover types were recognized: agriculture (A), garden (G), dense forest (DF), good rangeland (GR), poor forest (PF), waterway (W), mixture of garden and agriculture (MGA), mixture of agriculture with dry farming (MADF), mixture of agriculture with poor garden (MAPG), dry farming (DF), fallow (F), dense rangeland (DR), very poor forest (VPF), mixture of waterway and vegetation (MWV), mixture of moderate forest and agriculture (MMFA), mixture of moderate rangeland and agriculture (MMRA), mixture of poor rangeland and fallow (MPRF), mixture of low forest and fallow (MLFF), woodland (WL), moderate forest (MF), moderate rangeland (MR), poor rangeland (PR), bare soil and rock (BSR), urban and residential (UR), mixture of very poor forest (MVPF), and rangeland (R) (Fig. 3j).

Rainfall is a major source of recharge for the groundwater. In this research, 15 years (2000–2015) of mean annual rainfall data from four rain-gauge stations of the study area were used. The rainfall map (Fig. 3k) with five categories (300–400, 400–500, 500–600, 600–700, and 700–800 mm) was generated using an inverse distance weighting method due to a lower root mean square error (RMSE; Khosravi et al., 2016a, b). A soil map at a scale of 1 : 50 000 for the study area was provided by the Iranian Water Resources Department (IWRD). The soil types are soil rock outcrops/Entisols, rock outcrops/Inceptisols, Inceptisols, Inceptisols/Vertisols, and badlands (Fig. 3l).

Lithology for the study area, at a scale of 1 : 100 000, was provided by the Iranian Department of Geology Survey (IDGS). Accordingly, 30 classes were used: OMq, PeEf, PlQc, K1bl, Plc, pd, TRKubl, TRJvm, MPlfgp, OMql, Plbk, E2c, TRKurl, Qft2, MuPlaj, KEpd-gu, Kgu, Qft1, Ekn, KPeam, PeEtz, Kbgp, EMas-sb, Mgs, TRJlr, Klsol, JKbl, Kur, OMas, and Mmn (Fig. 3m). Finally, using ArcGIS 10.2 software, all the aforementioned groundwater conditioning factors were converted to a raster format with a grid size of 30 m × 30 m for modeling purposes.

3.2 Spatial relationship between spring location and conditioning factors

To assess the spatial relationship between spring location and conditioning factors, in this research, step-wise assessment ratio analysis (SWARA) (Keršuliene et al., 2010), a multi-criteria decision-making (MCDM) analysis, was used. SWARA has received great attention in various fields in the last 5 years (Alimardani et al., 2013; Hong et al., 2018). The working principles of SWARA are briefly described as follows.

Phase one. First, the experts will define the problem-solving criteria. By using the practical knowledge of the experts, the priority for the criteria is determined and these criteria are organized in descending order.

Phase two. The following trends are employed for the estimation of the weight of the criteria.

Starting from the second criterion, the respondent explains the relative importance of the criterion j in relation to the $(j-1)$ criterion, and for each particular criterion as well. As Keršuliene mentioned, this process specifies the comparative importance of the average value, S_j, as follows (Keršuliene et al., 2010):

$$S_j = \frac{\sum_1^n A_i}{n}, \tag{1}$$

where n is the number of experts, A_i explicates the offered ranks for each factor by the experts, and j stands for the number of the factor.

Subsequently, the coefficient K_j is determined as follows:

$$K_j = \begin{cases} 1 & j = 1 \\ S_j + 1 & j > 1 \end{cases}. \tag{2}$$

Recalculation of weight Q_j is done as follows:

$$Q_j = \frac{X_{j-1}}{K_j}. \tag{3}$$

The relative weights of the evaluation criteria are calculated by the following equation:

$$W_j = \frac{Q_j}{\sum_{j=1}^m Q_j}, \tag{4}$$

where W_j shows the relative weight of jth criterion, and m is the total number of criteria.

3.3 Groundwater spring prediction modeling

As mentioned earlier, in this research, five new metaheuristic optimization algorithms (IWO, DE, FA, PSO, and BA) were investigated in order to optimize the parameters of ANFIS. This section briefly presents the theoretical background of these algorithms and ANFIS.

3.3.1 Adaptive neuro-fuzzy inference system

An adaptive neuro-fuzzy inference system (ANFIS) was proposed by Jang (1993) to solve nonlinear and complex problems in one framework. ANFIS converts input data to fuzzy inputs by using a membership function; there are different membership functions that describe system behavior (Jang, 1993). ANFIS is applied to the Takagi–Sugeno–Kang (TSK) fuzzy model with two "if-then" rules, both with two inputs, x_1 and x_2, and one output, f (Takagi and Sugeno, 1985), as follows:

Rule 1 : if x_2 is A_2 and x_2 is B_2,

$$\text{then } f_2 = p_2 x_2 + q_2 x_2 + r_2, \tag{5}$$

Rule 2 : if x_1 is A_1 and x_2 is B_1,

$$\text{then } f_1 = p_1 x_1 + q_1 x_2 + r_1. \tag{6}$$

Jang's ANFIS consists of a feed-forward neural network with six distinct layers. Detailed description of ANFIS can be found in Jang (1993).

3.3.2 Metaheuristic optimization algorithms

The main goal of these algorithms is to find the optimal antecedent and the consequent parameters of the ANFIS model using IWO, DE, FA, PSO, and BA algorithms. Figure 4 illustrates a general methodological data flow of the ANFIS model.

Invasive weed optimization algorithm

Invasive weed optimization (IWO) mimics the colonizing behavior of weeds. Its design, by Mehrabian and Lucas (2006), is based on the way that weeds find proper space for growth and reproduction. One characteristic of this algorithm is its simple structure; the number of input parameters is low and it has strong robustness. Furthermore, it is easy to understand and the same merit causes it to be used for solving difficult nonlinear optimization problems (Ghasemi et al., 2014; Naidu and Ojha, 2015; Zhou et al., 2015). This algorithm consists of four parts: initialization, reproduction, spatial dispersal, competitive exclusion, and termination condition.

Differential evolution algorithm

Differential evolution (DE) is an evolutionary algorithm for finding global optimal answers for problems with continuous space (Das et al., 2009). This algorithm starts by producing a random population in which each individual of the population is a solution to the problem. Vector $X_i^G = \left(x_{1,i}^G, x_{2,i}^G, x_{3,i}^G, \ldots, x_{D,i}^G \right)$ shows each individual of the population, $i = \{0, 1, 2, \ldots, NP\}$ is a number denoting each individual, in which D stands for the search dimension, or in other words, is a component problem, and $G = \{0, 1, 2, \ldots, G_{\max}\}$ generation time, where $G_{\max}$ is the total number of generations.

By assuming the maximum and minimum of every dimension of searching space, there are $X_L = \left(x_{1,L} x_{2,L}, \ldots, x_{D,L} \right)$ and $X_U = \left(x_{2,U}, \ldots, x_{D,U} \right)$, respectively; initial population is defined by the following (Storn and Price, 1997):

$$x_{j,i}^0 = x_{j,L} + \text{rand}\,(0,1) \times \left(x_{j,U} - x_{j,L} \right), \tag{7}$$

where rand $(0, 1)$ is a uniformly distributed random number in [0, 1]. Detailed description of DE can be found in Chen et al. (2017).

Firefly algorithm

The firefly algorithm (FA) is as a metaheuristic algorithm, proposed by Yang (2010), that is originated from the flashing and communication behavior of fireflies. Like other swarm intelligence algorithms, of which components are known as solutions to the problems, in this algorithm, each firefly is a solution and its light intensity is the objective function value. In general, the FA follows three idealized rules as given below: (1) all firefly species are unisex, with each of them attracting other fireflies without considering their gender (Amiri et al., 2013); (2) attractiveness of a firefly is related to its light intensity, and thus, from two flashing firefly species, one with lower light intensity moves toward the other one with higher light intensity; (3) light intensity of a firefly is defined as an objective function value and must be optimized.

Particle swarm optimization algorithm

The article swarm optimization (PSO) algorithm has been inspired by the way birds use their collective intelligence for finding the best way to get food (Kennedy and Eberhart, 1995). Each bird implemented in this algorithm acts as a particle that is in fact a representative of a solution to the problem. These particles find the optimum answers for the problem by searching in "n" dimensional space, and "n" is the number of the problem's parameters. For this purpose, particles were scattered randomly in space at the beginning of algorithm execution. Detailed description of PSO can be found in Kennedy (2011).

Bees algorithm

The bees algorithm (BA), which was introduced by Pham et al. (2005), is inspired by the foraging behavior of bees'

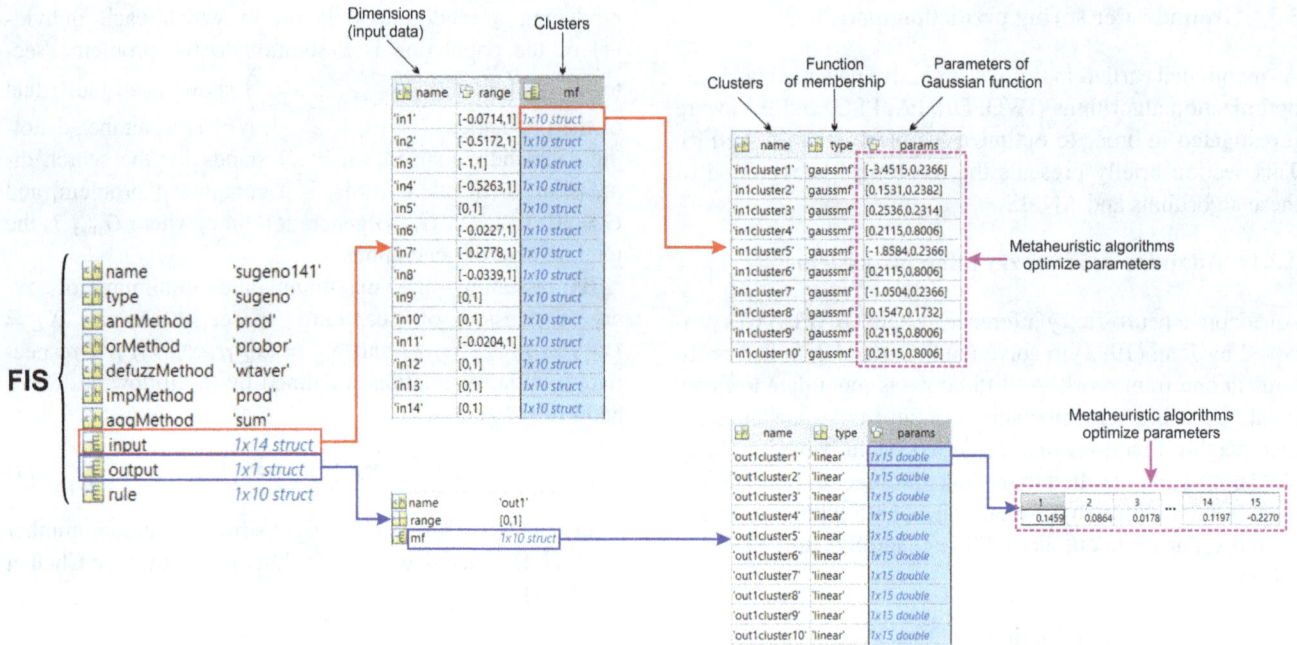

Figure 4. General methodological flow of ANFIS.

colonies in search of food sources located near the hive. In the initial setup, evenly distributed scout bees are scattered randomly in different directions to identify flower patches. After that, scout bees come back to the hive and start a specific dance called the waggle dance. This dance is for communicating with others in order to share the information of discovered flower patches. The information indicates direction, distance, and nectar quality of the flower patches, and helps the colony to have proper evaluation of all flower patches. After evaluation, scout bees come back to the location of discovered flower patches with other bees, named recruit bees. Dependent on the distance and the amount of nectar, a different number of recruit bees is assigned to each flower patch. In other words, those flower patches with better nectar quality are designated more recruit bees. Recruit bees then evaluate the quality of flower patches when performing the harvest process, and leave the flower patches with a low quality. Conversely, if the flower patch quality is good, it will be announced during the next waggle dance.

3.4 Performance assessment of models

According to Chung and Fabbari (2003), without validation, the results (achieved maps) of the models do not have any scientific significance. Prediction capability of these five spatial groundwater models must be evaluated using both success-rate curves and prediction-rate curves (Hong et al., 2015). Success-rate curves show how suitable the built model is for groundwater potential assessment (Gaprindashvili et al., 2014). Success-rate curves have been constructed using groundwater potential maps and the number of spring lo-

cations used in the training dataset (Tien Bui et al., 2012). Prediction-rate curves constructed using the testing dataset demonstrate how good the model is and evaluate the prediction power of the models. Therefore, it can be used for model prediction capabilities (Brenning, 2005). The area under the curve (AUC) of success and prediction rate is the basis for assessing the accuracy of the groundwater potential models quantitatively (Khosravi et al., 2016a, b; Pham et al., 2017b). The AUC value varies from 0.5 to 1; the higher the AUC, the better the prediction capability of models (Tien Bui and Hoang, 2018; Ngoc-Thach et al., 2018).

In addition, the mean squared error (MSE) was further used (Bui et al., 2016) as follows:

$$MSE = \frac{\sum_{i=1}^{n}(O_i - E_i)^2}{N},\qquad(8)$$

where O_i and E_i are observation (target) and prediction (output) values in both training and testing datasets, and N is the total samples in the training or the testing dataset.

3.5 Inferential statistics

3.5.1 Friedman test

Nonparametric statistical procedures such as the Friedman test (Friedman, 1937) can be used regardless of statistical assumptions (Derrac et al., 2011) and do not presuppose the data to be normally distributed. The main aim of this test is to find whether there is a significant difference between the models' performance or not. In other words, multiple comparisons are performed to detect significant differences between the behaviors of two or more models (Beasley and

Table 1. Multicollinearity analysis of conditioning factors.

No.	Groundwater conditioning factor	Collinearity statistics	
		TOL	VIF
1	Slope degree	0.231	2.401
2	Slope aspect	0.206	4.270
3	Altitude	0.801	2.097
4	Plan curvature	0.513	1.446
5	SPI	0.410	1.689
6	TWI	0.541	2.113
7	TRI	0.328	1.939
8	Distance from fault	0.408	2.25
9	Distance from river	0.212	3.126
11	Land use/land cover	0.296	3.891
12	Rainfall	0.298	1.686
13	Soil order	0.205	4.039
10	Geology (unit)	0.215	4.150

Zumbo, 2003). The null hypothesis (H0) is that there are no differences among the performances of the groundwater potential models. The higher the p value, the higher the probability that the null hypothesis is not true, since if the p value is less than the significance level ($\alpha = 0.05$), the null hypothesis will be rejected.

3.5.2 Wilcoxon signed-rank test

Because the Friedman test only illustrates whether there is any difference between the models or not, this test does not provide pairwise comparisons among compared models. Therefore, another nonparametric statistical test named the Wilcoxon signed-rank test has been applied. To evaluate the significance of differences between the performances of groundwater potential models, the p value and z value have been used.

4 Result and analysis

4.1 Multicollinearity diagnosis

Results of the multicollinearity analysis in this study are shown in Table 1. The analysis revealed that as the VIF is less than 5 and TOL is greater than 0.1, no multicollinearity problems exist among conditioning factors.

4.2 Determination of the most important parameters

The most common method, the information gain ratio (IGR), was applied to identify the most important conditioning factors. Result shows that all 13 conditioning factors affect groundwater occurrence. The land use/land cover factor has the most important impact on groundwater (IGR = 0.502), followed by lithology (IGR = 0.465), rainfall (IGR = 0.421),

TWI (IGR = 0.400), soil (IGR = 0.370), TRI (IGR = 0.337), slope degree (IGR = 0.317), altitude (IGR = 0.287), distance to river (IGR = 0.139), aspect (IGR = 0.066), plan curvature (IGR = 0.0548), distance to fault (IGR = 0.0482), and SPI (IGR = 0.0323).

4.3 Spatial relationship between springs and conditioning factors using the SWARA method

The spatial correlation between groundwater springs and the conditioning factors is shown in Table 2. Regarding the slope, the class of $0–5.5°$ shows the highest probability (0.45) of groundwater spring occurrence. As the slope degree increases, the probability of spring occurrence is reduced. In the case of slope aspect, the east aspect (0.44) has the most impact on spring occurrence. According to calculated results, in terms of altitude, springs are the most abundant in altitudes of 1703–2068 m (0.6). The SWARA model is high in flat areas (0.4), followed by areas with concave (0.38) and convex (0.2) curvature. For SPI, the highest SWARA value is found for the classes of 583 969–1 330 153 (0.46). In the case of the TWI, the SWARA values decrease when the TWI reduces. There is an inverse relationship between TRI and SWARA values, and as the TRI increases, the SWARA value reduces.

Regarding distance from the fault, a distance less than 2000 m has the highest impact on spring occurrence, and with an increase in distance (greater than 2000 m), the probability of spring occurrence is reduced. Regarding distance to river, it can be seen that the class of 0–200 m has the highest correlation with spring occurrence (0.46) and there is an inverse relationship between spring occurrence and SWARA values. In the case of land use, the highest SWARA values are shown for garden areas (0.219), followed by a mixture of garden and agriculture (0.17) and agricultural areas (0.12), whereas the lowest SWARA is for bare soil and rock (0.00063). Rainfall between 500 and 600 mm has the highest SWARA value (0.61). The Inceptisols class has the highest SWARA values (0.5), followed by rock outcrops/Entisols (0.39), rock outcrops/Inceptisols (0.056), Inceptisols/Vertisols (0.028), and badlands (0.014). The highest probability is shown for the highly porous and very good water reservoir karstic Oligo-Miocene and Cretaceous pure carbonate formation (OMq and K1bl), the young and poorly consolidated highly porous detrital rock units (PeEf and Plq), and the unconsolidated Quaternary alluvium (PlQc).

4.4 Application of ANFIS ensemble models and their assessment

In the current study, hybrids of the ANFIS model and five metaheuristic algorithms were designed, constructed, and implemented in MATLAB 8.0 software. These models were built using the training dataset. Weights gained by the SWARA method for each conditioning factor were fed as in-

Table 2. Spatial correlation between the conditioning factors and spring locations using the SWARA method.

Factors	Classes	Comparative importance of average value Kj	Coefficient Kj = Sj + 1	wj = (X(j − 1))/kj	Weight wj/ sigma wj
Slope (degree)	0–5.55		1.000	1.000	0.454
	5.55–12.11	0.300	1.300	0.769	0.349
	12.11–19.43	1.500	2.500	0.308	0.140
	19.43–28.77	2.000	3.000	0.103	0.047
	28.77–64.37	3.500	4.500	0.023	0.010
Slope aspect	East		1.000	1.000	0.448
	North	1.000	2.000	0.500	0.224
	West	0.300	1.300	0.385	0.172
	South	0.100	1.100	0.350	0.156
	Flat	0.8	1.05	0.31	0.121
Altitude (m)	1703–2068		1.000	1.000	0.608
	1385–1703	2.200	3.200	0.313	0.190
	2068–3175	0.800	1.800	0.174	0.106
	531–1070	1.000	2.000	0.087	0.053
	1070–1385	0.200	1.200	0.072	0.044
Plan curvature	Flat		1.000	1.000	0.408
	Concave	0.050	1.050	0.952	0.388
	Convex	0.900	1.900	0.501	0.204
SPI	583969.72–1330153.27		1.000	1.000	0.466
	227099.33–583969.72	1.000	2.000	0.500	0.233
	48664.14–227099.33	0.200	1.200	0.417	0.194
	0–48664.14	1.000	2.000	0.208	0.097
	1330153.27–4136452.25	10.000	11.000	0.019	0.009
TWI	6.64–7.92		1.000	1.000	0.471
	5.60–6.64	0.700	1.700	0.588	0.277
	7.92–11.97	1.300	2.300	0.256	0.120
	4.63–5.60	0.100	1.100	0.233	0.110
	2.12–4.63	4.000	5.000	0.047	0.022
TRI	0–5.59		1.000	1.000	0.544
	5.59–12.66	0.800	1.800	0.556	0.302
	12.66–20.62	1.500	2.500	0.222	0.121
	20.62–30.93	3.000	4.000	0.056	0.030
	30.93–75.13	10.000	11.000	0.005	0.003
Distance from fault (m)	0–200		1.000	1.000	0.242
	200–500	0.050	1.050	0.952	0.231
	500–1000	0.100	1.100	0.866	0.210
	1000–2000	0.050	1.050	0.825	0.200
	> 2000	0.700	1.700	0.485	0.118
Distance from river (m)	0–200		1.000	1.000	0.464
	200–500	1.900	2.900	0.345	0.160
	500–1000	0.050	1.050	0.328	0.152
	1000–2000	0.300	1.300	0.253	0.117
	> 2000	0.100	1.100	0.230	0.107
Land use/ land cover	Garden		1.000	1.000	0.219
	Mixture of garden and agriculture	0.282	1.282	0.780	0.171
	Agriculture	0.340	1.340	0.582	0.128
	Mixture of poor rangeland and fallow	0.419	1.419	0.410	0.090
	Fallow	0.233	1.233	0.333	0.073
	Mixture of moderate rangeland and agriculture	0.294	1.294	0.257	0.056
	Mixture of very poor forest	0.124	1.124	0.229	0.050
	Mixture of waterway and vegetation	0.549	1.549	0.148	0.032
	Moderate forest	0.205	1.205	0.122	0.027
	Mixture of agriculture with dry farming	0.064	1.064	0.115	0.025

Table 2. Continued.

Factors	Classes	Comparative importance of average value Kj	Coefficient Kj = Sj + 1	wj = (X(j − 1))/kj	Weight wj/ sigma wj
	Woodland	0.030	1.030	0.112	0.024
	Good rangeland	0.043	1.043	0.107	0.023
	Rangeland	0.333	1.333	0.080	0.018
	Poor rangeland	0.030	1.030	0.078	0.017
	Poor forest	0.210	1.210	0.065	0.014
	Moderate rangeland	0.281	1.281	0.050	0.011
	Bare soil and rock	0.237	1.237	0.041	0.009
	Dense rangeland	0.278	1.278	0.032	0.007
	Dense forest	10.000	11.000	0.003	0.001
	Waterway	0.000	1.000	0.003	0.001
	Mixture of agriculture with poor garden	0.000	1.000	0.003	0.001
	Very poor forest	0.000	1.000	0.003	0.001
	Mixture of moderate forest and agriculture	0.000	1.000	0.003	0.001
	Mixture of low forest and fallow	0.000	1.000	0.003	0.001
	Urban and residential	0.000	1.000	0.003	0.001
Rainfall (mm)	600–700		1.000	1.000	0.617
	700–800	2.200	3.200	0.313	0.193
	800–900	0.600	1.600	0.195	0.121
	500–600	1.500	2.500	0.078	0.048
	400–500	1.300	2.300	0.034	0.021
Soil order	Rock outcrops/Entisols		1.000	1.000	0.509
	Rock outcrops/Inceptisols	0.300	1.300	0.769	0.392
	Inceptisols	5.900	6.900	0.111	0.057
	Inceptisols/Vertisols	1.000	2.000	0.056	0.028
	Badlands	1.000	2.000	0.028	0.014
Lithology (unit)	OMq		1.000	1.000	0.133
	PeEf	0.309	1.309	0.764	0.101
	PlQc	0.253	1.253	0.610	0.081
	K1bl	0.113	1.113	0.548	0.073
	Plc	0.014	1.014	0.541	0.072
	pd	0.059	1.059	0.511	0.068
	TRKubl	0.223	1.223	0.417	0.055
	TRJvm	0.027	1.027	0.406	0.054
	MPlfgp	0.048	1.048	0.388	0.051
	OMql	0.015	1.015	0.382	0.051
	Plbk	0.081	1.081	0.353	0.047
	E2c	0.291	1.291	0.274	0.036
	TRKurl	0.059	1.059	0.258	0.034
	Qft2	0.335	1.335	0.194	0.026
	MuPlaj	0.100	1.100	0.176	0.023
	KEpd-gu	0.080	1.080	0.163	0.022
	Kgu	0.566	1.566	0.104	0.014
	Qft1	0.064	1.064	0.098	0.013
	Ekn	0.109	1.109	0.088	0.012
	KPeam	0.027	1.027	0.086	0.011
	PeEtz	0.328	1.328	0.065	0.009
	Kbgp	0.445	1.445	0.045	0.006
	EMas-sb	0.310	1.310	0.034	0.005
	Mgs	0.626	1.626	0.021	0.003
	TRJlr	10.000	11.000	0.002	0.000
	Klsol	0.000	1.000	0.002	0.000
	JKbl	0.000	1.000	0.002	0.000
	Kur	0.000	1.000	0.002	0.000
	OMas	0.000	1.000	0.002	0.000
	Mmn	0.000	1.000	0.002	0.000

put into the training dataset. The results are shown in Figs. 5 and 6.

As can be seen in Fig. 5, the MSEs of the ANFIS–IWO model, the ANFIS–DE model, the ANFIS–FA model, the ANFIS–PSO model, and the ANFIS–BA model using the training dataset are 0.066, 0.066, 0.066, 0.049, and 0.09, respectively. This indicates that the ANFIS–PSO model has the highest performance, whereas the ANFIS–BA model presents the lowest one. The prediction performance of the five models using the validation dataset is shown in Fig. 6. MSEs of the ANFIS–IWO model, the ANFIS–DE model, the ANFIS–FA model, the ANFIS–PSO model, and the ANFIS–BA model are 0.060, 0.060, 0.060, 0.045, and 0.09, respectively. Therefore, it could be concluded that the ANFIS–PSO model and ANFIS–BA model have the highest and lowest prediction performances, respectively.

However, it should be noticed that, in addition to accuracy, the execution speed of the five models was found to be of significance. To measure this, the running time for 1000 iterations was estimated. The result is shown in Fig. 7. It could be seen that the running time of the ANFIS–IWO model, the ANFIS–DE model, the ANFIS–FA model, the ANFIS–PSO model, and the ANFIS–BA model was 8036, 547, 22 111, 1050, and 6993 s, respectively. It can be concluded that the ANFIS–DE model had the lowest running time and the ANFIS–FA model had the maximum time.

On the other hand, it is possible to test how each model achieves convergence in learning. Using the cost function values, a convergence graph for all five models was constructed and shown in Fig. 8. The results show that cost function values of the ANFIS–DE model and the ANFIS–BA model were stable from 30 and 95 iterations, indicating a rapid convergence of the models, while the ANFIS–PSO model, the ANFIS–IWO model, and the ANFIS–FA model showed a convergence after 650, 650, and 360 iterations, respectively. This indicates a slow convergence.

4.5 Generation of groundwater spring potential maps using ANFIS hybrid models

Once the five models were successfully trained and validated, these models were used to compute groundwater spring indices for all the pixels of the study areas. Then, these indices were exported from MATLAB into ArcGIS10.2 software for generating groundwater spring potential maps. Ultimately, the achieved maps could be sorted into five classes: very low, low, moderate, high, and very high (Fig. 9a–e).

Many methods can be used for determining thresholds for the five classes: manual, equal interval, geometric interval, quantile, natural break, and standard deviation. Selection of a method depends on the characteristics of the data and the distribution of the groundwater spring indexes in a histogram (Ayalew and Yamagishi, 2005). If the indexes have a positive or negative skewness, the quantile or natural break classification is suitable for index classification (Akgun, 2012). In this

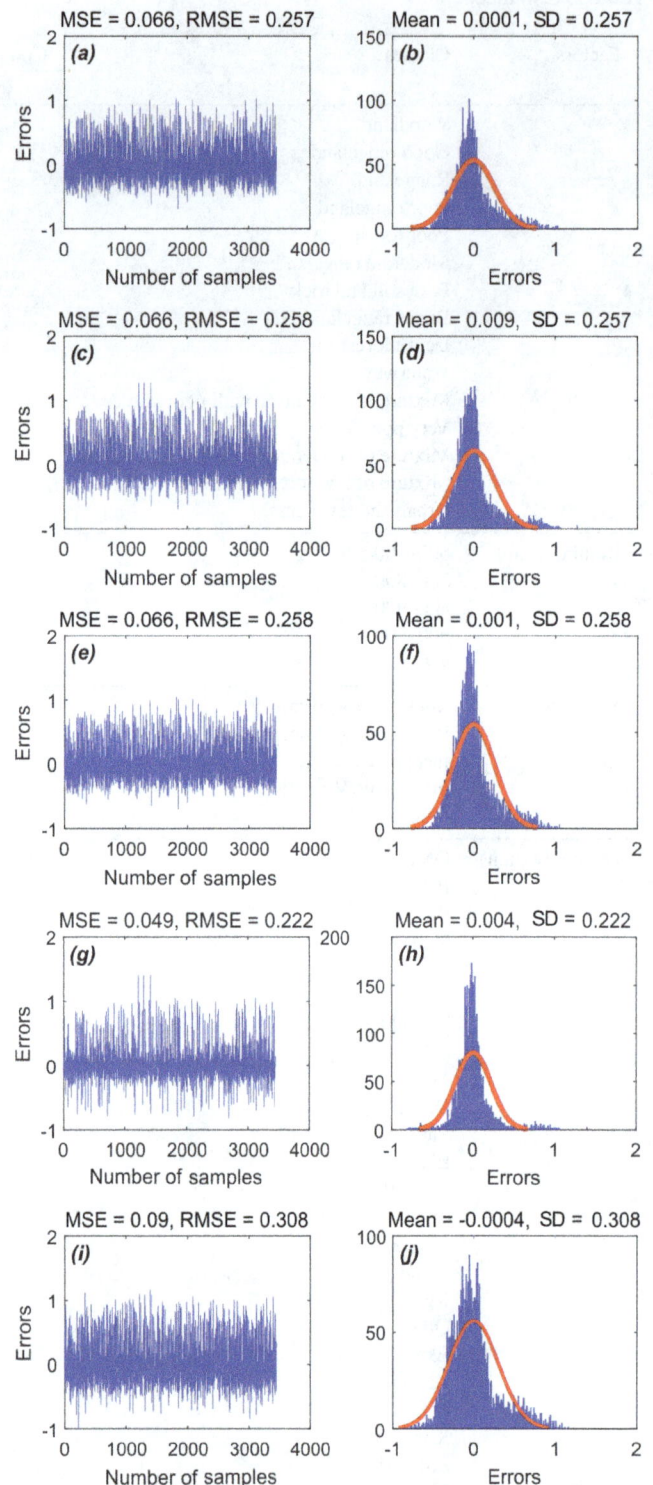

Figure 5. MSE and RMSE values of the five models using the training dataset of **(a)** ANFIS–IWO, **(c)** ANFIS–DE, **(e)** ANFIS–FA, **(g)** ANFIS–PSO, and **(i)** ANFIS–BA. Frequency errors of the five models using the train dataset of **(b)** ANFIS–IWO, **(d)** ANFIS–DE, **(f)** ANFIS–FA, **(h)** ANFIS–PSO, and **(j)** ANFIS–BA.

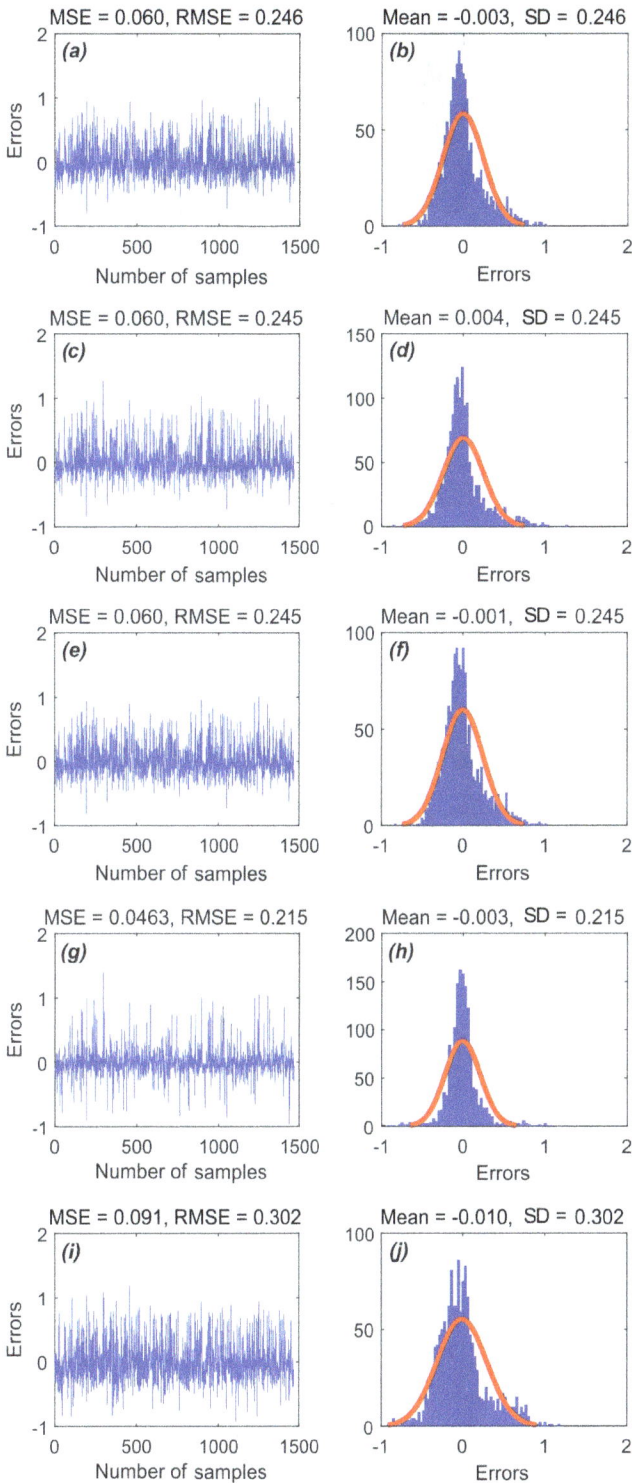

Figure 6. MSE and RMSE values of the five models using the validation dataset of (**a**) ANFIS–IWO, (**c**) ANFIS–DE, (**e**) ANFIS–FA, (**g**) ANFIS–PSO, and (**i**) ANFIS–BA. Frequency errors of the five models using the validation dataset of (**b**) ANFIS–IWO, (**d**) ANFIS–DE, (**f**) ANFIS–FA, (**h**) ANFIS–PSO, and (**j**) ANFIS–BA.

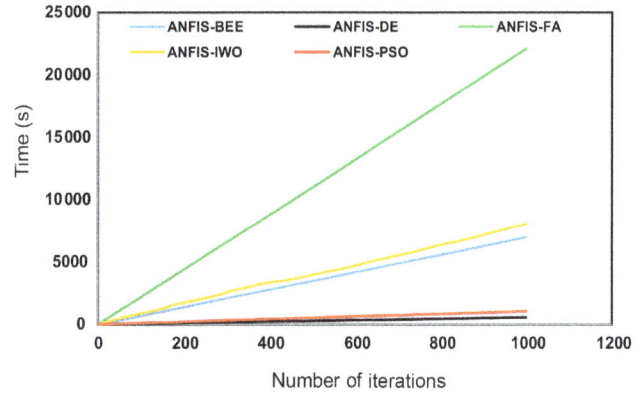

Figure 7. Processing time used for training the five models.

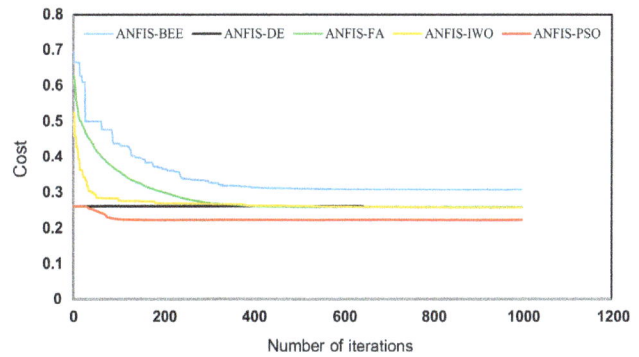

Figure 8. Convergence plot of the models.

research, the histogram was checked and the results revealed that the quantile method was better than other methods for index classification.

4.6 Validation and comparisons of the groundwater spring potential map

The prediction ability and reliability of the five achieved maps have been evaluated using both the training and the validation datasets. The results of the success rate revealed that the ANFIS–DE model had the highest AUC value (0.883), followed by the ANFIS–IWO model (0.882), the ANFIS–FA model (0.882), the ANFIS–PSO model (0.871), and the ANFIS–BA model (0.852) (Fig. 10a). Regarding the prediction rate, all five models had a good prediction capability but the ANFIS–DE model has the highest prediction rate (0.873), followed by the ANFIS-IWO model (0.873) and the ANFIS–FA model (0.873), the ANFIS–PSO model (0.865), and the ANFIS–BA model (0.839)(Fig. 10b).

4.7 Statistical tests

The result of the Friedman test (Table 3) revealed that as significance and chi-square values were less than 0.05 and greater than 3.84, respectively, the null hypothesis was rejected. The result also indicated that there was a statistically

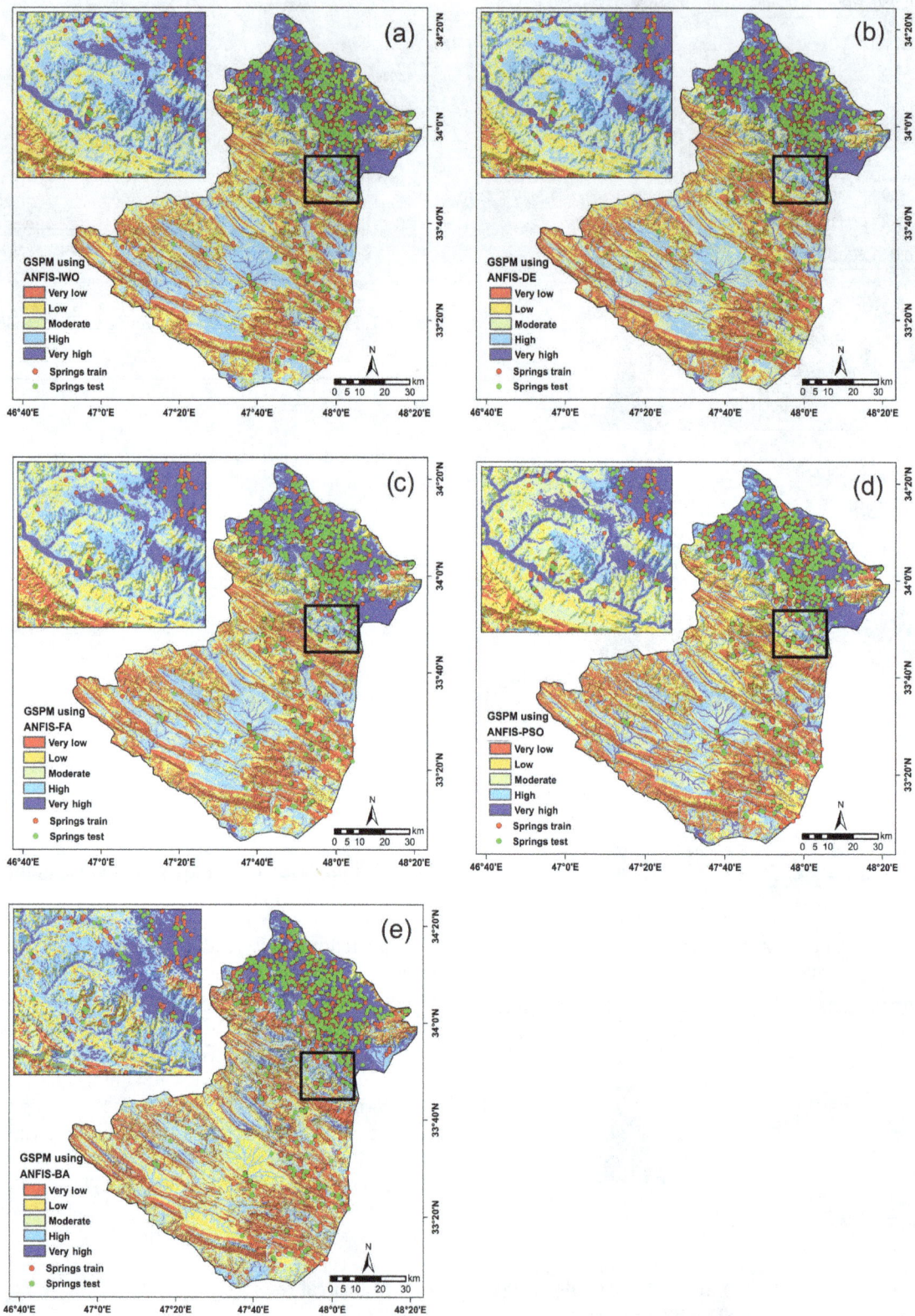

Figure 9. Groundwater spring potential map using **(a)** the ANFIS–IWO model, **(b)** the ANFIS–DE model, **(c)** the ANFIS–FA model, **(d)** the ANFIS–PSO model, and **(e)** the ANFIS–BA model.

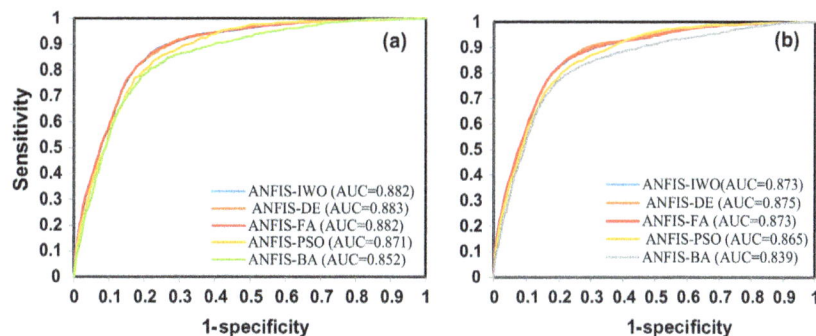

Figure 10. (a) Success rate and (b) prediction rate of the five models.

Table 3. Results of Friedman test.

No.	Performed models	Mean rank	Chi-square	Significance
1	ANFIS–DE	3.04		
2	ANFIS–IWO	3.13		
3	ANFIS–FA	2.98	64.84	0.00
4	ANFIS–PSO	2.72		
5	ANFIS–BA	3.12		

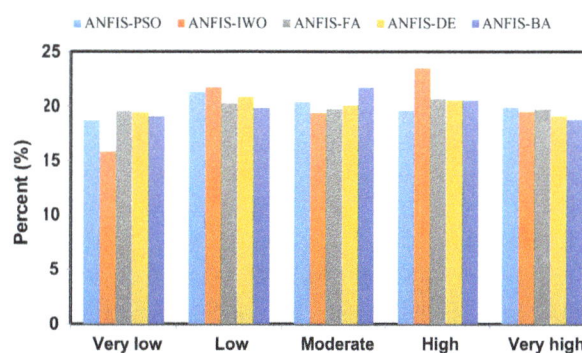

Figure 11. Percentage areas of different groundwater spring potential classes for five models.

significant difference between prediction capabilities of the five models.

The results of the Wilcoxon signed-rank test showed that both p values and z values were far from the standard values of 0.05 and from -1.96 to $+1.96$, respectively, except for the ANFIS–FA model vs. the ANFIS–DE model and the ANFIS–PSO model vs. the ANFIS–DE model (Table 4). This indicates that there are statistically significant differences between models' performance, except for ANFIS–FA vs. ANFIS–DE and ANFIS–PSO vs. ANFIS–DE.

4.8 Percentage area

The percentage area of each class of final map resulting from the five hybrid models is shown in Fig. 11. According to the results of the ANFIS–DE, the most accurate model in groundwater spring potential mapping, the percentage areas of very low, low, moderate, high, and very high groundwater spring potential are about 19.06, 19.88, 21.72, 20.55, and 18.78 % of the study area, respectively.

5 Discussion

5.1 The impact of conditioning factor classes on GSPM

Assessment of conditioning factor is a necessary step in finding the correlation analysis between the groundwater spring and the conditioning factors. It should be noted that no universal guideline is available regarding the number and size

of the classes as well as the selection of the conditioning factors. They were selected mostly based on characteristics of the study area and previous similar studies (Xu et al., 2013). As the slope increases, the probability of water infiltration reduces and runoff generation will increase. Thus, the steeper the slope, the lower the spring occurrence probability. According to the results of the SWARA method, the springs tend to occur at middle altitudes or on mountain slopes. Land in the flat curvature class retains rainfall which then infiltrates. Therefore, the amount of groundwater in these areas is higher than for concave or convex curvature. The east aspect has more springs than other aspects. These results are in accordance with Pourtaghi and Pourghasemi (2014), who reported that most springs occurred at elevations of 1600–1900 m with an east slope aspect (using the FR method).

TWI shows the amount of wetness, and it is obvious that the more the TWI, the higher the groundwater springs probability occurrence is. The terrain roughness index (TRI), or topographic roughness or terrain ruggedness, calculates the sum of change in elevation between a grid cell and its neighborhood, and the lesser the roughness, the higher the spring potential mapping. The SPI shows that the erosive power of the water and the mountainous area is higher than the plain area. So, as the SPI increases, the spring potential occurrence increases. Rivers are one of the most important sources

Table 4. Results of the Wilcoxon signed-rank test.

No.	Pairwise comparison	z value	p value	Significance
1	ANFIS–DE vs. ANFIS–BA	−3.97	0.00	Yes
2	ANFIS–FA vs. ANFIS–BA	−2.37	0.017	Yes
3	ANFIS–IWO vs. ANFIS–BA	−2.35	0.018	Yes
4	ANFIS–PSO vs. ANFIS–BA	−3.04	0.002	Yes
5	ANFIS–FA vs. ANFIS–DE	−1.32	0.185	No
6	ANFIS–IWO vs. ANFIS–DE	−3.96	0.00	Yes
7	ANFIS–PSO vs. ANFIS–DE	−0.841	0.41	No
8	ANFIS–IWO vs. ANFIS–FA	−3.19	0.001	Yes
9	ANFIS–PSO vs. ANFIS–FA	−1.90	0.057	Yes
10	ANFIS–PSO vs. ANFIS–IWO	−2.44	0.015	Yes

of groundwater recharge, and the nearer to river, the higher the probability of spring occurrence. Also, as the rainfall increases, the groundwater spring incidence increases, but in the current study, some other conditioning factors affected the spring occurrence.

Most of the springs were located in the garden land use/land cover category. Therefore, it can be stated that the gardens have been established near the springs. Pliocene–Quaternary formation is newer on a geologic timescale, and Quaternary formation has a higher potential for groundwater spring incidence due to high permeability. A fault causes discontinuity in a volume of rock. Thus, the nearer to the fault, the higher the spring occurrence probability will be. Inceptisols soils are relatively new and are characterized as only having a weak appearance of horizons. They are the most abundant on the Earth (https://www.britannica.com/science/Inceptisol, last access: 1 May 2018) and are mostly formed from colluvial and alluvial materials. So, due to high permeability and high rainfall infiltration, they have a high potential for spring occurrence. In the case of lithological units, there are four suitable rock types in a water reservoir based on physical phenomena such as porosity and permeability that consist of (1) unconsolidated sands and gravels, (2) sandstones, (3) limestones, and (4) basaltic lava flows. In this study area lithological units include sedimentary rocks, mostly carbonate and detrital rocks, with cover of alluvium and minor soil.

5.2 Advantages/disadvantages of the models and performance analysis

The highest accuracy based on the MSE in both the training and validation datasets is found for the ANFIS–PSO model. However, based on the AUC of the success rate and the prediction rate, the ANFIS–DE model has the highest performance. The problem with the MSE comes from the fact that it is based on the error assessment. But the models should be acted upon holistically based on abilities. The AUC is based on the true positive (TP), true negative (TN), false positive (FP), and false negative (FN) (Pham et al., 2018b), and

therefore is more accurate than the RMSE for comparison (Termeh et al., 2018).

ANFIS has the potential to capture the benefits of both a neural network and fuzzy logic in a single framework and can be considered to be a robust model. ANFIS had some advantages including the following: (1) a much better learning ability, (2) a need for fewer adjustable parameters than those required in other neural network structure, (3) the allowance of a better integration with other control design methods through its networks, and (4) more flexibility (Vahidnia et al., 2010).

Despite several advantages of ANFIS, determination of the membership function is the biggest disadvantage of this model. Finding the optimal parameter for a neural fuzzy model in a membership function is difficult; therefore, the best parameter should make use of other optimization models. This problem was addressed in this paper by being solved using five metaheuristic algorithms, namely IWO, DE, FA, PSO, and BA.

In the current study, the results showed that DE algorithm optimized the parameter for the neural fuzzy model better than the four other algorithms. The main DE algorithm's advantage is its simplicity as it consists of only three parameters called N (size of population), F (mutation parameter), and C (crossover parameter) for controlling the search process (Tvrdýk, 2006). Advantages of the DE algorithm can be explained as follows: (1) ability to handle non-differentiable, nonlinear, and multimodal cost functions, (2) parallelizability to cope with computation-intensive cost functions, (3) good convergence properties, and (4) random sampling and combination of vectors in the present population for creating vectors for the next generation.

Finally, it should be noted that each algorithm has some advantages or disadvantages according to the optimization problems which can be summarized as follows.

Some of the advantages of IWO include the manner of reproduction, spatial dispersal, and competitive exclusion (Mehrabian and Lucas, 2006) as well as the fact that seeds and their parents are ranked together and that those with better fitness survive and become reproductive (Ahmed et al., 2014). This algorithm can benefit from combined advantages of retaining the dominant poles and the error minimization (Abu-Al-Nadi et al., 2013).

The bees algorithm does not employ any probability approach, but it utilizes fitness evaluation to drive the search (Yuce et al., 2013). This algorithm uses a set of parameters that makes it powerful, including the number of scout bees in the selected patches, the number of best patches in the selected patches, the number of elite patches in the selected best patches, the number of recruited bees in the elite patches, the number of recruited bees in the non-elite best patches, the size of neighborhood for each patch, the number of iterations and the difference between the value of first and last iterations.

The firefly algorithm's advantages are summarized as (1) the handling of highly nonlinear, multimodal optimization problems efficiently, (2) non-utilization of velocities, (3) its ability to be integrated with other optimization techniques as a flexible method, and finally (4) no need of a good initial solution to begin its iteration process.

Advantages of the particle swarm optimization (PSO) algorithm can be summarized as follows: (1) particles update themselves with the internal velocity; (2) particles have a memory, which is important for the algorithm; (3) the "best" particle gives out the information to others; (4) it often produces quality solutions more rapidly than alternative methods; and (5) it automatically searches for the optimum solution in the solution space (Wan, 2013).

As a result, there is no algorithm which works perfectly for all optimization problems, and each algorithm has a different performance accuracy based on different data. New algorithms, therefore, should be applied and tested, and finally the most powerful algorithm should be selected as the conclusion of the research demands.

5.3 Previous works and future work proposal

Some research has been carried out on groundwater well or spring potential mapping using bivariate statistical models (Nampak et al., 2014; Guru et al., 2017; Al-Manmi and Rauf, 2016), random forests (Rahmati et al., 2016), and boosted trees for regression and classification (Naghibi et al., 2016). The ANFIS–metaheuristic hybrid models have not been used in groundwater potential mapping. However, these hybrid models have proven to be efficient in flood susceptibility mapping (Bui et al., 2016; Termeh et al., 2018) and landslide susceptibility mapping (Chen et al., 2017). Bui et al. (2016) ensemble the ANFIS using two optimization models, namely a genetic algorithm (GA) and particle swarm optimization (PSO), for the identification of flood-prone areas in Vietnam. Termeh et al. (2018) used ANFIS–ant colony optimization, ANFIS–GA, and ANFIS–PSO in flood susceptibility mapping of the Jahrom basin, and stated that ANFIS–PSO had higher prediction capabilities than the other two models. Chen et al. (2017) applied three hybrid

models, namely ANFIS–GA, ANFIS–differential evolution (DE), and ANFIS–PSO, for identifying the areas prone to landslides in Hanyuan County, China. The results showed that ANFIS–DE had a higher performance (AUC = 0.84), followed by ANFIS–GA (AUC = 0.82) and ANFIS–PSO (AUC = 0.78).

In general, the results of the present study, as well as previous research, find that by applying hybrid models, better results could be achieved for spatial prediction modeling including groundwater potential mapping. The ensembles of ANFIS and metaheuristic algorithms can be applied for any spatial prediction modeling such as groundwater potential mapping, flood susceptibility mapping, landslide susceptibility assessment, gully occurrence susceptibility mapping, and other endeavors at a regional scale and in other areas.

For future work, it is recommended that (1) the water quality of the Koohdasht–Nourabad plain be investigated and the water quality of areas with high potential be determined for different aspects such as drinking and agricultural and industrial activities, and (2) the groundwater vulnerability assessment be applied by some common methods including the DRASTIC model for which the zones with a high potential for groundwater occurrence should be protected against pollution.

6   Conclusion

Groundwater is the most important natural resource in the world and about 25 % of all freshwater is estimated to be groundwater. Thus, groundwater potential mapping has been considered to be one of the most effective methods for the management of groundwater resources for better exploitation. The main results of the present study can be summarized by the following points.

1. The results showed that although all models had good results, the ANFIS–DE had the highest prediction power (0.875), followed by ANFIS–IWO and ANFIS–FA (0.873), ANFIS–PSO (0.865), and ANFIS–BA (0.839).

2. According to the results of the SWARA method, most springs existed in altitudes of 1703–2068 m, with a flat curvature, east aspect, TWI of 6.6–7.9, TRI of 0–8.7, SPI of 583 969–1 330 153, Inceptisols soil type, a slope of 0–5.5°, 0–200 m distance from river, 500–1000 m distance from fault, rainfall between 500 and 600 mm, and a garden land cover category, and in a Pliocene–Quaternary lithological age with an OMq lithology unit.

3. Based on the information gain ratio, the most important factors affecting groundwater occurrence are land

use/land cover, lithology, rainfall, and TWI. The least important factors are plan curvature, distance to fault, and SPI.

4. Based on the ANFIS–DE model, a total of 39.33 % of the study area has a high and very high groundwater potential, situated in the north.

The results of the current study are helpful for the Iran Water Resources Management Company (IWRMC) for sustainable management of groundwater resources. Overall, the maps resulting from these hybrid artificial intelligence algorithms can be applied for the better management of groundwater resources in the study area.

Author contributions. KK collected and processed the data, and wrote the manuscript. MP designed the experiment, performed the analysis, and wrote the manuscript. DTB checked the analysis and results, and reviewed, edited, and wrote the manuscript.

Competing interests. The authors declare that they have no conflict of interest.

Acknowledgements. We would like to thank Bjørn Kristofersen at the University of South-Eastern Norway for checking the English of the manuscript and also two reviewers and Dimitri Solomatine, editor of the Journal of Hydrology and Earth System Science, for their positive and helpful comments.

Edited by: Dimitri Solomatine

References

Abu-Al-Nadi, D. I., Alsmadi, O. M., Abo-Hammour, Z. S., Hawa, M. F., and Rahhal, J. S.: Invasive weed optimization for model order reduction of linear MIMO systems, Appl. Math. Model., 37, 4570–4577, 2013.

Adiat, K., Nawawi, M., and Abdullah, K.: Assessing the accuracy of GIS-based elementary multi criteria decision analysis as a spatial prediction tool – A case of predicting potential zones of sustainable groundwater resources, J. Hydrol., 440, 75–89, 2012.

Ahmed, A., Al-Amin, R., and Amin, R.: Design of static synchronous series compensator based damping controller employing invasive weed optimization algorithm, SpringerPlus, 3, 394, https://doi.org/10.1186/2193-1801-3-394, 2014.

Akgun, A.: A comparison of landslide susceptibility maps produced by logistic regression, multi-criteria decision, and likelihood ratio methods: a case study at Ýzmir, Turkey, Landslides, 9, 93–106, https://doi.org/10.1007/s10346-011-0283-7, 2012.

Al-Manmi, D. A. M. and Rauf, L. F.: Groundwater potential mapping using remote sensing and GIS-based, in Halabja City, Kurdistan, Iraq, Arab. J. Geosci., 9, 357, https://doi.org/10.1007/s12517-016-2385-y, 2016.

Alimardani, M., Hashemkhani Zolfani, S., Aghdaie, M. H., and Tamoðaitienë, J.: A novel hybrid SWARA and VIKOR methodology for supplier selection in an agile environment, Technol. Econ. Dev. Eco-, 19, 533–548, 2013.

Amiri, B., Hossain, L., Crawford, J. W., and Wigand, R. T.: Community detection in complex networks: Multi–objective enhanced firefly algorithm, Knowl.-Based Syst., 46, 1–11, 2013.

Ayalew, L. and Yamagishi, H.: The application of GIS-based logistic regression for landslide susceptibility mapping in the Kakuda-Yahiko Mountains, Central Japan, Geomorphology, 65, 15–31, 2005.

Beasley, T. M. and Zumbo, B. D.: Comparison of aligned Friedman rank and parametric methods for testing interactions in split-plot designs, Comput. Stat. Data An., 42, 569–593, 2003.

Berhanu, B., Seleshi, Y., and Melesse, A. M.: Surface Water and Groundwater Resources of Ethiopia: Potentials and Challenges of Water Resources Development, in: Nile River Basin, edited by: Melesse, A., Abtew, W., and Setegn, S., Springer, Cham, 97–117, 2014.

Brenning, A.: Spatial prediction models for landslide hazards: review, comparison and evaluation, Nat. Hazards Earth Syst. Sci., 5, 853–862, https://doi.org/10.5194/nhess-5-853-2005, 2005.

Bui, D. T., Lofman, O., Revhaug, I., and Dick, O.: Landslide susceptibility analysis in the Hoa Binh province of Vietnam using statistical index and logistic regression, Nat. Hazards, 59, 1413, https://doi.org/10.1007/s11069-011-9844-2, 2011.

Bui, D. T., Pradhan, B., Revhaug, I., Nguyen, D. B., Pham, H. V., and Bui, Q. N.: A novel hybrid evidential belief function-based fuzzy logic model in spatial prediction of rainfall-induced shallow landslides in the Lang Son city area (Vietnam), Geomat. Nat. Haz. Risk, 6, 243–271, 2015.

Bui, D. T., Pradhan, B., Nampak, H., Bui, Q.-T., Tran, Q.-A., and Nguyen, Q.-P.: Hybrid artificial intelligence approach based on neural fuzzy inference model and metaheuristic optimization for flood susceptibilitgy modeling in a high-frequency tropical cyclone area using GIS, J. Hydrol., 540, 317–330, 2016.

Bui, K.-T. T., Tien Bui, D., Zou, J., Van Doan, C., and Revhaug, I.: A novel hybrid artificial intelligent approach based on neural fuzzy inference model and particle swarm optimization for horizontal displacement modeling of hydropower dam, Neural Comput. Appl., 29, 1495–1506, 2018.

Chen, W., Panahi, M., and Pourghasemi, H. R.: Performance evaluation of GIS-based new ensemble data mining techniques of adaptive neuro-fuzzy inference system (ANFIS) with genetic algorithm (GA), differential evolution (DE), and particle swarm optimization (PSO) for landslide spatial modelling, CATENA, 157, 310–324, 2017.

Chung, C.-J. F. and Fabbri, A. G.: Validation of spatial prediction models for landslide hazard mapping, Nat. Hazards, 30, 451–472, 2003.

Das, S., Abraham, A., Chakraborty, U. K., and Konar, A.: Differential evolution using a neighborhood-based mutation operator, IEEE T. Evolut. Comput., 13, 526–553, 2009.

Derrac, J., García, S., Molina, D., and Herrera, F.: A practical tutorial on the use of nonparametric statistical tests as a methodology for comparing evolutionary and swarm intelligence algorithms, Swarm Evol. Comput., 1, 3–18, 2011.

Emamgholizadeh, S., Moslemi, K., and Karami, G.: Prediction the groundwater level of bastam plain (Iran) by artificial neural network (ANN) and adaptive neuro-fuzzy inference system (ANFIS), Water Resour. Manag., 28, 5433–5446, 2014.

Ercin, A. E. and Hoekstra, A. Y.: Water footprint scenarios for 2050: A global analysis, Environ. Int., 64, 71–82, https://doi.org/10.1016/j.envint.2013.11.019, 2014.

Fashae, O. A., Tijani, M. N., Talabi, A. O., and Adedeji, O. I.: Delineation of groundwater potential zones in the crystalline basement terrain of SW-Nigeria: an integrated GIS and remote sensing approach, Applied Water Science, 4, 19–38, 2014.

Fitts, C. R.: Groundwater science, Academic press, Oxford, UK, 2002.

Friedman, M.: The Use of Ranks to Avoid the Assumption of Normality Implicit in the Analysis of Variance, J. Am. Stat. Assoc., 32, 675–701, https://doi.org/10.1080/01621459.1937.10503522, 1937.

Gaprindashvili, G., Guo, J., Daorueang, P., Xin, T., and Rahimy, P.: A new statistic approach towards landslide hazard risk assessment, International Journal of Geosciences, 5, 38–49, https://doi.org/10.4236/ijg.2014.51006, 2014.

Ghasemi, M., Ghavidel, S., Akbari, E., and Vahed, A. A.: Solving non-linear, non-smooth and non-convex optimal power flow problems using chaotic invasive weed optimization algorithms based on chaos, Energy, 73, 340–353, 2014.

Guru, B., Seshan, K., and Bera, S.: Frequency ratio model for groundwater potential mapping and its sustainable management in cold desert, India, Journal of King Saud University – Science, 29, 333–347, 2017.

Hong, H., Pradhan, B., Xu, C., and Bui, D. T.: Spatial prediction of landslide hazard at the Yihuang area (China) using two-class kernel logistic regression, alternating decision tree and support vector machines, Catena, 133, 266–281, 2015.

Hong, H., Panahi, M., Shirzadi, A., Ma, T., Liu, J., Zhu, A.-X., Chen, W., Kougias, I., and Kazakis, N.: Flood susceptibility assessment in Hengfeng area coupling adaptive neuro-fuzzy inference system with genetic algorithm and differential evolution, Sci. Total Environ., 621, 1124–1141, 2018.

Israil, M., Al-Hadithi, M., and Singhal, D.: Application of a resistivity survey and geographical information system (GIS) analysis for hydrogeological zoning of a piedmont area, Himalayan foothill region, India, Hydrogeol. J., 14, 753–759, 2006.

Jang, J.-S.: ANFIS: adaptive-network-based fuzzy inference system, IEEE T. Syst. Man. Cyb., 23, 665–685, 1993.

Jha, M. K., Chowdhury, A., Chowdary, V., and Peiffer, S.: Groundwater management and development by integrated remote sensing and geographic information systems: prospects and constraints, Water Resour. Manag., 21, 427–467, 2007.

Jha, M. K., Chowdary, V., and Chowdhury, A.: Groundwater assessment in Salboni Block, West Bengal (India) using remote sensing, geographical information system and multi-criteria decision analysis techniques, Hydrogeol. J., 18, 1713–1728, 2010.

Kaliraj, S., Chandrasekar, N., and Magesh, N.: Identification of potential groundwater recharge zones in Vaigai upper basin, Tamil Nadu, using GIS-based analytical hierarchical process (AHP) technique, Arab. J. Geosci., 7, 1385–1401, 2014.

Kennedy, J.: Particle swarm optimization, in: Encyclopedia of machine learning, Springer, New York, USA, 760–766, 2011.

Kennedy, J. and Eberhart, R.: Particle swarm optimization, in: Proceedings of ICNN'95 – International Conference on Neural Networks, Perth, WA, Australia, 27 November–1 December 1995, IEEE, https://doi.org/10.1109/ICNN.1995.488968, 1995.

Keršuliene, V., Zavadskas, E. K., and Turskis, Z.: Selection of rational dispute resolution method by applying new step-wise weight assessment ratio analysis (SWARA), J. Bus. Econ. Manag., 11, 243–258, 2010.

Khosravi, K., Nohani, E., Maroufinia, E., and Pourghasemi, H. R.: A GIS-based flood susceptibility assessment and its mapping in Iran: a comparison between frequency ratio and weights-of-evidence bivariate statistical models with multi-criteria decision-making technique, Nat. Hazards, 83, 947–987, 2016a.

Khosravi, K., Pourghasemi, H. R., Chapi, K., and Bahri, M.: Flash flood susceptibility analysis and its mapping using different bivariate models in Iran: a comparison between Shannon's entropy, statistical index, and weighting factor models, Environ. Monit. Assess., 188, 656, https://doi.org/10.1007/s10661-016-5665-9, 2016b.

Khosravi, K., Pham, B. T., Chapi. K., Shirzadi, A., Shahabi, H., Revhaug, I., Prakash, I., and Tien Bui, D.: A comparative assessment of decision trees algorithms for flash flood susceptibility modeling at Haraz watershed, northern Iran, Sci. Total Environ., 627, 744–755, https://doi.org/10.1016/j.scitotenv.2018.01.266, 2018.

Li, Y.-F., Xie, M., and Goh, T.-N.: Adaptive ridge regression system for software cost estimating on multi-collinear datasets, J. Syst. Software, 83, 2332-2343, 2010.

Lohani, A., Kumar, R., and Singh, R.: Hydrological time series modeling: A comparison between adaptive neuro-fuzzy, neural network and autoregressive techniques, J. Hydrol., 442, 23-35, 2012.

Maiti, S., and Tiwari, R.: A comparative study of artificial neural networks, Bayesian neural networks and adaptive neuro-fuzzy inference system in groundwater level prediction, Environ. Earth Sci., 71, 3147–3160, 2014.

Mehrabian, A. R., and Lucas, C.: A novel numerical optimization algorithm inspired from weed colonization, Ecol. Inform., 1, 355–366, 2006.

Mohanty, S., Jha, M. K., Raul, S., Panda, R., and Sudheer, K.: Using artificial neural network approach for simultaneous forecasting of weekly groundwater levels at multiple sites, Water Resour. Manag., 29, 5521–5532, 2015.

Mukherjee, S.: Targeting saline aquifer by remote sensing and geophysical methods in a part of Hamirpur-Kanpur, India, Hydrogeol. J., 19, 53-64, 1996.

Naghibi, S. A., Pourghasemi, H. R., Pourtaghi, Z. S., and Rezaei, A.: Groundwater qanat potential mapping using frequency ratio and Shannon's entropy models in the Moghan watershed, Iran, Earth Sci. Inform., 8, 171–186, 2015.

Naghibi, S. A., Pourghasemi, H. R., and Dixon, B.: GIS-based groundwater potential mapping using boosted regression tree, classification and regression tree, and random forest machine learning models in Iran, Environ. Monit. Assess., 188, 44, https://doi.org/10.1007/s10661-015-5049-6, 2016.

Naidu, Y. R. and Ojha, A.: A hybrid version of invasive weed optimization with quadratic approximation, Soft Comput., 19, 3581–3598, 2015.

Nampak, H., Pradhan, B., and Manap, M. A.: Application of GIS based data driven evidential belief function model to predict groundwater potential zonation, J. Hydrol., 513, 283–300, 2014.

Ngoc-Thach, N., Ngo, D. B.-T., Xuan-Canh, P., Hong-Thi, N., Thi, B. H., NhatDuc, H., and Dieu, T. B.: Spatial pattern assessment of tropical forest fire danger at Thuan Chau area (Vietnam) using GIS-based advanced machine learning algorithms: A comparative study, Ecol. Inform., 46, 74–85, 2018.

Nosrati, K., and Van Den Eeckhaut, M.: Assessment of groundwater quality using multivariate statistical techniques in Hashtgerd Plain, Iran, Environm. Earth Sci., 65, 331–344, 2012.

Nourani, V., Alami, M. T., and Vousoughi, F. D.: Hybrid of SOM-Clustering Method and Wavelet-ANFIS Approach to Model and Infill Missing Groundwater Level Data, J. Hydrol. Eng., 21, 05016018, https://doi.org/10.1061/(ASCE)HE.1943-5584.0001398, 2016.

O'brien, R. M.: A caution regarding rules of thumb for variance inflation factors, Qual. Quant., 41, 673–690, 2007.

Oh, H.-J., Kim, Y.-S., Choi, J.-K., Park, E., and Lee, S.: GIS mapping of regional probabilistic groundwater potential in the area of Pohang City, Korea, J. Hydrol., 399, 158–172, 2011.

Ozdemir, A.: Using a binary logistic regression method and GIS for evaluating and mapping the groundwater spring potential in the Sultan Mountains (Aksehir, Turkey), J. Hydrol., 405, 123–136, 2011a.

Ozdemir, A.: GIS-based groundwater spring potential mapping in the Sultan Mountains (Konya, Turkey) using frequency ratio, weights of evidence and logistic regression methods and their comparison, J. Hydrol., 411, 290–308, 2011b.

Pham, B. T., Bui, D. T., Pourghasemi, H. R., Indra, P., and Dholakia, M.: Landslide susceptibility assesssment in the Uttarakhand area (India) using GIS: a comparison study of prediction capability of naïve bayes, multilayer perceptron neural networks, and functional trees methods, Theor. Appl. Climatol., 128, 255–273, 2017a.

Pham, B. T., Khosravi, K., and Prakash, I.: Application and comparison of decision tree-based machine learning methods in landside susceptibility assessment at Pauri Garhwal Area, Uttarakhand, India, Environmental Processes, 4, 711–730, 2017b.

Pham, B. T., Hoang, T.-A., Nguyen, D.-M., and Bui, D. T.: Prediction of shear strength of soft soil using machine learning methods, Catena, 166, 181–191, 2018a.

Pham, B. T., Prakash, I., and Tien Bui, D.: Spatial prediction of landslides using a hybrid machine learning approach based on Random Subspace and Classification and Regression Trees, Geomorphology, 303, 256–270, 2018b.

Pham, D., Ghanbarzadeh, A., Koc, E., Otri, S., Rahim, S., and Zaidi, M.: The bees algorithm. Technical note, Manufacturing Engineering Centre, Cardiff University, UK, 1–57, 2005.

Pourghasemi, H. R. and Beheshtirad, M.: Assessment of a data-driven evidential belief function model and GIS for groundwater potential mapping in the Koohrang Watershed, Iran, Geocarto Int., 30, 662–685, 2015.

Pourghasemi, H. R., Pradhan, B., and Gokceoglu, C.: Application of fuzzy logic and analytical hierarchy process (AHP) to landslide susceptibility mapping at Haraz watershed, Iran, Nat. Hazards, 63, 965–996, 2012.

Pourghasemi, H. R., Moradi, H., and Aghda, S. F.: Landslide susceptibility mapping by binary logistic regression, analytical hi-erarchy process, and statistical index models and assessment of their performances, Nat. Hazards, 69, 749–779, 2013a.

Pourghasemi, H. R., Pradhan, B., Gokceoglu, C., Mohammadi, M., and Moradi, H. R.: Application of weights-of-evidence and certainty factor models and their comparison in landslide susceptibility mapping at Haraz watershed, Iran, Arab. J. Geosci., 6, 2351–2365, 2013b.

Pourtaghi, Z. S. and Pourghasemi, H. R.: GIS-based groundwater spring potential assessment and mapping in the Birjand Township, southern Khorasan Province, Iran, Hydrogeol. J., 22, 643–662, 2014.

Rahmati, O., Samani, A. N., Mahdavi, M., Pourghasemi, H. R., and Zeinivand, H.: Groundwater potential mapping at Kurdistan region of Iran using analytic hierarchy process and GIS, Arab. J. Geosci., 8, 7059–7071, 2015.

Rahmati, O., Pourghasemi, H. R., and Melesse, A. M.: Application of GIS-based data driven random forest and maximum entropy models for groundwater potential mapping: a case study at Mehran Region, Iran, Catena, 137, 360–372, 2016.

Richey, A. S., Thomas, B. F., Lo, M. H., Reager, J. T., Famiglietti, J. S., Voss, K., Swenson, S., and Rodell, M.: Quantifying renewable groundwater stress with GRACE, Water Resour. Res., 51, 5217–5238, 2015.

Sander, P., Chesley, M. M., and Minor, T. B.: Groundwater assessment using remote sensing and GIS in a rural groundwater project in Ghana: lessons learned, Hydrogeol. J., 4, 40–49, 1996.

Siebert, S., Henrich, V., Frenken, K., and Burke, J.: Update of the digital global map of irrigation areas to version 5, Rheinische Friedrich-Wilhelms-Universität, Bonn, Germany and Food and Agriculture Organization of the United Nations, Rome, Italy, 2013.

Singh, A. K. and Prakash, S. R.: An integrated approach of remote sensing, geophysics and GIS to evaluation of groundwater potentiality of Ojhala sub-watershed, Mirjapur district, UP, India, Asian conference on GIS, GPS, aerial photography and remote sensing, Bangkok, Thailand, 7–9 August 2002.

Storn, R. and Price, K.: Differential evolution – a simple and efficient heuristic for global optimization over continuous spaces, J. Global Optim., 11, 341–359, 1997.

Sun, Y., Wendi, D., Kim, D. E., and Liong, S.-Y.: Technical note: Application of artificial neural networks in groundwater table forecasting – a case study in a Singapore swamp forest, Hydrol. Earth Syst. Sci., 20, 1405–1412, https://doi.org/10.5194/hess-20-1405-2016, 2016.

Takagi, T. and Sugeno, M.: Fuzzy identification of systems and its applications to modeling and control, IEEE T. Syst. Man. Cyb., SMC-15, 116–132, 1985.

Tehrany, M. S., Pradhan, B., and Jebur, M. N.: Spatial prediction of flood susceptible areas using rule based decision tree (DT) and a novel ensemble bivariate and multivariate statistical models in GIS, J. Hydrol., 504, 69–79, 2013.

Tehrany, M. S., Pradhan, B., and Jebur, M. N.: Flood susceptibility mapping using a novel ensemble weights-of-evidence and support vector machine models in GIS, J. Hydrol., 512, 332–343, 2014.

Termeh, S. V. R., Kornejady, A., Pourghasemi, H. R., and Keesstra, S.: Flood susceptibility mapping using novel ensembles of adaptive neuro fuzzy inference system and metaheuristic algorithms, Sci. Total Environ., 615, 438–451, 2018.

Tien Bui, D. and Hoang, N.-D.: GIS-based spatial prediction of tropical forest fire danger using a new hybrid machine learning method, Ecol. Inform., 48, 104–116, 2018.

Tien Bui, D., Pradhan, B., Lofman, O., Revhaug, I., and Dick, O. B.: Spatial prediction of landslide hazards in Hoa Binh province (Vietnam): a comparative assessment of the efficacy of evidential belief functions and fuzzy logic models, Catena, 96, 28–40, 2012.

Tien Bui, D., Bui, Q.-T., Nguyen, Q.-P., Pradhan, B., Nampak, H., and Trinh, P. T.: A hybrid artificial intelligence approach using GIS-based neural-fuzzy inference system and particle swarm optimization for forest fire susceptibility modeling at a tropical area, Agr. Forest Meteorol., 233, 32–44, 2017.

Todd, D. K. and Mays, L. W.: Groundwater Hydrology, 2nd edn., Wiley, New York, 1980.

Tvrdýk, J.: Competitive differential evolution and genetic algorithm in GA-DS toolbox, Technical Computing Prague, Praha, Humusoft, 99–106, 2006.

Umar, Z., Pradhan, B., Ahmad, A., Jebur, M. N., and Tehrany, M. S.: Earthquake induced landslide susceptibility mapping using an integrated ensemble frequency ratio and logistic regression models in West Sumatera Province, Indonesia, Catena, 118, 124–135, 2014.

Vahidnia, M. H., Alesheikh, A. A., Alimohammadi, A., and Hosseinali, F.: A GIS-based neuro-fuzzy procedure for integrating knowledge and data in landslide susceptibility mapping, Comput. Geosci., 36, 1101–1114, 2010.

Waikar, M. and Nilawar, A. P.: Identification of groundwater potential zone using remote sensing and GIS technique, International Journal of Innovative Research in Science, Engineering and Technology, 3, 1264–1274, 2014.

Wan, S.: Entropy-based particle swarm optimization with clustering analysis on landslide susceptibility mapping, Environ. Earth Sci., 68, 1349–1366, 2013.

Xu, C., Dai, F., Xu, X., and Lee, Y. H.: GIS-based support vector machine modeling of earthquake-triggered landslide susceptibility in the Jianjiang River watershed, China, Geomorphology, 145, 70–80, 2012.

Xu, C., Xu, X., Dai, F., Wu, Z., He, H., Shi, F., Wu, X., and Xu, S.: Application of an incomplete landslide inventory, logistic regression model and its validation for landslide susceptibility mapping related to the May 12, 2008 Wenchuan earthquake of China, Nat. Hazards, 68, 883–900, 2013.

Yang, X.-S.: Nature-inspired metaheuristic algorithms, Luniver press, University of Cambridge, UK, 2010.

Yuce, B., Packianather, M. S., Mastrocinque, E., Pham, D. T., and Lambiase, A.: Honey bees inspired optimization method: the bees algorithm, Insects, 4, 646–662, 2013.

Zare, M. and Koch, M.: Groundwater level fluctuations simulation and prediction by ANFIS- and hybrid Wavelet-ANFIS/Fuzzy C-Means (FCM) clustering models: Application to the Miandarband plain, J. Hydro-environ. Res., 18, 63–76, 2018.

Zhou, Y., Luo, Q., Chen, H., He, A., and Wu, J.: A discrete invasive weed optimization algorithm for solving traveling salesman problem, Neurocomputing, 151, 1227–1236, 2015.

9

Hydrogeochemical controls on brook trout spawning habitats in a coastal stream

Martin A. Briggs[1], **Judson W. Harvey**[2], **Stephen T. Hurley**[3], **Donald O. Rosenberry**[4], **Timothy McCobb**[5], **Dale Werkema**[6], **and John W. Lane Jr.**[1]

[1]U.S. Geological Survey, Hydrogeophysics Branch, 11 Sherman Place, Unit 5015, Storrs, CT 06269, USA
[2]U.S. Geological Survey, Water Cycle Branch, M.S. 430, Reston, VA 20192, USA
[3]Massachusetts Division of Fisheries and Wildlife, 195 Bournedale Road, Buzzards Bay, MA 02532, USA
[4]U.S. Geological Survey, National Research Program, M.S. 406, Bldg. 25, DFC, Lakewood, CO 80225, USA
[5]U.S. Geological Survey, 10 Bearfoot Road, Northborough, MA 01532, USA
[6]U.S. Environmental Protection Agency, Office of Research and Development, National Exposure Research Laboratory, Exposure Methods & Measurement Division, Environmental Chemistry Branch, Las Vegas, NV 89119 USA

Correspondence: Martin A. Briggs (mbriggs@usgs.gov)

Abstract. Brook trout (*Salvelinus fontinalis*) spawn in fall and overwintering egg development can benefit from stable, relatively warm temperatures in groundwater-seepage zones. However, eggs are also sensitive to dissolved oxygen concentration, which may be reduced in discharging groundwater (i.e., seepage). We investigated a 2 km reach of the coastal Quashnet River in Cape Cod, Massachusetts, USA, to relate preferred fish spawning habitats to geology, geomorphology, and discharging groundwater geochemistry. Thermal reconnaissance methods were used to locate zones of rapid groundwater discharge, which were predominantly found along the central channel of a wider stream valley section. Pore-water chemistry and temporal vertical groundwater flux were measured at a subset of these zones during field campaigns over several seasons. Seepage zones in open-valley sub-reaches generally showed suboxic conditions and higher dissolved solutes compared to the underlying glacial outwash aquifer. These discharge zones were cross-referenced with preferred brook trout redds and evaluated during 10 years of observation, all of which were associated with discrete alcove features in steep cutbanks, where stream meander bends intersect the glacial valley walls. Seepage in these repeat spawning zones was generally stronger and more variable than in open-valley sites, with higher dissolved oxygen and reduced solute concentrations. The combined evidence indicates that regional groundwater discharge along the broader valley bottom is predominantly suboxic due to the influence of near-stream organic deposits; trout show no obvious preference for these zones when spawning. However, the meander bends that cut into sandy deposits near the valley walls generate strong oxic seepage zones that are utilized routinely for redd construction and the overwintering of trout eggs. Stable water isotopic data support the conclusion that repeat spawning zones are located directly on preferential discharges of more localized groundwater. In similar coastal systems with extensive valley peat deposits, the specific use of groundwater-discharge points by brook trout may be limited to morphologies such as cutbanks, where groundwater flow paths do not encounter substantial buried organic material and remain oxygen-rich.

1 Introduction

The heat tracing of water can be used to map a distribution of spatially focused, or "preferential", groundwater-discharge zones throughout surface water systems at times of contrast between the surface and groundwater temperature. The measurement of the water temperature from the reach to watershed scale is now possible using thermal infrared and fiber-optic distributed temperature sensing (FO-DTS) methodology (Dugdale, 2016; Hare et al., 2015; Steel et al., 2017).

Remote infrared data collection throughout the river corridor has been enabled by handheld cameras, piloted aircraft, and the rapidly evolving capabilities of unmanned aerial systems. Researchers are capitalizing on the ongoing refinement of these technologies to identify zones of focused groundwater seepage to streams in order to map potential discrete preferential cold-water fish habitats such as summer thermal refugia (Dugdale et al., 2015). However, surface thermal surveys alone do not indicate groundwater flow path dynamics or the suitability of an interface aquatic habitat (Briggs et al., 2018a).

For example, dissolved oxygen (DO) concentration must be sufficiently high for cold groundwater seepage to provide support for fish life processes at the direct point of discharge to surface water (Ebersole et al., 2003), which is not apparent from thermal analysis alone. During warm summer periods in systems with suboxic groundwater, cold-water fish species such as salmonids can face a tradeoff between occupying discrete zones of preferred water temperatures with near-lethal DO levels and stream sections that are too warm for long-term survival (Matthews and Berg, 1997). The use of groundwater upwelling zones as thermal refugia is further complicated by competition with aggressive invasive species (to the northeastern USA) such as brown trout, which compete with native trout for resources (Hitt et al., 2017). Streams at higher elevations may support the persistence of reach-scale cold-water habitats where point-scale thermal refugia are not needed under current climatic conditions, serving as vital "climate refugia" against rising air temperatures (Isaak et al., 2015). In systems with reliably cold channel water in summer, which can also exist at low elevations when heavily influenced by discharging groundwater, salmonid fish may directly use groundwater-seepage zones for spawning rather than thermal refuge.

Brook trout (*Salvelinus fontinalis*) are a species of char that are native to eastern North America, from Georgia to Québec (MacCrimmon and Campbell, 1969). Populations have been stressed by warming temperatures and reduced water quality, particularly in low-elevation areas (Hudy et al., 2008). Stream network-scale tracking of fish has indicated that the brook trout directly utilize stream confluence mixing zones and preferential groundwater discharge to survive warm summer periods (Baird and Krueger, 2003; Petty et al., 2012; Snook et al., 2016). Additionally, brook trout spawn in the fall, and eggs deposited in redds develop over the winter before hatching in spring (Cunjak and Power, 1986). Oxygen use by the shallow buried embryos increases over the period of development (Crisp, 1981); therefore, DO concentration is a critical parameter of the pore water in which the eggs are bathed. Several studies have demonstrated the importance of hyporheic downwelling in increasing shallow oxygen concentrations, including for salmonid redds, where deeper stream-bed pore water is generally reduced in DO (e.g., Buffington and Tonina 2009; Cardenas et al., 2016; Harvey et al., 2013). Fine sediments can reduce the efficacy of hyporheic DO exchange in spawn zones (Obruca and Hauer, 2016) and are actively cleared by trout during the spawning process (Montgomery et al., 1996).

The importance of hyporheic exchange to salmonid spawning may be limited in the lowland streams that are expected to harbor native cold-water species in the 21st century, namely those with strong groundwater influence. Groundwater upwelling reduces the penetration of the hyporheic flow from surface water (Cardenas and Wilson, 2006) and may shut down hyporheic flushing in redds (Cardenas et al., 2016). While hyporheic exchange introduces oxygenated channel water into the shallow stream bed, the downward advection of heat associated with near-freezing surface water in winter will also cool stream-bed sediments (Geist et al., 2002), potentially impairing egg development. Coaster brook trout, a life-history variant of native brook trout exhibiting potadromous migrations within the Great Lakes, have been shown to specifically prefer groundwater-discharge zones for building redds (Van Grinsven et al., 2012). The development of trout in winter has been found to positively correlate with warmer stream water temperatures as influenced by groundwater seepage (French et al., 2017). Therefore, spatially discrete groundwater-discharge zones with adequate DO may form preferred brook trout spawning habitats (Curry et al., 1995).

Multiscale physical and biogeochemical factors influence temperature and DO concentrations along groundwater flow paths. In river valleys, discharge to the surface water of locally recharged groundwater is expected to emanate from more shallow, lateral flow paths controlled by the local topography (Modica, 1999; Winter et al., 1998). Shallow groundwater flow paths, particularly those within approximately 5 m of the land surface, will be more sensitive to annual air temperature patterns and long-term warming trends due to strong vertical conductive heat exchanges (Kurylyk et al., 2015b). The distance of seeps from upgradient groundwater recharge zones will also affect seepage temperature dynamics and associated aquatic ecosystems due to future changes in surface and recharge temperatures (Burns et al., 2017). Therefore, characterizing the hydrogeochemical attributes of discharging groundwater flow paths is critical in understanding the thermal stability of current and future point-scale preferential brook trout habitats (Briggs et al., 2018a). The complimentary methodology of geophysical remote sensing, geochemical sampling, and vertical bed temperature time series can indicate the physical and chemical properties of groundwater flow paths that source preferential discharge zones utilized routinely by fish for spawning.

Coarse-grained mineral-dominated aquifers with little fine particulate organic matter and low dissolved organic carbon supplies tend to result in generally oxic groundwater conditions (Back et al., 1993). The sandy surficial aquifer of Cape Cod, where our investigation took place, is a classic example of a mineral soil-dominated flow system (Frimpter and Gay, 1979). The flow of groundwater through near-stream organic deposits, however, can result in inverted redox gradients toward the upwelling interface, such that groundwater discharged to surface water is reduced in DO (Seitzinger et al., 2006). In sandy glacial terrain with superimposed peatland deposits, the specific flow patterns of groundwater to surface water in relation to buried peat will influence the groundwater-discharge biogeochemistry. Krause et al. (2013) found that stream-bed groundwater seepage was strongly reduced in DO in zones with peat deposits, likely due to an increase in both near-stream residence time and localized sources of dissolved organic carbon.

Interdisciplinary collaborations between physical and biological scientists are useful to better understand how cold-water species utilize the stream habitat influenced by groundwater discharge and the larger landscape-scale controls on discharge characteristics. While previous hydrogeological research in the coastal stream used for this study had focused on locating and quantifying discrete groundwater discharge (e.g., "cold anomalies", Hare et al., 2015; Rosenberry et al., 2016), here we endeavor to understand the hydraulic and biogeochemical controls on seepage zone distribution utilized directly by native brook trout. In this groundwater-dominated stream (e.g., likely climate refugia), brook trout do not need to occupy discrete inflows for summer thermal refugia but do favor certain upwelling zones for fall spawning. We compare over a decade of visual survey and electronic fish passive integrated transponder (PIT)-tag dropout data regarding repeat brook trout spawning locations to a comprehensive physical and chemical characterization of groundwater-seepage zones across 2 km of stream in order to do the following:

1. identify repeat brook trout spawning locations and determine if they are directly associated with the preferential discharge of groundwater through interface sediments, and

2. develop a hydrogeochemical characterization of trout-preferred groundwater-discharge zones that can aid in their identification in other less-studied systems and potential inclusion in stream habitat restoration efforts.

2 Site description and previous hydrogeologic characterization

Cape Cod is a peninsula in southeastern coastal Massachusetts, USA, composed primarily of highly permeable unconsolidated glacial moraine and outwash deposits. The largest of the Cape Cod sole-source aquifers occupies a western (landward) section of the peninsula (LeBlanc et al., 1986) and is incised by several linear valleys that drain groundwater south to the Atlantic Ocean via baseflow-dominated streams. Strong groundwater discharge to one such stream, the Quashnet River, supports a relatively stable flow regime that has averaged 0.49 ± 0.15 (SD) $m^3 s^{-1}$ from 1986 to 2015

(Rosenberry et al., 2016). The lower Quashnet River emerges from a narrow sand and gravel valley into a broader area with well-defined lateral floodplains. Historical cranberry farming practices, abandoned in the 1950s, have modified the stream corridor (Barlow and Hess, 1993). Primary modifications included the straightening of the main channel (reducing natural sinuosity), installation of flood-control structures, incision of shallow groundwater drainage ditches in the lateral peatland floodplain, and widespread application of sand to the floodplain surface. The current bank-full width of the main channel averages approximately 4 m.

The Quashnet River has long been recognized as a critical habitat for a naturally reproducing population of native sea-run brook trout (Mullan, 1958) with a genetically distinct population (Annett et al., 2012). Efforts to restore trout habitats by the group Trout Unlimited and others have been ongoing for over 40 years (Barlow and Hess, 1993). These efforts include the removal of flood-control structures, the planting of trees along the main channel, and the addition of wood structures to stabilize banks and provide cover from airborne predators. Furthermore, the Commonwealth of Massachusetts purchased 12.5 ha in 1956 and an additional 146 ha along the lower Quashnet River in 1987 and 1988 to protect the area from development. The Massachusetts Division of Fisheries and Wildlife has been monitoring trout populations since 1988 and their movement since 2007.

The groundwater influence on stream temperature is pronounced, particularly over the 2 km reach above the U.S. Geological Survey gage no. 011058837, below which the stream stage is tidally affected. Ambient regional groundwater temperature is approximately 11 °C (Briggs et al., 2014), and strong conductive and advective exchange with the proximal aquifer maintains the surface water temperature well below the lethal threshold for brook trout (maximum weekly average temperature > 23.3 °C, Wehrly et al., 2007). Therefore, point-scale thermal refugia are not a current concern in this system, as the stream supports a system-scale cold-water habitat that is likely to persist into the future and serve as warming "climate refugia" (Briggs et al., 2018a). In winter, seepage zones can be located as relatively warm anomalies, increasing and buffering surface water temperatures from ambient atmospheric influence.

Previous work has measured relatively large net gains in streamflow over the lower Quashnet River (Barlow and Hess, 1993; Rosenberry et al., 2016), which are attributed to groundwater discharge through direct stream-bed seepage and the harvesting of groundwater from the floodplain platform via relic agricultural drainage ditches. Deployments of fiber-optic temperature sensing (FO-DTS) cables along the thalweg stream-bed interface indicate that the greatest density of focused seepage zones occurs along the broader valley area, approximately 1 km upstream of the U.S. Geological Survey gage (Fig. 1). This zone coincides with the largest gains in net streamflow (Hare et al., 2015). Based on the stream-bed interface temperature data presented by

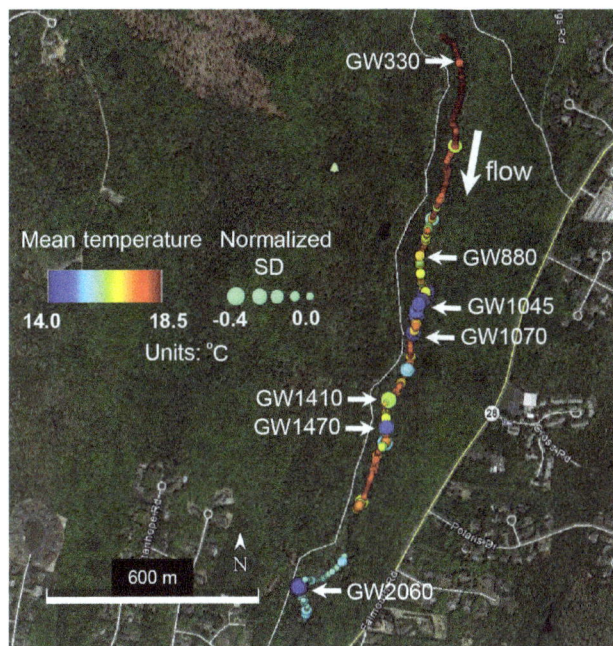

Figure 1. Fiber-optic distributed temperature data collected along the stream channel sediment–water interface over two days in July 2013 are summarized here using mean temperature (color) and temperature standard deviation normalized to known non-seepage locations (size). Locations of reduced mean temperature and the standard deviation of temperature can indicate zones of preferential groundwater upwelling. A subset of these apparent upwelling zones (labeled "GW" followed by the distance from upper reach boundary in meters) with varied thermal statistics was chosen for direct pore-water sampling and quantitative seepage measurements. This figure was modified from Rosenberry et al. (2016).

Rosenberry et al. (2016), Fig. 1 shows how temperature-sensitive fiber optic cables have been used to pinpoint possible groundwater-discharge zones based on an anomalously cold mean temperature and/or reduced thermal variance. A focused evaluation of FO-DTS anomalies with physical seepage meters and vertical temperature profilers confirmed localized, meter-scale seepage zonation along the streambed where discrete colder zones indicated through heat tracing showed approximately 5 times the groundwater-discharge rate of adjacent sandy bed locations only meters away (Rosenberry et al., 2016). The active heating of wrapped FO-DTS cables deployed vertically within an open-valley stream-bed seepage zone indicated the true vertical flow to at least 0.6 m into the bed sediments (Briggs et al., 2016), an expected characteristic of a more regional groundwater discharge (Winter et al., 1998), rather than that of a flow driven by the valley topography local to the river. Hyporheic exchange in the lower Quashnet River system is superimposed on the general upward hydraulic gradient to the stream, therefore being reduced to a thin, shallow hyporheic exchange zone (e.g., < 0.1 m depth) along the thalweg by these com-

peting pressures (Briggs et al., 2014). Vertically compressed hyporheic zones such as these have been simulated for similar stream systems (e.g., Cardenas and Wilson, 2006).

3 Methods

A combination of fish tagging and visual spawning observations, heat tracing, geophysical surveys, and focused pore-water sampling was used to investigate the interplay between the locations of preferential brook trout spawning and the local hydrogeology. For consistency between varied methods and years of data collection, all sample locations are spatially referenced as downstream channel distances from the fish ladder river crossing at the upper end of the study reach (Fig. 2).

3.1 Observations regarding repeat spawning locations

Observations of discrete repeat brook trout spawning locations were made opportunistically as part of an ongoing PIT tagging study of the native reproducing population of the Quashnet River. Large-scale trout movements are continuously monitored in the lower Quashnet River at three stationary fish counting sites (Fig. 2a). However, the spatial resolution of these counting sites, separated by hundreds of meters, is not adequate in studying how brook trout utilize specific decimeter- to meter-scale zones of groundwater discharge. For this finer scale characterization, dropped fish tags have also been located through roving surveys using a handheld portable PIT antenna (Biomark, Inc.), which have been conducted in spring and fall since 2007. The dropout of PIT tags from the fish body is a process that is more likely to happen during spawning behavior in salmonids, so dropped tags were electronically and spatially mapped to reveal discrete zones of repeat spawning. Although these roving surveys do not yield the temporal continuity of the instream counting gates, the clustering of dropped tags can be mapped at the sub-meter scale, presumably directly at trout redds. In addition, spawning brook trout were located visually during annual fall data collection events by Massachusetts Fish and Wildlife Staff, with redd development behavior captured in one seepage feature by an underwater video in 2015 using a GoPro Hero camera (San Mateo, CA). We refer to the three most prominent sites of brook trout spawning within the study reach as Spawn 1 (113 m), Spawn 2 (146 m), and Spawn 3 (2062 m), from upstream to downstream, respectively (Fig. 2).

3.2 Spatial mapping of preferential groundwater discharge

To augment existing stream-bed interface thermal surveys for preferential groundwater discharge (e.g., Rosenberry et al., 2016; Fig. 1) and to investigate the bank dependence of the discharge location, ruggedized fiber-optic cables suit-

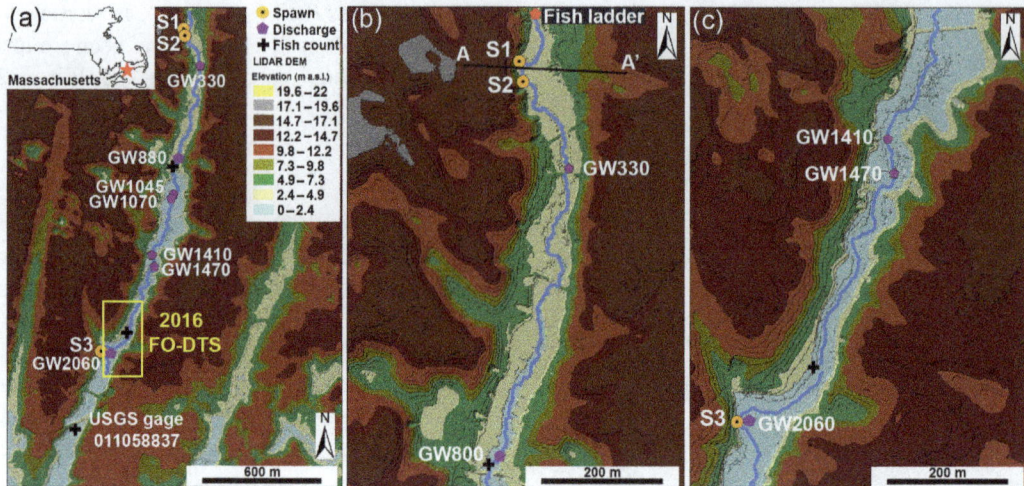

Figure 2. Lidar elevation data show the linear valley terrain of the Quashnet River study reach, as shown in panel **(a)** with Spawn (S1, S2, S3) locations and major open-valley seepage zones identified. The enlarged view of panel **(b)** shows the more narrow upper valley zone where Spawn 1 and 2 are located at the base of a steep cutbank and the topographic transecting point of Fig. 9 (A–A') is noted. Finally, panel **(c)** displays the lower open-valley reach where Spawn 3 is located along a major cutbank.

able for stream use were deployed in the river along the base of each bank from 1700 to 2160 m on 10 to 12 June 2016 (Fig. 2a). Two separate cables weighted with stainless steel armoring were installed directly along the foot of each bank on top of the stream-bed interface. Single-ended measurements made at the 1.01 m linear spatial sampling scale were integrated over 5 min intervals on each channel by an Oryx FO-DTS control unit (Sensornet Ltd.). During the same period, data were also collected along a high-resolution wrapped fiber-optic array for a dataset described in Kurylyk et al. (2017) but not shown here; this experimental setup resulted in measurements for each channel of four instrument channels, which were recorded at 20 min intervals. The calibration for dynamic instrument drift was performed automatically using approximately 30 m of cable for each channel, submerged in a continuously mixed ice bath and monitored with an independent Oryx T-100 thermistor.

3.3 Quantification of vertical groundwater discharge rates

Once preferential discharge locations are located along the stream bed with FO-DTS, actual vertical discharge rates can be assessed using a variety of methodologies (Kalbus et al., 2006). Temporal patterns in the groundwater-discharge flux rate can indicate source flow path hydrodynamics and can be derived from a bed-temperature time series using vertical temperature signal transport characteristics, as reviewed by Rau et al. (2013). Custom "1DTempProfilers" designed specifically for the quantification of groundwater discharge (Briggs et al., 2014) were used to monitor the stream-bed

temperature over time along a shallow vertical profile. Profilers were deployed within a subset of the thermal anomalies previously identified with FO-DTS. The profiler deployment locations were chosen to represent a range of preferential groundwater-discharge rates and characteristics based on the on the observed FO-DTS temperature anomalies, e.g., anomalies of the varied mean temperature and buffering effect (Fig. 1) located at 330, 880, 1045, 1070, 1410, 1470, and 2060 m. These groundwater-discharge locations are referred to with the prefix "GW" followed by the meter mark for the remainder of the paper, such that the major streambed seep 330 m downstream of the fish ladder is referred to as "GW330". Data were collected at various locations from 11 June to 13 July 2014, 21 August to 13 September 2015, and 5 June to 9 July 2016. These deployments included the installation of 1DTempProfilers at the near-bank and channel sides of observed repeat spawning zones.

Individual thermal data loggers (iButton Thermochron DS1922L, Maxim Integrated) were waterproofed with silicone caulk and inserted horizontally into short slotted-steel pipes (0.025 m diameter). The shallow thermal profilers were driven vertically into the stream bed so that sensors were positioned at some combination of 0.01, 0.04, 0.07, and 0.11 m depths. Data were collected at temporal intervals of 0.5 h in 2014, 0.5 h in 2015, and 1 h in 2016. Rosenberry et al. (2016) found that when a subset of the 2014 stream-bed temperature data presented here were analyzed using the diurnal signal amplitude attenuation models employed by VFLUX2 (Irvine et al., 2015), a near 1 : 1 relation was found in comparison to physical seepage meter measurements of groundwater discharge ranging from 0.5 to 3 m d^{-1}. A similar diurnal signal-based stream-bed thermal parameter estimation is used here.

3.4 Stream-bed groundwater discharge and spawning zone pore-water characterization

Subsurface water samples were collected for chemical analysis at seven major open-valley seepage locations and three repeat spawn locations. Geochemical data collection occurred in 2014 and 2016 along with the 1DTempProfiler deployments, while stable water isotope data were collected in August 2017. For geochemical sampling, 0.0095 m (nominal) stainless steel drive points were inserted to depths of 0.3, 0.6, and/or 0.9 m and Masterflex Norprene tubing was attached to the drive point. A peristaltic pump was used to extract pore-water samples until they were free of obvious turbidity (typically requiring 3 min of pumping), after which the pumping rate was slowed and the groundwater samples were collected by pumping into 60 mL high-density polyethylene (HDPE) syringe barrels. First an unfiltered sample for specific conductivity was pushed from the syringe into a 30 mL HDPE Nalgene sample bottle. Second, a filtered sample for anion analysis was collected after attaching a 0.2 μm pore size (25 mm diameter) Pall polyethersulfone filter to the syringe. Lastly, the pumping rate was slowed again and an overflow cup was attached to the Norprene sample tubing and was held upright until it overflowed, at which point the DO was measured by a field colorimetric test using the manufacturer's evacuated reagent vials (Chemetrics V-2000). DO concentrations were read twice and the test was repeated using an alternative vial kit if results were near the concentration range limit or out of range. The collected samples were kept cool and out of the light and analyzed for Cl^- upon return to the laboratory using standard ion chromatographic techniques.

In addition to the drive point samples, pore-water samples were also collected in June 2016 from shallow depths 0.015, 0.04, 0.08 and 0.15 m below the stream-bed surface at locations GW1045 and Spawn 1, 2, 3 using MINIPOINT samplers (e.g., Harvey and Fuller, 1998). Water was pumped simultaneously from all depths using a multi-head pump that withdrew small-volume samples (15 mL) at low flow rates (1.5 mL min^{-1}) to minimize the disturbance of natural subsurface fluxes and chemical gradients. Pumped lines terminated at press-on luer fittings that were pushed onto 0.2 μm pore size (25 mm diameter) Pall polyethersulfone filters. Samples for specific conductivity were collected, whereas filtered samples were collected for anions in pre-labeled 20 mL LDPE plastic scintillation vials with Polyseal™ caps. Sample lines were then attached to overflow cups and dissolved oxygen concentrations were measured as described above.

During a follow-up field effort in August 2017, stream-bed pore-water samples were collected at the Spawn sites and at GW1045, GW1140 (approximately 70 m downstream of GW1070), and GW1470. Additionally, two large hillslope springs were identified along the edge of the riparian zone,

upstream of Spawn 1, using a handheld thermal infrared camera (FLIR T640, FLIR Systems, Inc.). These exposed springs were sampled to identify a localized hillslope groundwater signature that would not be impacted by valley-floor peat deposits. Samples were drawn from push-point piezometers installed 0.2–0.44 m below the sediment interface, with deeper samples collected in the hillslope springs to avoid surface organic material. Pore water was evaluated for specific conductivity (SpC), DO, and stable water isotopes. Isotope samples were analyzed by the U.S. Geological Survey Stable Isotope Laboratory using dual-inlet isotope-ratio mass spectrometry. A substantial fraction of regional Cape Cod shallow groundwater exchanges with the numerous groundwater flow-through lakes as it discharges to the coast (Walter and Masterson, 2002). It is therefore assumed that the regional Cape Cod groundwater isotopic signature is likely to indicate evaporative processes (LeBlanc et al., 2008), offering a contrasting signal from locally recharged hillslope groundwater (no substantial evaporation). The local deuterium excess of contemporary water can indicate groundwater that has been influenced by evaporation in lakes and is therefore in disequilibrium with local meteoric water. Deuterium excess was determined here as $D_{xs} = \delta^2 H - 8 \cdot \delta^{18} O$ (Dansgaard, 1964).

As mentioned previously, historic cranberry farming practices extensively modified the Quashnet River valley, including the incision of drainage ditches into the floodplain. Some ditches extend from the valley wall to the main channel, whereas others are shorter or cut at angles. In addition to characterization of pore water, 34 major drainage ditches (observed flowing water) and a stream thalweg profile were spot-checked for specific conductivity on 16 June 2014 using the SmarTroll probe (YSI). At a subset of these ditch locations, filtered grab samples were collected and analyzed in the laboratory for Cl^- in a similar manner as the mini and drive point samples described above. In June 2016, the dataset was augmented for five ditch confluence locations upstream of Spawn 1.

3.5 Visualizing stream-bed sediment structure

Ground penetrating radars (GPR) have been successfully applied to several surface water and groundwater exchange studies to characterize underlying peat and sandy deposits (e.g., Lowry et al., 2009; Comas et al., 2011) due to strong expected differences in matrix porosity (water content), which can exceed 70 % in peat (Rezanezhad et al., 2016). An upstream to downstream GPR profile was collected on 7 July 2016 using a MALA HDR GX160 shielded antenna (MALA GPR, Sweden), towed down the stream center channel by hand with a small inflatable watercraft. The locations of major seep and spawning sites were specifically marked on the digital GPR record during data collection. The GPR data were processed using Reflexw software (Sandmeier, Germany) to convert reflection time to interface depth.

Figure 3. Several representative images of specific spawn zones and groundwater-discharge zones were collected in February 2016. The cutbank alcove at Spawn 1 is shown in **(a)**, while the open-valley seepage zone GW1045 is shown in **(b)**, and fresh cutbank slumping and visible seepage at Spawn 3 is shown in **(c)**. Underwater imagery collected at the Spawn 1 zone in fall 2015 is displayed in **(d)**, showing several fish clustered directly at the base of the cutbank where pore-water samples were obtained.

4   Results

The hydrogeochemical characterization of observed repeat trout spawning zones and other major stream-bed groundwater-discharge zones are contrasted below.

4.1   Observations regarding repeat spawning locations

Out of the dozens of preferential groundwater-discharge zones geolocated along the Quashnet River in this and previous work (e.g., Fig. 1), brook trout appear to consistently utilize only three discrete stream-bed locations for repeat spawning activity. These locations coincide with steep cutbanks where the river channel approaches the sand and gravel valley wall (Fig. 2b, c). Specifically, trout were found to occupy small "scalloped" alcove-bank features (Fig. 3a) that may be formed by groundwater sapping of fines and the subsequent slumping of sandy bank materials. In winter 2016, fresh slumping and direct seepage from the newly exposed sand wall was observed at Spawn 3 (Fig. 3c); a larger slump event had filled approximately one-third of the scalloped alcove at Spawn 2 by June 2016. Brook trout were observed clustered along the inner bank area at the Spawn 1 location in fall 2015 (Fig. 3d), and this spawning behavior was captured using an underwater video (Supplement).

Dropout PIT tags have been found repeatedly in each of the three preferential spawn zones. Seven dropout PIT tags were located in the Spawn 3 zone in March 2017, by far the most dropped tags found in any one location since the track-ing program began in 2007. The only other obvious scalloped bank features along the 2 km study reach are located at GW1045 (Fig. 3b). Compared to the trout spawning zone alcoves along the valley-wall cutbanks (e.g., Fig. 3a), this open-valley seepage alcove was overgrown with watercress and thick (tens of centimeters), loose deposits of organic material.

4.2   Spatial mapping of preferential groundwater discharge

As shown in Fig. 1, previously collected FO-DTS data were used to guide data collection at a subset of representative preferential stream-bed groundwater discharges. Additionally, paired FO-DTS cables were deployed at the base of both stream banks through a lower reach section in 2016 (Fig. 2c), revealing differing thermal anomaly patterns (Fig. 4; Briggs et al., 2018b). The cable along the downstream-right bank captures a large, 8 m long cooler zone at Spawn 3 (Fig. 4b), and this seepage signature is spatially reduced but visible along the opposing bank (Fig. 4a). Other thermal anomalies observed along one bank show little or no signature along the other. Air temperature dropped noticeably over the final 1.5 days of deployment, and smaller cool anomalies that appeared on warm days were no longer captured by the stream-bed FO-DTS deployment, though the Spawn 3 signature is still visible along both cables.

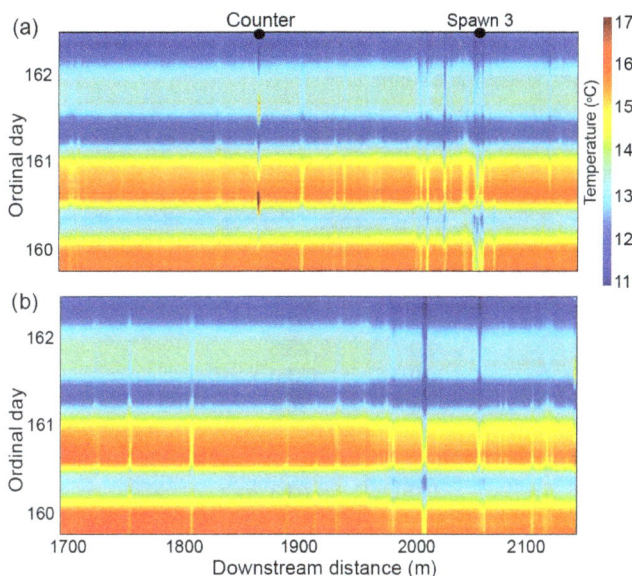

Figure 4. Fiber-optic-distributed temperature data collected from the approximate channel distance of 1700 to 2160 m along **(a)** the downstream right bank through the Spawn 3 meander bend area (see Fig. 2a for location), and **(b)** the downstream left bank along the same stream reach. The persistent vertical bands of relatively cool temperatures indicate discrete groundwater discharge. Some larger zones display a thermal signature on both bank cables, while smaller discharges may be specific to one bank.

4.3 Quantification of vertical groundwater-discharge rates

Ambient stream-bed temperature signal data can be used to measure stream-bed thermal conduction parameters (Luce et al., 2013), which is particularly important when applying heat-based methods to quantify upward vertical fluid flux (Rosenberry et al., 2016), compared to downward fluid-flux models that generally show less sensitivity to stream-bed thermal parameters. Diurnal signal-based thermal diffusivity measurements derived from a pair of 1DTempProfilers inserted in sandy channel sediments for a month in 2014 have the same geometric mean value of $0.11\,\mathrm{m^2\,d^{-1}}$, and this value is used here to model vertical groundwater discharge for all locations and data collection periods (Briggs et al., 2018b). Sub-daily groundwater-discharge fluxes evaluated over similar spring and early summer time periods in 2014 and 2016 show relatively stable patterns at open-valley seepage zones, generally $< 1\,\mathrm{m\,d^{-1}}$ (Fig. 6). At Spawn 1 and 3 seepage is stronger (2 to $3.5\,\mathrm{m\,d^{-1}}$) and more variable than at open-valley zones. The Darcy-based horizontal seepage estimate through the Spawn 3 bank, made using the bank piezometer, is $2.3\,\mathrm{m\,d^{-1}}$, which is similar to the temperature-based seepage rates at the Spawn 3 interface (Fig. 6), and indicates lateral discharge through the cutbank wall from a more localized groundwater flow path. The Spawn 2 zone shows a reduced and more stable discharge

rate during summer 2016, and is likely impacted by a large bank slump into this zone that occurred during the winter of 2016, partially filling the alcove. Seepage patterns collected at Spawn 1 and 2 in late-summer 2015 show greater temporal stability, even though the stream stage at the downstream U.S. Geological Survey gage showed substantial variation. Discharge rates along the inner bank wall of the scalloped bank spawn zones were consistently higher than at bed areas located just a few meters away toward the channel.

4.4 Stream-bed groundwater discharge and spawning zone pore-water characterization

Based on previous characterization, the Cape Cod sand and gravel aquifer generally has high DO concentrations (9–$11\,\mathrm{mg\,L^{-1}}$), relatively dilute specific conductance (SpC, $62\,\mu\mathrm{S\,cm^{-1}}$), and dilute chloride concentrations (Cl$^-$, $9.3\,\mathrm{mg\,L^{-1}}$) at depths ranging between 12 and 20 m (Savoie et al., 2012). The groundwater that discharges to the Quashnet River, however, is often strongly variable in all three of these parameters (Harvey et al., 2018). In June 2014, drive point data were primarily collected in open-valley seepage zones identified with FO-DTS (Fig. 1); these locations are suboxic to anoxic at 0.3 and 0.6 m stream-bed depths (Table 1). The highest stream-bed seepage DO is found at GW330 in the tighter upstream valley section ($4.6\,\mathrm{mg\,L^{-1}}$ at both depths) and Spawn 3, where DO is 9.0 and $7.6\,\mathrm{mg\,L^{-1}}$ at 0.3 and 0.6 m depths, respectively (Table 1). SpC is also variable, but lowest and similar to the regional signal at GW330 and Spawn 3. Note that SpC and Cl$^-$ are used here to indicate aquifer flow path hydrogeochemical properties and not unsuitable spawn habitats based on chemical concentration, as their range is well within general brook trout tolerances.

Drive point data collected at the 0.3 m depth in June 2016, primarily around spawn zones, generally show high DO and relatively low SpC at the interior of Spawn zones 1 and 3 near the cutbank (Table 1). Data collected a few meters toward the main channel from these near-bank spawn locations are reduced in DO with increased SpC. The Spawn 2 data were collected at the toe of the recent large sediment slump that had partially filled the alcove, and DO data are suboxic at 0.3 m ($3.9\,\mathrm{mg\,L^{-1}}$) but more oxygen-rich at 0.9 m depth ($7.2\,\mathrm{mg\,L^{-1}}$), indicating the potential for shallow stream-bed respiration that removes oxygen from discharging groundwater (assuming vertical flow) in the slumped material. In contrast to the spawn zones, the major open-valley seepage location GW1045 is nearly anoxic at all depths with SpC similar to the 2014 stream water profile grab samples ($n = 8$, $101.4 \pm 1.7\,\mu\mathrm{S\,cm^{-1}}$). Little difference was observed between near-bank and channel positions at GW1045 (both are suboxic) even though a large scalloped seepage bank feature was observed (Fig. 3b).

The drainage-ditch grab samples generally show Cl$^-$ concentrations that are lower than the average 2014 channel grab samples ($n = 10$, $19 \pm 0.4\,\mathrm{mg\,L^{-1}}$), though the two most

Table 1. This table lists 2014 and 2016 drive point pore-water chemistry data collected in major stream-bed groundwater-discharge zones located with fiber-optic heat tracing and in zones of observed repeat trout spawning directly along the bank and farther toward the stream center channel. The italicized values indicate sample depths that differ from others in the same column.

Open valley groundwater discharges	0.3 m depth		0.6 m depth	
	DO $mg\,L^{-1}$	SpC $\mu S\,cm^{-1}$	DO $mg\,L^{-1}$	SpC $\mu S\,cm^{-1}$
GW330	4.6	53.8	4.6	61.3
GW880	1.4	97.7	3.4	65.1
GW1045	0.1	78.8	0.0	82.5
GW1045 (bank)	0.16	105.5	0.39	104.0
GW1045 (channel)	0.31	99.1	0.18	96.4
GW1070	0.2	100.0	0.2	89.8
GW1410	0.0	77.7	0.0	79.0
GW1470	0.1	69.1	0.0	64.3
GW2060	1.4	75.0	0.5	79.4
mean	0.9	84.1	1.0	80.2
Spawning locations (channel)	**0.3 m depth**		**0.9 m depth**	
Spawn 1 channel	4.41	143.9	5.68	143.2
Spawn 2 channel	5.25	139.3	n/a	n/a
Spawn 3 channel	1.76	82.1	2.68	79.9
mean	3.8	121.8	4.2	111.6
Spawning locations (bank)	**0.3 m depth**		**0.9 m depth**	
Spawn 1 bank	7.28	70.6	9.76	55.9
Spawn 2 bank	3.89	70.8	7.17	57.6
Spawn 3 bank (2016)	9.11	60.4	4.91	71.9
Spawn 3 bank (2014)	9.0	56.4	7.6 *(0.6 m)*	60.9 *(0.6 m)*
mean	7.3	64.6	7.4	61.6

n/a: not applicable.

upstream ditches are similar to stream water, and 2 open-valley ditches are appreciably higher in Cl^- (Fig. 7a). Spawn zones 1, 2, and 3 approximate the lowest Cl^- concentrations observed in drainage ditches, and Spawn 3 has a similar concentration to the adjacent 2016 stream-bank piezometer in both the 2014 and 2016 data. An analogous pattern is shown in the more widespread SpC data, with many drainage ditches and all spawn zones having concentrations around $60\,\mu S\,cm^{-1}$. However, several ditches cluster around the stream water average or higher, particularly in the open-valley area.

The shallow, shallow pore-water samples collected with the MINIPOINT system in discrete intervals show that stream-bed SpC is appreciably lower than stream water, even at the 0.02 m depth, at all near-bank spawn zones (Fig. 8a). Conversely, the shallow channel sediments at Spawn 1 and open-valley seepage at GW1045 approximate the stream water value for SpC. DO is high and stable along the shallow profiles (to 0.14 m) at the interior of Spawn zones 1 and 3 but suboxic at the Spawn 1 channel sample and Spawn 2 zones and essentially anoxic along the bank at GW1045. Center channel pore-water samples at GW1045 show moderate oxy-

gen enrichment at 0.02 m ($4.6\,mg\,L^{-1}$), which may result from hyporheic mixing, as deeper intervals along the same profile are nearly anoxic.

The underwater video collected here in the fall of 2015 indicates Quashnet River brook trout clustered tightly around an approximate $1\,m^2$ bed area in Spawn 1 (Fig. 3d, Supplement), directly at the base of the sandy cutbank. During the June 2016 collection of pore-water data, drive points were installed precisely in this area. A chemical analysis of 0.3 m deep pore water shows a strong gradient from the near-bank Spawn 1 zone to the outer alcove area, with specific conductance rising dramatically (70.6 to $143.9\,\mu S\,cm^{-1}$) and DO falling (7.28 to $4.41\,mg\,L^{-1}$) (Table 1). Spawn 3 shows a similar pattern from the near-bank zone toward the main channel (60.4 to $82.1\,\mu S\,cm^{-1}$ SpC; 9.11 to $1.76\,mg\,L^{-1}$ DO). Spawn 2, although complicated by the large slump during the previous winter, shows an increase in SpC from 70.6 to $139.3\,\mu S\,cm^{-1}$ from the inner to outer alcove. Conversely, pore water collected at 0.3, 0.6, and 0.9 m depths in the open-valley seepage alcove at GW1045 (pictured in Fig. 3b) are functionally anoxic with elevated SpC

Table 2. This table lists 2017 drive point pore-water chemistry and stable water isotope data collected in a subset of major stream-bed groundwater-seepage zones, zones of observed repeat trout spawning, and from springs located above the waterline along the same hillslope as the meander cutbanks of Spawn 1 and Spawn 2.

Location	Sample depth (m)	SpC ($\mu S\,cm^{-1}$)	DO ($mg\,L^{-1}$)	$\delta^2 H$ (‰)	$\delta^{18}O$ (‰)	D_{xs} $\delta^2 H - 8 \cdot \delta^{18}O$
Hillslope 1	40	74.82	5.004	−51.38	−8.2	14.22
Hillslope 2	44	60.59	9.318	−51.81	−8.73	18.03
Spawn 1	20	72.45	6.853	−48.9	−7.9	14.3
Spawn 2	20	51.75	5.419	−48.2	−7.95	15.4
Spawn 3	20	42.62	9.054	−44.32	−7.33	14.32
GW 1045	20	109.8	0.043	−34.03	−4.93	5.41
GW 1140	20	103.4	0.043	−32.56	−4.8	5.84
GW 1470	20	97.68	0.04	−33.05	−4.72	4.71

compared to inner spawn zones and have little gradient from the bank to the channel.

Pore-water data collected in August 2017 indicate that all three Spawn sites are similar to emergent hillslope springs, characterized by relatively high DO and low SpC, compared to major open-valley stream-bed seepage zones that are anoxic with higher SpC (Table 2). Additionally, the stable isotopic signatures of the hillslope and Spawn zones are similar, but are contrasted by the lower deuterium excess metric determined for the open-valley seepages. This indicates that groundwater discharging through the stream bed away from the hillslope shows the evaporative signature of groundwater flow-through lakes and can therefore be considered regional discharge, compared to locally recharged hillslope groundwater apparently favored by trout for spawning.

4.5 Visualizing stream-bed sediment structure

Radar data were collected over most of the study reach length depicted in Fig. 2a, and although spatial reference data were not collected for each sample point due to integrated global positioning system failure, Spawn and groundwater-discharge zones of interest were precisely marked in the record (Fig. 5). The GPR data collected along the thalweg adjacent to Spawn 1 and 2 indicate that a contiguous thin layer of material underlies the sandy stream bed that may be peat deposited over deeper sands and gravels (Fig. 5a). The GPR profile through open-valley groundwater-discharge locations GW1045 and GW1070 shows the strongest radar signal reflectors of anywhere along the open-valley section (Fig. 5b). These discontinuous geologic structures are interpreted as layered sand and gravel, interspersed with thicker peat deposits. Otherwise, discontinuous reflections indicative of sediment-type interfaces of variable depths are observed near the downstream open-valley seepage zones and strongly attenuated GPR signals indicate thick lenses of buried peat with high water content (Fig. 5b,c).

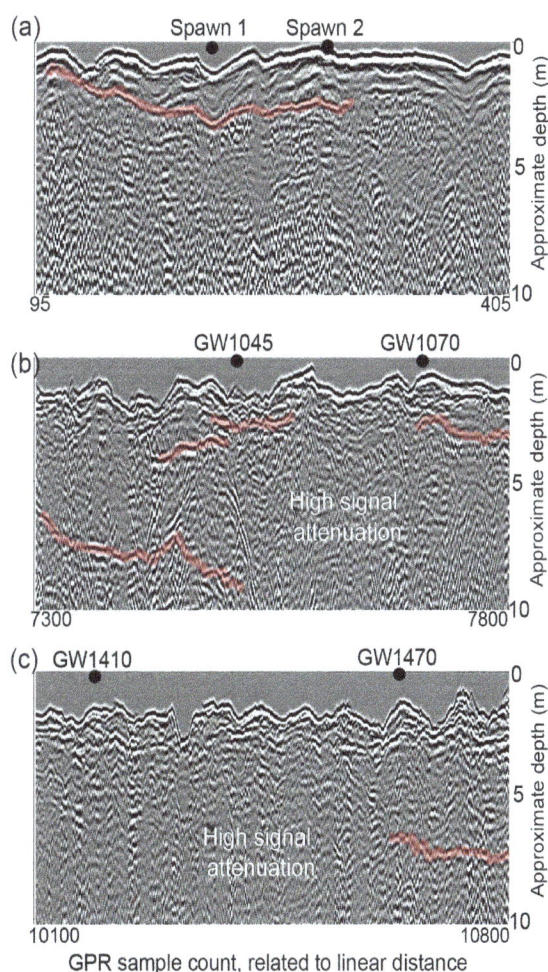

Figure 5. These images show ground-penetrating radar profiles collected down the center of the river channel to indicate peat, sand, and gravel layering in the stream bed. Stronger apparent radar reflectors are highlighted in red and likely indicate sediment layer boundaries (e.g., sand and gravel vs. peat). Spawn- and groundwater-discharge locations were directly marked in the radar data stream during collection and are shown for each sub-reach panel.

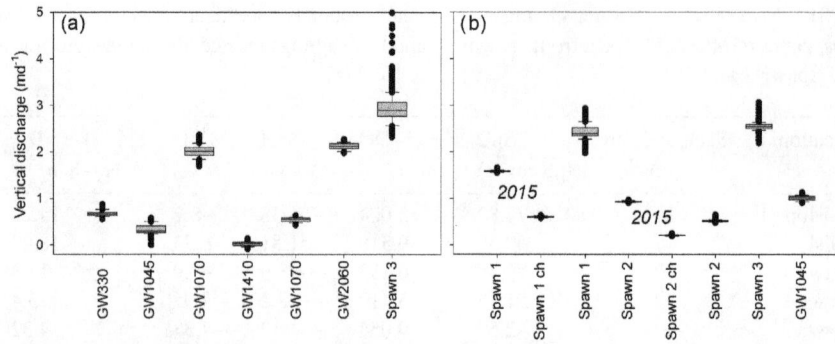

Figure 6. Summarizing box plots of sub-daily vertical groundwater-discharge rates modeled for the open-valley groundwater discharge and Spawn 3 bank locations for the 11 June to 13 July 2014 period are shown in panel (**a**). Additionally, panel (**b**) displays discharge rates collected in Spawn and GW1045 locations directly against the cutbanks and farther out towards the channel (indicated by "ch") for the 21 August to 13 September 2015 and 5 June to 9 July 2016 periods.

5 Discussion

Heat tracing reconnaissance technologies, such as FO-DTS and thermal infrared, offer an efficient means to comprehensively characterize preferential groundwater-discharge points at the reach to watershed scale (Briggs and Hare, 2018). Using the groundwater-fed Quashnet River as an example, Rosenberry et al. (2016) showed that cold stream-bed interface anomalies in summer indeed correspond to discrete zones of particularly high groundwater discharge through stream-bed sediments. This spatial characterization of discharge points alone is not sufficient to understand the physical and chemical drivers of a niche habitat, but can efficiently guide additional data collection, as was done here. Compared to more randomly distributed stream-bed field parameter surveys and larger spatial scale evaluations of net groundwater discharge made with differential gaging, the comprehensive spatial mapping of groundwater discharge using heat is a great advance in the context of understanding groundwater-dependent ecosystems. However, in fast flowing streams, FO-DTS cable placement on the stream bed will likely impact which specific groundwater-discharge zones are captured with FO-DTS, as shown here by applying cables along opposite banks through the Spawn 3 area (Fig. 4). The largest seepage zones may have a spatial footprint that encompasses the stream-bed area from bank to bank (e.g., the Spawn 3 cold anomaly), but a subset of more discrete seepage zones are bound to be missed with a single linear cable deployment. We did not capture Spawn zones 1 and 2 in early FO-DTS field efforts (Fig. 1), but fish tracking indicated their importance in regards to trout spawning behavior. Therefore, in studies of niche stream habitats as influenced by preferential groundwater discharge, a combination of heat tracing and biological observation may be needed to both identify major discharge points and discern which points are directly used by the biota of interest (e.g., brook trout).

In a study of the regional Cape Cod aquifer condition, Frimpter and Gay (1979) state that groundwater is typically near DO saturation, except in the case of the downgradient of peat or river bottom sediments, where consumption of DO allows the mobilization of natural iron and manganese. Visible observations along the open-valley section, in addition to stream-bed sediment coring (Briggs et al., 2014), revealed the widespread coating of shallow stream-bed sediment grains with metal oxides, consistent with the conceptual model of organic material influence on near-surface groundwater (Fig. 9). Aquifer recharge passing through upgradient groundwater flow-through kettle lakes (e.g., Stoliker et al., 2016) may also serve to decrease the DO content of the regional flow paths that discharge vertically through the bed of the Quashnet River, although we hypothesize that localized peat deposits may be the primary control on both seepage zone distribution and chemistry.

Out of the dozens of preferential groundwater-discharge zones located along the lower Quashnet with heat tracing, most were suboxic to anoxic (Table 1). Brook trout consistently prefer three areas for fall spawning, all along meander bend cutbanks into the sand and gravel valley wall. Zones of locally enhanced seepage, likely controlled by subtle differences in sediment hydraulic conductivity, can lead to the groundwater sapping of fines, reduction in bank stability, and consequent slumping of bank material into the river; this process was observed in real time at the Spawn 3 meander in February 2016 (Fig. 3c). Slumping effectively forms *seepage-driven* alcoves outside of the main flow and are more suitable for redd placement, along with forming a more favorable coarse sand and gravel substrate (Bowerman et al., 2014; Hausle and Coble, 1976; Raleigh, 1982).

In other systems, trout have been observed to occupy microhabitat around and within groundwater-discharge zones, even being segregated by fish size and desirable temperature range (e.g., Fig. 2.4.1.2 in Torgersen et al., 2012). Here, real-time observation and visual imagery show trout clustering tightly against the bank in Spawn 3 (Fig. 3d, Supplement) where pore water was found to be more oxygen rich and lower in SpC. The month-long time series of vertical

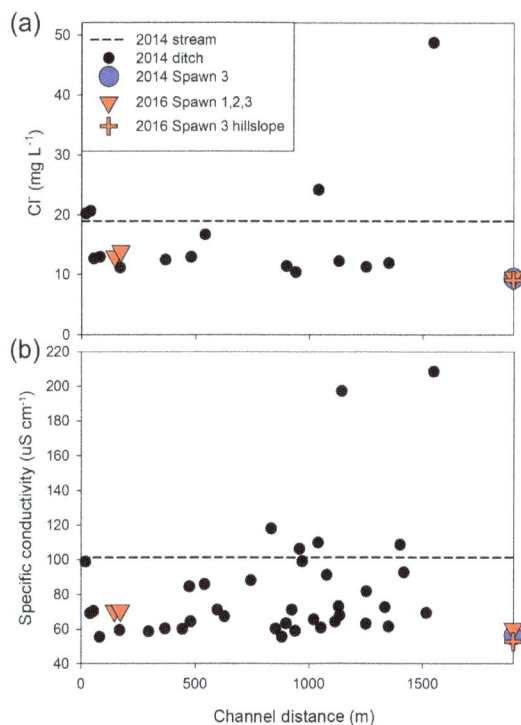

Figure 7. Drainage ditch chemistry throughout the lower Quashnet, showing **(a)** Cl$^-$ and **(b)** specific conductance that was collected in June 2014, just above the confluence with the main channel. Data are plotted as the distance from the upper flood control structure in the narrow valley reach and are compared to groundwater-seepage data collected in preferential spawning locations and a hillslope piezometer.

groundwater-discharge rates are reduced considerably from the near-bank to the near-channel areas at all spawning zones (Fig. 6), indicating in part a reduction in stream-bed hydraulic conductivity as influenced by peat deposits under the main channel and as observed in GPR data (Fig. 5). The evidence of higher near-bank vertical groundwater flux rates and DO combined with lower SpC indicates limited interaction between the shallow groundwater flow paths and peat against the meander bend cutbanks. As observed in other systems, it appears that even short travel distances through organic deposits toward the center channel at Spawn 1 and 2 may be sufficient in increasing total dissolved solids, depleting DO (e.g., Levy et al., 2016), and rendering upwelling zones undesirable for redd construction. Therefore, near-surface channel sediments may need to be specifically characterized in preferential groundwater-discharge zones, as net chemical reactivity over the last ~ 1 m of transport may dominate net chemical change of the discharging groundwater.

The alcove seepage features utilized by trout in this study are apparently similar to the numerous cold-water alcove patches observed in another stream system by Ebersole et al. (2003). In that study of preferential salmonid habitats, alcoves were often located where streams converged on valley walls and were the most abundant type of discrete cold-water habitat type identified. Conversely, valley-wall alcoves were the least common type of seep morphology observed along the Quashnet River. It is likely that the artificial reduction in channel sinuosity along the Quashnet River by farming practices has reduced the number of natural higher-quality spawning locations.

Other bank and alcove features with strong groundwater discharge found along the open-valley section (Fig. 3b) were highly influenced by organic material deposition and did not apparently support spawning habitats. Our research indicates that in lowland systems with organic-rich floodplain sediments, valley-wall alcoves alone create a favorable brook trout spawning habitat via local mineral soildominated groundwater-discharge flow paths, as shown in conceptual Fig. 9. This finding might help inform future ecologically based stream restoration practices in using the natural landscape to predict desirable preferential groundwaterdischarge points, as was recently done by Hare et al. (2017) to inform the engineering of a large-scale cranberry bog restoration.

The pore-water SpC, Cl$^-$, and DO data alone do not definitively show that seepage at the cutbank spawn sites is derived from more localized groundwater recharge, as opposed to regional groundwater that is unadulterated by buried peat lenses. However, the hydrodynamic data derived from long-term vertical temperature profiling in seepage zones does offer additional insight. In general, groundwaterdischarge rates are more variable at cutbank spawn zones than in the open-valley stream-bed zones (Fig. 6), and this variability may be tied to shorter-term changes in local river stage and/or water table depth, impacting the local hydraulic gradient. The relatively stable patterns of openvalley groundwater discharge may be controlled by the regional gradient, where the flow path length term dominates the Darcy relation and is therefore relatively insensitive to local changes in river stage and water table fluctuations. Furthermore, the stable water isotope data display evaporative signatures at the open-valley stream-bed discharge sites, indicating regional groundwater that has passed through one or more upgradient flow-through lakes (Table 2). In contrast, the Spawn sites all show isotope signals that fall along the local meteoric waterline and therefore likely represent recharge to the hillslopes more local to the river. These localized groundwater flow systems would be expected to be less influenced by regional groundwater contamination, which is widespread in the regional Cape Cod aquifer (Walter and Masterson, 2002).

Groundwater drainage-ditch data collected along the river corridor indicate that low SpC/Cl$^-$ conditions exist for the majority of ditches throughout the lower Quashnet River riparian areas (Fig. 7). The hillslope piezometer in sand and gravel at the down valley wall has a similar chemical signature along with high DO. This similarity further indicates that low-SpC groundwater discharges even to the lower por-

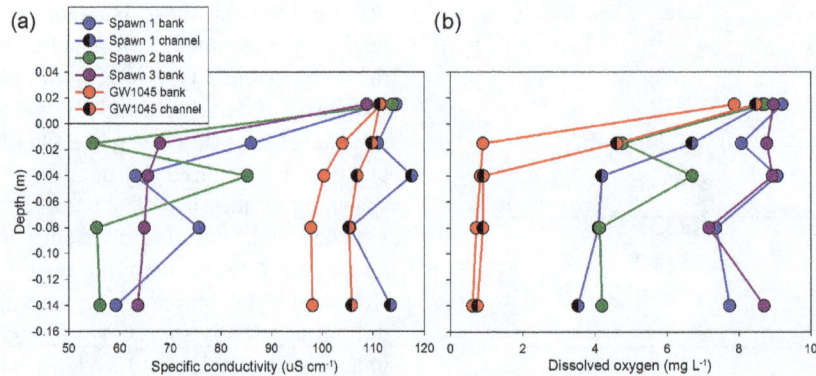

Figure 8. Minipoint pore-water chemistry data showing high spatial resolution profiles of (**a**) specific conductance and (**b**) dissolved oxygen, collected in June 2016 at the major seepage alcoves. Triangle symbols indicate data collected farther toward the thalweg from the respective alcove bank, and all profiles include a local stream water sample taken just above the stream-bed interface.

tion of the river corridor but is chemically modified by travel through near-stream organics. The relic drainage ditches allow the discharging groundwater to effectively short-circuit the valley floor peat deposits and remain high in DO, similar to the natural valley-wall springs and cutbank alcoves. Future restoration strategies that seek to actively enhance groundwater discharge (e.g., Kurylyk et al., 2015a) may consider capitalizing on this short circuit behavior, possibly by auguring through buried stream-bed peat or through the movement of the stream channel toward the valley wall to create more desirable brook trout aquatic habitat.

6   Conclusions

The three repeatedly utilized discrete spawning zone locations that have been identified for over a decade of observation have coupled strongly discharging groundwater with high DO concentration. A conceptual diagram of the hydrogeochemical setting of spawn zones vs. other non-favorable stream-bed locations of groundwater discharge is shown in Fig. 9. Spawn zones are located exclusively in side alcoves of the channel created by bank slumps along meanders, where the river cuts into steep hillslopes along the glacial sands and gravel valley wall. In the alcoves at the base of the cutbanks, hillslope groundwater with high DO concentrations is discharged through the stream bed without appreciable loss of oxygen. Just a few meters away toward the main channel, however, groundwater consistently discharges at lower rates, reduces in DO, and increases in SpC. The lowest oxygen concentrations in groundwater are associated with water emerging from the stream bed adjacent to the wide riparian areas that flank the Quashnet in the open-valley section of the study reach, even though groundwater-discharge rates were also relatively high. In the open valley, where the stream is not near the valley walls, proximity to the stream bank does not seem to control seepage chemistry, and GPR data indicated thick zones of discontinuous stream-bed peat. In this

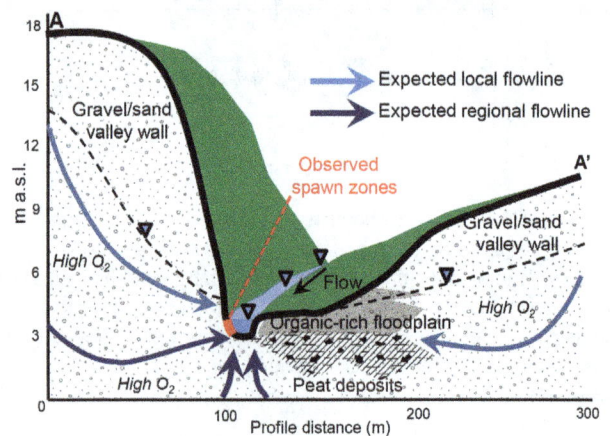

Figure 9. This conceptual model shows how valley-wall cutbank discharge zones are likely sourced by locally recharged hillslope groundwater that avoids substantial interaction with valley-floor organic material. The discharging groundwater remains oxygen-rich, therefore supporting trout spawning activity along discrete streambed sections at the meter scale. The topographic profile shown here (A–A') is derived from airborne lidar data and is oriented perpendicular to the stream at the Spawn 1 zone, as geolocated in Fig. 2b.

and other groundwater-dominated streams that are expected to serve as climate refugia for future native trout populations, hyporheic exchange will be limited by a strong upward hydraulic gradient. Therefore, preferential spawning habitat in such lowland valley systems may be primarily supported by discrete zones of oxic groundwater upwelling at the meter to sub-meter scale, as has been indicated by previous work (e.g., Curry et al., 1995).

In systems where all groundwater discharge is universally anoxic, preferential salmonid spawning zonation may be controlled by points of downwelling hyporheic water where shallow sediments remain high in DO (Buffington and Tonina, 2009; Cardenas et al., 2016). However, these hyporheic areas will deliver cold surface water to shallow sediments during winter, which may impair the overwintering of brook trout eggs (French et al., 2017). Here and in many other coastal systems, groundwater temperature is expected to range from approximately 10–12 °C, which is an ideal range for brook trout egg development (Raleigh, 1982). Points of oxic groundwater upwelling devoid of near-stream buried organics, combined with a recirculating side alcove and favorable sand and gravel sediments, may provide an ideal and unique and preferential spawning habitat for native trout.

Stream surface or stream-bed interface heat tracing of groundwater discharge offers an efficient means to locate discrete seepage zones but offers only limited insight into source groundwater flow path hydraulics and geochemistry. A combined toolkit that also includes spatially informed (using heat tracing) geochemical and isotope sampling and geophysical imaging can be used to trace groundwater flow paths back into the source aquifer, and develop a robust hydrogeochemical characterization. Additionally, as digital elevation models become more refined and combined with infrared data derived from unmanned aerial systems, the remote identification of relatively small features such as the seepage alcoves described here should be possible. A comprehensive and process-based characterization of a niche stream habitat can be used to guide a stream ecological restoration design that directly incorporates the local preferential groundwater-discharge template.

Author contributions. All authors contributed to the analysis of field data and the development of this paper.

Competing interests. The authors declare that they have no conflict of interest.

Acknowledgements. Comments from anonymous reviewers and U.S. Geological Survey (USGS) reviews by Nathaniel Hitt and Paul Barlow are gratefully acknowledged. The U.S. Environmental Protection Agency (USEPA) through its Office of Research and Development partially funded and collaborated in the research described here under agreement number DW-14-92381701 to the USGS. The USGS authors were supported by the following USGS entities: the Office of Groundwater, Water Availability and Use Science Program, National Water Quality Program, and the Toxic Substances Hydrology Program. Field and laboratory assistance from Allison Swartz, Jay Choi, Jenny Lewis, Yao Du, Danielle Hare, Courtney Scruggs, Rayna Mitzman, David Rey, Geoff Delin, Eric White, MassWildlife Southeast District Staff, Jennifer Salas, and volunteers from Trout Unlimited is greatly appreciated. The paper has been subjected to Agency review and approved for publication. The views expressed in this article are those of the authors and do not necessarily represent the views or policies of the USEPA. Any use of trade, firm, or product names is for descriptive purposes only and does not imply endorsement by the U.S. government.

Edited by: Alberto Guadagnini

References

Annett, B., Gerlach, G., King, T. L., and Whiteley, A. R.: Conservation Genetics of Remnant Coastal Brook Trout Populations at the Southern Limit of Their Distribution: Population Structure and Effects of Stocking, T. Am. Fish. Soc., 141, 1399–1410, https://doi.org/10.1080/00028487.2012.694831, 2012.

Back, W., Baedecker, M. J., and Wood, W. W.: Scales in chemical hydrogeology: a historical perspective, in: Regional Ground-Water Quality, edited by: Alley, W. M., Van Nostrand Reinhold, New York, 111–128, 1993.

Baird, O. E. and Krueger, C. C.: Behavioral thermoregulation of brook and rainbow trout: comparison of summer habitat use in an Adirondack River, New York, T. Am. Fish. Soc., 132, 1194–1206, 2003.

Barlow, P. M. and Hess, K. M.: Simulated Hydrologic Responses of the Quashnet River Stream-Aquifer System to Proposed Ground-Water Withdrawals, Cape Cod, Massachusetts, USGS, Water-Resources Investigations Report, 93-4064, 51 pp., 1993.

Bowerman, T., Neilson, B. T., and Budy, P.: Effects of fine sediment, hyporheic flow, and spawning site characteristics on survival and development of bull trout embryos, Can. J. Fish. Aquat. Sci., 71, 1059–1071, 2014.

Briggs, M. A. and Hare, D. K.: Explicit consideration of preferential groundwater discharges as surface water ecosystem control points, Hydrol. Process., 2, 2435–2440, https://doi.org/10.1002/hyp.13178, 2018.

Briggs, M. A., Lautz, L. K., Buckley, S. F., and Lane, J. W.: Practical limitations on the use of diurnal temperature signals to quantify groundwater upwelling, J. Hydrol., 519, 1739–1751, https://doi.org/10.1016/j.jhydrol.2014.09.030, 2014.

Briggs, M. A., Buckley, S. F., Bagtzoglou, A. C., Werkema, D., and Lane, J. W.: Actively heated high-resolution fiber-optic distributed temperature sensing to quantify flow dynamics in zones of strong groundwater upwelling, Water Resour. Res., 52, 5179–5194, https://doi.org/10.1002/2015WR018219, 2016.

Briggs, M. A., Johnson, Z. C., Snyder, C. D., Hitt, N. P., Kurylyk, B. L., Lautz, L., Irvine, D. J., Hurley, S. T., and Lane, J. W.: Inferring watershed hydraulics and cold-water habitat persistence using multi-year air and stream temperature signals, Sci. Total Environ., 636, 1117–1127, https://doi.org/10.1016/j.scitotenv.2018.04.344, 2018a.

Briggs, M. A., Scruggs, C. R., Hurley, S. T., and White, E. A.: Temperature and geophysical data collected along the Quashnet River, Mashpee/Falmouth MA, U.S. Geological Survey data release, https://doi.org/10.5066/F7PN93QF, 2018b.

Buffington, J. M. and Tonina, D.: A three-dimensional model for analyzing the effects of salmon redds on hyporheic exchange and egg pocket habitat A three-dimensional model for analyzing the effects of salmon redds on hyporheic exchange and egg pocket habitat, Can. J. Fish. Aquat. Sci., 66, 2157–2173, https://doi.org/10.1139/F09-146, 2009.

Burns, E. R., Zhu, Y., Zhan, H., Manga, M., Williams, C. F., Ingebritsen, S. E., and Dunham, J.: Thermal effect of climate change on groundwater-fed ecosystems, Water Resour. Res., 53, 3341–3351, https://doi.org/10.1002/2016WR020007, 2017.

Cardenas, M. B. and Wilson, J. L.: The influence of ambient groundwater discharge on exchange zones induced by current-bedform interactions, J. Hydrol., 331, 103–109, 2006.

Cardenas, M. B., Ford, A. E., Kaufman, M. H., Kessler, A. J., and Cook, P. L. M.: Hyporheic flow and dissolved oxygen distribution in fish nests: the effects of open channel velocity, permeability patterns, and groundwater upwelling, J. Geophys. Res.-Biogeosci., 121, 3113–3130, https://doi.org/10.1002/2016JG003381, 2016.

Comas, X., Slater, L., and Reeve, A. S.: Pool patterning in a northern peatland: Geophysical evidence for the role of postglacial landforms, J. Hydrol., 399, 173–184, https://doi.org/10.1016/j.jhydrol.2010.12.031, 2011.

Crisp, D. T.: A desk study of the relationship between temperature and hatching time for the eggs of five species of salmonid species, Freshwater Biol., 11, 361–368, https://doi.org/10.1111/j.1365-2427.1981.tb01267.x, 1981.

Cunjak, R. A. and Power, G.: Seasonal changes in the physiology of brook trout, Salvelinus fontinalis (Mitchill), in a sub-Arctic river system, J. Fish Biol., 29, 279–288, 1986.

Curry, R., Noakes, D. L. G., and Morgan, G. E.: Groundwater and the incubation and emergence of brook trout (Salvelinus fontinalis), Can. J. Fish. Aquat. Sci., 52, 1741–1749, 1995.

Dansgaard, W.: Stable isotopes in precipitation, Tellus, 16, 436–468, https://doi.org/10.3402/tellusa.v16i4.8993, 1964.

Dugdale, S. J.: A practitioner's guide to thermal infrared remote sensing of rivers and streams: recent advances, precautions and considerations, WIREs Water, 3, 251–268, https://doi.org/10.1002/wat2.1135, 2016.

Dugdale, S. J., Bergeron, N. E., and St-Hilaire, A.: Spatial distribution of thermal refuges analysed in relation to riverscape hydromorphology using airborne thermal infrared imagery, Remote Sens. Environ., 160, 43–55, https://doi.org/10.1016/j.rse.2014.12.021, 2015.

Ebersole, J. L., Liss, W. J., and Frissell, C. A.: Cold water patches in warm streams: physicochemical characteristics and the influence of shading, J. Am. Water Resour. As., 59860, 355–368, 2003.

French, W. E., Vondracek, B., Ferrington, L. C., Finlay, J. C., and Dieterman, D. J.: Brown trout (Salmo trutta) growth and condition along a winter thermal gradient in temperate streams, Can. J. Fish. Aquat. Sci., 74, 56–64, https://doi.org/10.1139/cjfas-2016-0005, 2017.

Frimpter, M. H. and Gay, F. B.: Chemical quality of ground water on Cape Cod, Massachusetts, U.S. Geological Survey, Water-Resources Investigations Report 79-65, 1979.

Geist, D. R., Hanrahan, T. P., Arntzen, E. V, Mcmichael, G. A., Murray, C. J., and Chien, Y.: Physicochemical Characteristics of the Hyporheic Zone Affect Redd Site Selection by Chum Salmon and Fall Chinook Salmon in the Columbia River, N. Am. J. Fish. Manage., 22, 1077–1085, 2002.

Hare, D. K., Briggs, M. A., Rosenberry, D. O., Boutt, D. F., and Lane, J. W.: A comparison of thermal infrared to fiber-optic distributed temperature sensing for evaluation of groundwater discharge to surface water, J. Hydrol., 530, 153–166, https://doi.org/10.1016/j.jhydrol.2015.09.059, 2015.

Hare, D. K., Boutt, D. F., Clement, W. P., Hatch, C. E., Davenport, G., and Hackman, A.: Hydrogeological controls on spatial patterns of groundwater discharge in peatlands, Hydrol. Earth Syst. Sci., 21, 6031–6048, https://doi.org/10.5194/hess-21-6031-2017, 2017.

Harvey, J. W. and Fuller, C. C.: Effect of enhanced manganese oxidation in the hyporheic zone on basin-scale geochemical mass balance, Water Resour. Res., 34, 623–636, 1998.

Harvey, J. W., Böhlke, J. K., Voytek, M. A., Scott, D., and Tobias, C. R.: Hyporheic zone denitrification: Controls on effective reaction depth and contribution to whole-stream mass balance, Water Resour. Res., 49, 6298–6316, https://doi.org/10.1002/wrcr.20492, 2013.

Harvey, J. W., Briggs, M. A., Buskirk, B., Swartz, A., Lewis, J., and Du, Y.: Surface water and groundwater water chemistry data collected along the Quashnet River, Mashpee/Falmouth, MA, U.S. Geological Survey data release, https://doi.org/10.5066/F7M044MF, 2018.

Hausle, D. A. and Coble, D. W.: Influence of sand in redds on survival and emergence of brook trout (Salvelinus fontinalis), T. Am. Fish. Soc., 105, 57–63, 1976.

Hitt, N. P., Snook, E. L., and Massie, D. L.: Brook trout use of thermal refugia and foraging habitat influenced by brown trout, Can. J. Fish. Aquat. Sci., 74, 406–418, https://doi.org/10.1139/cjfas-2016-0255, 2017.

Hudy, M., Thieling, T. M., Gillespie, N., and Smith, E. P.: Distribution, status, and land use characteristics of subwatersheds within the native range of brook trout in the Eastern United States, N. Am. J. Fish. Manage., 28, 1069–1085, 2008.

Irvine, D. J., Lautz, L. K., Briggs, M. A., Gordon, R. P., and Mckenzie, J. M.: Experimental evaluation of the applicability of phase, amplitude, and combined methods to determine water flux and thermal diffusivity from temperature time series using VFLUX 2, J. Hydrol., 531, 728–737, 2015.

Isaak, D. J., Young, M. K., Nagel, D. E., Horan, D. L., and Groce, M. C.: The cold-water climate shield: Delineating refugia for preserving salmonid fishes through the 21st century, Glob. Change Biol., 21, 2540–2553, https://doi.org/10.1111/gcb.12879, 2015.

Kalbus, E., Reinstorf, F., and Schirmer, M.: Measuring methods for groundwater – surface water interactions: a review, Hydrol. Earth Syst. Sci., 10, 873–887, https://doi.org/10.5194/hess-10-873-2006, 2006.

Krause, S., Tecklenburg, C., Munz, M., and Naden, E.: Streambed nitrogen cycling beyond the hyporheic zone: Flow controls on horizontal patterns and depth distribution of nitrate and dissolved oxygen in the upwelling groundwater of a lowland river, J. Geophys. Res.-Biogeosci., 118, 54–67, https://doi.org/10.1029/2012JG002122, 2013.

Kurylyk, B. L., Macquarrie, K. T. B., Linnansaari, T., Cunjak, R. A., and Curry, R. A.: Preserving, augmenting, and creating cold-water thermal refugia in rivers: concepts derived from research on the Miramichi River, New Brunswick (Canada), Ecohydrology, 8, 1095–1108, https://doi.org/10.1002/eco.1566, 2015a.

Kurylyk, B. L., MacQuarrie, K. T. B., Caissie, D., and McKenzie, J. M.: Shallow groundwater thermal sensitivity to climate change and land cover disturbances: derivation of analytical expressions and implications for stream temperature modeling, Hydrol. Earth Syst. Sci., 19, 2469–2489, https://doi.org/10.5194/hess-19-2469-2015, 2015b.

Kurylyk, B. L., Irvine, D. J., Carrey, S., Briggs, M. A., Werkema, D., and Bonham, M.: Heat as a hydrologic tracer in shallow and deep heterogeneous media: analytical solution, spreadsheet tool, and field applications, Hydrol. Process., 31, 2648–2661, https://doi.org/10.1002/hyp.11216, 2017.

LeBlanc, B. D. R., Massey, A. J., Cochrane, J. J., King, J. H., Smith, K. P., and Survey, U. S. G.: Distribution and Migration of Ordnance-Related Compounds and Oxygen and Hydrogen Stable Isotopes in Ground Water near Snake Pond, Sandwich, Massachusetts, 2001–2006, U.S. Geological Survey Scientific Investigations Report 2008-5052, 19 pp., 2008.

LeBlanc, D. R., Guswa, J. H., Frimpter, M. H., and Londquist, C. J.: Ground-water resources of Cape Cod, Massachusetts, U.S. Geological Survey, Hydrologic Atlas, 692, 4 sheets, 1986.

Levy, Z. F., Siegel, D. I., Glaser, P. H., Samson, S. D., and Dasgupta, S. S.: Peat porewaters have contrasting geochemical fingerprints for groundwater recharge and discharge due to matrix diffusion in a large, northern bog-fen complex, J. Hydrol., 541, 941–951, https://doi.org/10.1016/j.jhydrol.2016.08.001, 2016.

Lowry, C. S., Fratta, D., and Anderson, M. P.: Ground penetrating radar and spring formation in a groundwater dominated peat wetland, J. Hydrol., 373, 68–79, https://doi.org/10.1016/j.jhydrol.2009.04.023, 2009.

Luce, C. H., Tonina, D., Gariglio, F., and Applebee, R.: Solutions for the diurnally forced advection-diffusion equation to estimate bulk fluid velocity and diffusivity in streambeds from temperature time series, Water Resour. Res., 49, 488–506, https://doi.org/10.1029/2012WR012380, 2013.

MacCrimmon, H. R. and Campbell, S. C.: World Distribution of Brook Trout, *Salaelinus fontinalis*, J. Fish. Res. Board Can., 26, 1699–1725, 1969.

Matthews, K. R. and Berg, N. H.: Rainbow trout responses to water temperature and dissolved oxygen stress in two southern California stream pools, J. Fish Biol., 59, 50–67, 1997.

Modica, E.: Source and age of ground-water seepage to streams, U.S. Geological Survey, Fact Sheet Fact Sheet 063-99, 1999.

Montgomery, D. R., Buffington, J. M., Peterson, N. P., SchuettHames, D., and Quinn, T. P.: Stream-bed scour, egg burial depths, and the influence of salmonid spawning on bed surface mobility and embryo survival, Can. J. Fish. Aquat. Sci., 53, 1061–1070, https://doi.org/10.1139/cjfas-53-5-1061, 1996.

Mullan, J. W.: The sea run or "Salter" brook trout (*Salvelinus fontinalis*) fishery of the coastal streams of Cape Cod, Massachusetts, Massachusetts Division of Fisheries and Game, Bulletin No. 17, 1958.

Obruca, W. and Hauer, C.: Physical laboratory analyses of intergravel flow through brown trout redds (*Salmo trutta fario*) in response to coarse sand infiltration, Earth Surf. Proc. Land., 42, 670–680, https://doi.org/10.1002/esp.4009, 2016.

Petty, J. T., Hansbarger, J. L., Huntsman, B. M., and Mazik, P. M.: Transactions of the American Fisheries Society Brook Trout Movement in Response to Temperature, Flow, and Thermal Refugia within a Complex Appalachian Riverscape Brook Trout Movement in Response to Temperature, Flow, and Thermal Refugia within a Compl, T. Am. Fish. Soc., 141, 1060–1073, https://doi.org/10.1080/00028487.2012.681102, 2012.

Raleigh, R. F.: Habitat suitability index models: Brook trout, U.S. Fish and Wildlife Service, FWS/OBS, 82/10.24, 1982.

Rau, G. C., Andersen, M. S., McCallum, A. M., Roshan, H., and Acworth, R. I.: Heat as a tracer to quantify water flow in near-surface sediments, Earth-Sci. Rev., 129, 40–58, https://doi.org/10.1016/j.earscirev.2013.10.015, 2014.

Rezanezhad, F., Price, J. S., Quinton, W. L., Lennartz, B., Milojevic, T., and Cappellen, P. Van: Structure of peat soils and implications for water storage, flow and solute transport?: A review update for geochemists, Chem. Geol., 429, 75–84, https://doi.org/10.1016/j.chemgeo.2016.03.010, 2016.

Rosenberry, D. O., Briggs, M. A., Delin, G., and Hare, D. K.: Combined use of thermal methods and seepage meters to efficiently locate, quantify, and monitor focused groundwater discharge to a sand-bed stream, Water Resour. Res., 52, 4486–4503, https://doi.org/10.1002/2016WR018808, 2016.

Seitzinger, S., Harrison, J. A., Böhlke, J. K., Bouwman, A. F., Lowrance, R., Peterson, B., Tobias, C., and Van Drecht, G.: Denitrification across landscapes and waterscapes: a synthesis, Ecol. Appl., 16, 2064–2090, https://doi.org/10.1890/1051-0761(2006)016[2064:DALAWA]2.0.CO;2, 2006.

Snook, E. L., Letcher, B. H., Dubreuil, T. L., Zydlewski, J., Donnell, M. J. O., Whiteley, A. R., Hurley, S. T., and Danylchuk, A. J.: Movement patterns of Brook Trout in a restored coastal stream system in southern Massachusetts, Ecol. Freshw. Fish, 26, 360–375, https://doi.org/10.1111/eff.12216, 2016.

Steel, E. A., Beechie, T. J., Torgersen, C. E., and Fullerton, A. H.: Envisioning, Quantifying, and Managing Thermal Regimes on River Networks, BioScience, 67, 506–522, https://doi.org/10.1093/biosci/bix047, 2017.

Stoliker, D. L., Repert, D. A., Smith, R. L., Song, B., LeBlanc, D. R., Mccobb, T. D., Conaway, C. H., Hyun, S. P., Koh, D., Moon, H. S., and Kent, D. B.: Hydrologic Controls on Nitrogen Cycling Processes and Functional Gene Abundance in Sediments of a Groundwater Flow-Through Lake, Environ. Sci. Technol., 50, 3649–3657, https://doi.org/10.1021/acs.est.5b06155, 2016.

Van Grinsven, M., Mayer, A., and Huckins, C.: Estimation of Streambed Groundwater Fluxes Associated with Coaster Brook Trout Spawning Habitat, Groundwater, 50, 432–441, https://doi.org/10.1111/j.1745-6584.2011.00856.x, 2012.

Walter, B. D. A. and Masterson, J. P.: Simulated Pond-Aquifer Interactions under Natural and Stressed Conditions near Snake Pond, Cape Cod, U.S. Geological Survey, Water-Resources Investigations Report 99-4174, 34 pp., 2002.

Wehrly, K. E., Wang, L., and Mitro, M.: Field-based estimates of thermal tolerance limits for trout: incorporating exposure time and temperature fluctuation, T. Am. Fish. Soc., 136, 365–374, 2007.

Winter, T. C., Harvey, J. W., Franke, O. L., and Alley, W. M.: Ground water and surface water; a single resource, U.S. Geological Survey, Circular, 1139, 79 pp., 1998.

10

Contributions of catchment and in-stream processes to suspended sediment transport in a dominantly groundwater-fed catchment

Yan Liu[1], Christiane Zarfl[1], Nandita B. Basu[2], Marc Schwientek[1], and Olaf A. Cirpka[1]

[1]Center for Applied Geoscience, University of Tübingen, 72074 Tübingen, Germany
[2]Department of Civil and Environmental Engineering, University of Waterloo, Waterloo, ON N2L 3G1, Canada

Correspondence: Olaf A. Cirpka (olaf.cirpka@uni-tuebingen.de)

Abstract. Suspended sediments impact stream water quality by increasing the turbidity and acting as a vector for strongly sorbing pollutants. Understanding their sources is of great importance to developing appropriate river management strategies. In this study, we present an integrated sediment transport model composed of a catchment-scale hydrological model to predict river discharge, a river-hydraulics model to obtain shear stresses in the channel, a sediment-generating model, and a river sediment-transport model. We use this framework to investigate the sediment contributions from catchment and in-stream processes in the Ammer catchment close to Tübingen in southwestern Germany. The model is calibrated to stream flow and suspended-sediment concentrations. We use the monthly mean suspended-sediment load to analyze seasonal variations of different processes. The contributions of catchment and in-stream processes to the total loads are demonstrated by model simulations under different flow conditions. The evaluation of shear stresses by the river-hydraulics model allows the identification of hotspots and hot moments of bed erosion for the main stem of the Ammer River. The results suggest that the contributions of suspended-sediment loads from urban areas and in-stream processes are higher in the summer months, while deposition has small variations with a slight increase in summer months. The sediment input from agricultural land and urban areas as well as bed and bank erosion increase with an increase in flow rates. Bed and bank erosion are negligible when flow is smaller than the corresponding thresholds of 1.5 and 2.5 times the mean discharge, respectively. The bed-erosion rate is higher during the summer months and varies along the main stem. Over the simulated time period, net sediment trapping is observed in the Ammer River. The present work is the basis to study particle-facilitated transport of pollutants in the system, helping to understand the fate and transport of sediments and sediment-bound pollutants.

1 Introduction

Suspended sediments are comprised of fine particulate matter (Bilotta and Brazier, 2008), which is an important component of the aquatic environment (Grabowski et al., 2011). Sediment transport plays significant roles in geomorphology, e.g., floodplain formation (Kaase and Kupfer, 2016), and transport of nutrients, such as particulate phosphorus and nitrogen (Haygarth et al., 2006; Slaets et al., 2014; Scanlon et al., 2004). Fine sediments are important for creating habitats for aquatic organisms (Amalfitano et al., 2017; Zhang et al., 2016). Conversely, high suspended-sediment concentrations can have negative impacts on water quality, especially, by facilitating transport of sediment-associated contaminants, such as heavy metals (Mukherjee, 2014; Peraza-Castro et al., 2016; Quinton and Catt, 2007) and hydrophobic organic pollutants such as polycyclic aromatic hydrocarbons (PAHs) (Rügner et al., 2014; Schwientek et al., 2013b; Dong et al., 2015, 2016), polychlorinated biphenyls (PCBs), and other persistent organic pollutants (Meyer and Wania, 2008; Quesada et al., 2014). Without understanding the transport of particulate matter, stream transport of strongly sorbing pollutants cannot be understood.

An efficient approach to estimate suspended-sediment loads is by rating curves, relating concentrations of suspended sediments to discharge. By this empirical approach, however, we cannot gain any information on the sources of suspended sediments, which is important for the assessment of particle-bound pollutants. Therefore, a model considering the various processes leading to the transport of suspended sediments in streams is needed. Numerous sediment-transport models have been developed during the past decades, including empirical and physically based models. Commonly used empirical models include the Universal Soil Loss Equation (USLE) (Wischmeier and Smith, 1978) and the Sediment Delivery Distributed (SEDD) model (Ferro and Porto, 2000). The USLE was designed to estimate soil loss on the plot scale. It is incapable to deal with heterogeneities along the transport pathways of soil particles and thus cannot be applied to entire subcatchments. The SEDD model considers morphological effects at annual and event scales. The two models cannot distinguish different in-stream processes. Among the models simulating physical processes, the Water Erosion Prediction Project (WEPP) (Flanagan and Nearing, 1995), the EUROpean Soil Erosion Model (EUROSEM) (Morgan et al., 1998), the Soil and Water Assessment Tool (SWAT) (Neitsch et al., 2011), the Storm Water Management Model (SWMM) (Rossman and Huber, 2016), the Hydrological Simulation Program Fortran (HSPF) model (Bicknell et al., 2001), and the Hydrologic Engineering Center's River Analysis System (HEC–RAS) (Brunner, 2016) are widely used. WEPP and EUROSEM are applied to simulate soil erosion from hillslopes on the timescale of single storm events. The two models do not have the capability of estimating urban particles. SWAT uses a modified USLE method to calculate soil erosion from catchments. SWMM aims at simulating runoff quantity and quality from primarily urban areas, including particle accumulation and wash-off in urban areas. HSPF considers pervious and impervious land surfaces. All of these models estimate sediment productions from the catchment and model the transport in the river channel with simplified descriptions of in-stream processes by simplifying the shape of cross sections. Various sediment-transport models for river channels exist that rely on detailed river hydraulics, particularly the bottom shear stress, which controls the onset of erosion and the transport capacity of a stream for a given grain diameter (Zhang and Yu, 2017; Siddiqui and Robert, 2010). HEC–RAS solves the full one-dimensional Saint Venant equation for any type of cross section, including cases with changes in the flow regime, which is beneficial to obtaining detailed information on river hydraulics.

In this study, we present a numerical modeling framework to understand the combined contributions from catchment and in-stream processes to suspended-sediment transport. The main objectives of this study were (i) to develop an integrated sediment-transport model taking sediment-generating processes (e.g., particle accumulation and particle wash-off), and river sediment-transport processes (e.g., bed erosion and bank erosion) into consideration, (ii) to understand annual load and seasonal variations of suspended sediments from different processes, (iii) to investigate how the contributions of suspended sediments from catchment and in-stream processes change under different flow conditions, and (iv) to identify hotspots and hot moments of bed erosion. The model is applied to a specific catchment introduced in the next section, implying that model components that control the behavior of suspended sediments in this catchment are given specific attention, whereas processes of less relevance are simplified in the model formulation. All model components are made available in the Supplement to facilitate modifications that may be needed when applying the framework to catchments with different controls.

2 Study area

2.1 The Ammer catchment, Germany

We applied the integrated sediment transport model to the Ammer catchment, located in southwestern Germany (Fig. 1). The River Ammer is a tributary to the River Neckar within the Rhine basin. It covers approximately $130\,\mathrm{km}^2$ and is dominated by agricultural land use that accounts for 67 % of the total area. The hydrogeology is dominated by the middle-Triassic Upper Muschelkalk limestone formation which forms the main karstified aquifer (Selle et al., 2013). In this catchment, annual precipitation is 700–800 mm. The Ammer River, approximately 12 km long, is the main stem, with a mean discharge of $\sim 1\,\mathrm{m}^3\,\mathrm{s}^{-1}$. It has two major tributaries, the Kochhart and Käsbach streams. Two wastewater treatment plants (WWTPs), Gäu-Ammer and Hailfingen, also contribute flow and suspended sediments to the Ammer River. During dry weather conditions, the discharge of WWTP Gäu-Ammer is $0.10–0.12\,\mathrm{m}^3\,\mathrm{s}^{-1}$, and the effluent turbidity is approximately 3 NTUs (nephelometric turbidity units). The WWTP in Hailfingen is comparatively small, with flow rates of $0.012–0.015\,\mathrm{m}^3\,\mathrm{s}^{-1}$, and its turbidity is in the same range as that of the WWTP Gäu-Ammer.

With the exception of a small stripe at the northeastern boundary of the study domain, highlighted by the forest land use in Fig. 1, the topography of the catchment is only slightly hilly (with mean slope of 4.2°), which agrees with the bed rock being a carbonate platform, partially overlain by upper Triassic mudstones and loess. Soils are dominated by luvisols on loess with mostly high probability of deep infiltration and low risk of soil erosion, according to the state geological survey of the state of Baden-Württemberg (LGRB, http://maps.lgrb-bw.de, last access: 21 June 2018).

Based on the digital elevation model (DEM) of the Ammer catchment, we delineated 14 subcatchments using the watershed delineation tool of the Better Assessment Science Integrating point & Nonpoint Sources (BASINS) model (see Fig. 1). Table 1 shows the proportions of different land-use types and the areas of each subcatchment.

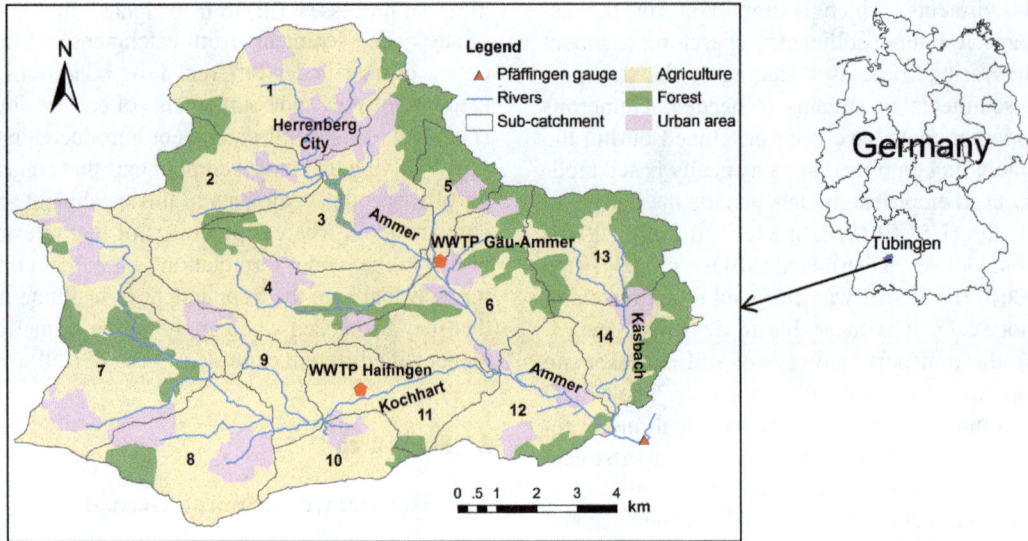

Figure 1. Location of the Ammer catchment and its subcatchments, rivers, and land uses. The numbers show identifiers of 14 subcatchments that are characterized in more detail in Table 1. Two red regular pentagons represent two WWTPs in the study domain. The red triangular indicates the gauge at the catchment outlet.

Table 1. Properties of the Ammer subcatchments.

ID of subcatchment	Area of subcatchment (km^2)	Urban Area (km^2)	Agriculture* (km^2)	Forest (km^2)
1	12.70	3.78	7.80	1.13
2	8.13	0.70	6.06	1.38
3	13.53	2.47	8.13	2.92
4	11.15	1.19	8.70	1.25
5	3.97	0.46	1.62	1.89
6	11.80	1.53	7.69	2.59
7	17.12	3.30	10.65	3.16
8	10.10	2.41	6.74	0.95
9	6.14	0.66	5.48	0.00
10	4.55	0.50	3.87	0.18
11	7.74	0.05	7.39	0.30
12	8.66	1.04	6.73	0.89
13	8.36	0.21	3.39	4.76
14	6.60	0.58	3.66	2.35
Area of land use (km^2)	130.54	18.87	87.92	23.75
Proportion of land use (%)	100	14.45	67.35	18.19

* The agricultural land in the Ammer catchment is dominated by nonirrigated arable land (80.2 % of the total agricultural areas), the crop of which is mainly cereals with annual rotation, and complex cultivation land (e.g., vegetables, 17.5 %). The rest (2.3 %) is principally agricultural area with natural vegetation. Therefore, we summarize the three types of arable land and use the same parameterization to estimate soil erosion.

2.2   Data sources

Hourly precipitation and air-temperature data are the driving forces of the hydrological model. We use hourly precipitation data of the weather station Herrenberg, operated by the German weather service DWD (CDC, 2017), whereas air temperatures are taken from the weather station Bondorf of the agrometeorological service Baden-Württemberg (BwAm, 2016). The generation and transport of sediments behave differently for different land uses and topography. We use the digital elevation model with 10 m resolution and land-use map of the state topographic service of Baden-Württemberg and Federal Agency for Cartography and Geodesy (BKG, 2009; LGRB, 2011; UBA, 2009). The river-hydraulics model requires bathymetric profiles of the River Ammer and its main tributaries. We use 230 profiles at

100 m spacing, obtained from the environmental protection agency of Baden-Württemberg (LUBW, 2010).

Only one gauging station is installed in the main channel of the Ammer River at the outlet of the studied catchment in Pfäffingen (red triangle in Fig. 1); here, hourly discharge and turbidity measurements are available, which we used for model calibration and validation. The water levels and turbidity data were measured by online probes (UIT GmbH, Dresden, Germany). The hydrograph was converted to discharge time series by rating curves, whereas the suspended sediment concentrations are derived from continuous turbidity measurements (Rügner et al., 2013). The linear relationship between suspended-sediment concentrations and turbidity with a conversion factor of 2.02 (mg L^{-1} NTU^{-1}) has been reported to be robust in the Ammer River (Rügner et al., 2013, 2014).

The simulation period covers the years 2013–2016. In this time, the maximum discharge reflected an event with a 2–10-year return period according to the long-time statistics of the gauging station (LUBW, http://www.hvz. baden-wuerttemberg.de/, last access: 21 June 2018).

3   Model setup

3.1   Model structure and assumptions

The integrated sediment-transport model consists of a catchment-scale hydrological model, a river-hydraulics model, a catchment sediment-generating model and a river sediment-transport model (Fig. 2). The catchment-scale hydrological model is used to estimate river discharge along the entire stream. The river-hydraulics model uses the discharge of the hydrological model and the river bathymetry to compute the river stage, cross-sectional area, velocity, and bottom shear stress, which are needed for the river-transport model. In this study we use HEC–RAS in quasi-steady-state mode. The catchment sediment-generating model is used for simulating particle accumulation in urban areas during dry weather periods, particle wash-off during storms, and erosion from rural areas during rain periods. The river sediment-transport model is used to simulate in-stream processes (advection, dispersion, and deposition, as well as bank and bed erosion). Wastewater treatment plants are treated as point inputs with constant discharge and sediment concentration during dry weather periods. Under low-flow conditions, when no soil erosion and urban particle wash-off occur and the suspended sediment concentrations in the streams are relatively small, we use a constant concentration to represent the sediment input under these conditions. Based on our prior knowledge of the Ammer catchment, soil erosion is very limited (the information supporting this statement will be discussed in Sect. 4.2), and thus a well-known approach and a simplified method are used to simulate particles from urban and rural areas, respectively. Mobilization of particles from differ-

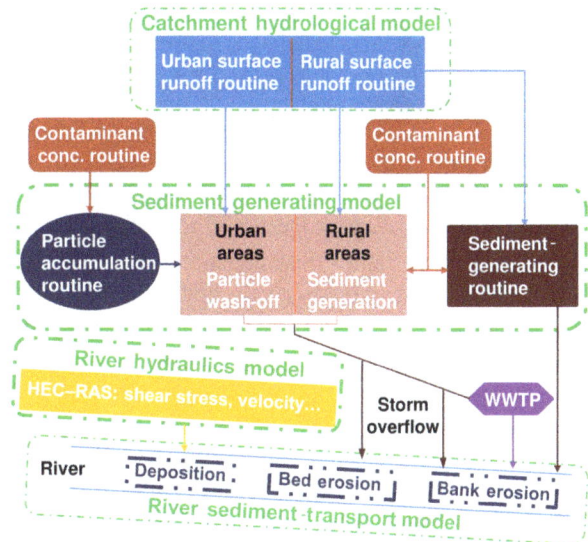

Figure 2. Integrated sediment transport model, consisting of a catchment-scale hydrological model, a river-hydraulic model, a sediment-generating model, and a river sediment-transport model.

ent sources depends on different processes; e.g., input of urban particles depends on the build-up and wash-off processes and rural particles rely on soil erosion, whereas bed and bank erosion are substantially affected by river hydraulics. Considering these processes enables us to diagnose the importance of different sediment sources well.

3.2   Catchment-scale hydrological model

The catchment-scale hydrological model is based on the HBV model (Hydrologiska Byråns Vattenbalansavdelning) (Lindström et al., 1997). However, we have added a quick recharge component and an urban surface runoff component to explain the special behavior of discharge in the Ammer catchment (see Sect. 2.1). The main Ammer springs are fed by groundwater from the karstified middle-Triassic Muschelkalk formation. The measured hydrograph indicates a rapid increase of base flow in sporadic events. We explain this behavior with a model that contains three storages of water in the subsurface: soil moisture in the top soils, a subsurface storage in the deeper unsaturated zone, and groundwater in the karstic aquifer. In our conceptual model, we assume water storage in the deep unsaturated zone, which spills over when a threshold value is reached, causing quick groundwater recharge to occur that then leads to a rapid increase of base flow. An urban surface runoff component is used to obtain surface runoff depths in urban areas in order to simulate particle wash-off from urban land surface. Details of the hydrological model are given in Appendix A. The temporal resolution of the hydrological model is 1 h. We use the catchment-scale hydrological model to simulate discharge contributions from the 14 subcatchments shown in Fig. 1 (detailed information see Sect. 2.1).

3.3 River-hydraulics model

In order to better understand in-stream processes, we feed the discharge data of the hydrological model into the river-hydraulics model HEC–RAS (Brunner, 2016), which solves the one-dimensional Saint Venant equations. The HEC–RAS model simulates hourly quasi-steady flow using the hourly discharge of the 14 subcatchments simulated by the hydrological model as change-of-discharge input. The locations where the discharge from 14 subcatchments enters into the main channel are set to the corresponding cross sections. The upstream boundary condition was set to time series of flow and the downstream one to normal depth. We have 258 measured cross sections and we used the built-in interpolation algorithm in HEC-RAS to obtain the additional cross sections, which results in totally 385 cross sections for the entire river network. The distances between computed cross sections range from 10 to 100 m depending on the changes of river bathymetry. The model requires river profiles in cross sections along the river channel and yields the water-filled cross-sectional area, the water depth, flow velocity, and shear stress, among other factors, as model output, which are needed in the river sediment-transport model. The detailed settings of HEC–RAS can be found in the Supplement.

3.4 Sediment-generating model

The land use is classified into urban and rural areas as well as forested areas. Impervious surfaces such as roads and roofs are regarded as urban areas, while rural areas consist of pervious surfaces such as gardens, parks, and agricultural land. The sediment-generating processes are different for the two types of land use. Sediment generation in forested areas is considered to be negligible. The sediment-generating model is used to obtain hourly sediments of urban and rural particles from the 14 subcatchments.

3.4.1 Urban areas

We use the urban-area algorithm of SWMM, which performs well on particle build-up and wash-off for urban land use (Wicke et al., 2012; Gong et al., 2016), to describe sediment generation from urban areas. The corresponding processes are described below.

1. *Particle accumulation.* An exponential function is used to simulate particle accumulation during dry periods under the assumption that particles in the urban areas have a capacity that is governed by the accumulation process during dry periods.

$$\frac{dM}{dt} = kM_{max}e^{-kt}, \tag{1}$$

in which M ($\mathrm{g\,m^{-2}}$) and M_{max} ($\mathrm{g\,m^{-2}}$) represent the particle build-up at the current time and the maximum build-up (particle mass per unit area), respectively; k ($\mathrm{s^{-1}}$) is the rate constant for particle accumulation, and t (s) denotes time since the last wash-off event. The maximum build-up depends on the location because the particle production (such as traffic density, population density, and industry density) and cleaning frequency (removing urban particles) differ in different urban areas. In our model it is obtained as uniform value for the entire catchment by calibration. The particle accumulation is restarted at the beginning of every accumulation period considering remaining particles after the flush period.

2. *Particle wash-off.* A power function is used to simulate particle wash-off during rain periods. The particle wash-off quantity is a function of surface runoff and the initial buildup of the corresponding rain period.

$$\frac{dM}{dt} = r_w = -k_w q^{n_w} M, \tag{2}$$

$$c_{sw} = -\frac{r_w}{q}, \tag{3}$$

in which r_w ($\mathrm{g\,m^{-2}\,s^{-1}}$), q ($\mathrm{m\,s^{-1}}$), and c_{sw} ($\mathrm{mg\,L^{-1}}$) are the rate of wash-off, the surface runoff velocity, and the concentration of washed suspended sediment, respectively; k_w ($\mathrm{s^{n_w-1}\,m^{-n_w}}$) and n_w (–) represent a wash-off coefficient and a wash-off exponent.

3.4.2 Rural areas

In contrast to urban areas, the supply of suspended sediments from rural areas can be seen as "infinite" because they mainly originate from eroded soils. Soil erosion is assumed to linearly depend on shear stress, provided that the shear stress generated by surface runoff is larger than a critical shear stress. The sediment generation from rural areas is based on the study of Patil et al. (2012).

$$\tau = \rho_w g R_{surface} \tan\theta, \tag{4}$$

$$y_h = \begin{cases} C_h (\tau - \tau_c) & \text{if } \tau > \tau_c, \\ 0 & \text{otherwise,} \end{cases} \tag{5}$$

$$c_{sed} = \frac{y_h}{q}, \tag{6}$$

in which τ ($\mathrm{N\,m^{-2}}$) is the mean shear stress generated by the average depth of surface runoff $R_{surface}$ (m), $\tan\theta$ (–) is the mean slope of the subcatchment, ρ_w ($\mathrm{kg\,m^{-3}}$) is the density of water, and g ($\mathrm{m\,s^{-2}}$) is the gravitational acceleration constant. The rural sediment load y_h ($\mathrm{kg\,m^{-2}\,s^{-1}}$) is directly proportional to the difference between the mean shear stress τ and the critical rural shear stress τ_c ($\mathrm{N\,m^{-2}}$). C_h ($\mathrm{s\,m^{-1}}$) is a proportionality constant, c_{sed} ($\mathrm{kg\,m^{-3}}$) is the concentration of sediment generated in rural areas, and q ($\mathrm{m\,s^{-1}}$) is, like

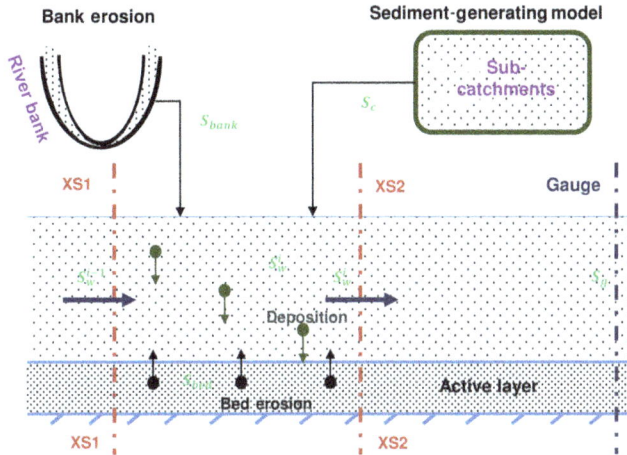

Figure 3. In-stream processes of the river suspended-sediment transport model considering deposition, bed erosion, bank erosion, and input from the catchment. XS1 and XS2 are the two cross sections bounding a cell in a finite-volume scheme. S_c and S_{bank} are sediments from the catchment and bank erosion. S_{bed} indicates the bed sediment mass. S_w^i stands for the concentration of suspended sediments in the ith cell. S_g is the suspended-sediment concentration at a river gauge.

above, the surface runoff velocity. This is a simplified approach to estimate the average sediment delivery from rural areas to streams. It does not explicitly consider all processes on the hillslope scale. In particular, we do not consider the dependence of the coefficients on the crop type and time-dependent phenology of the crops. Instead, all rural areas are treated the same. We justify this strong simplification by an overall low sediment input from rural areas discussed further below. In catchments with larger sediment load from rural areas, distinctions should be made.

3.5   River sediment-transport model

We consider two types of sediment: suspended sediment in the aqueous phase (mobile component) and bed sediment (immobile component). Figure 3 shows a schematic of the river sediment-transport model, which considers advection, dispersion, deposition, bank erosion, bed erosion, and lateral input of suspended sediments. We use this model to calculate the average concentration of the mobile component and the mass of the immobile component for every computation cell (formed by two cross sections) every hour.

1. *Mobile component.* We use a finite-volume discretization for suspended-sediment transport for the main channel, considering storage in the aqueous phase, advection, dispersion, bed and bank erosion, deposition,

and lateral inputs (tributaries and WWTPs):

$$\frac{\partial (c_w V)}{\partial t} = -\frac{\partial (c_w Q)}{\partial x}\Delta x + A D \frac{\partial^2 c_w}{\partial x^2}\Delta x \\ + (r_{bed} + r_{bank})\Delta x - r_d V + \sum c_{lat}^i Q_{lat}^i, \quad (7)$$

in which c_w $(mg\,L^{-1})$ is suspended-sediment concentration; V (m^3) is the cell volume; Δx (m) is the cell length; Q $(m^3\,s^{-1})$ and A (m^2) are the flow rate and cross sectional area; D $(m^2\,s^{-1})$ is the dispersion coefficient; c_{lat}^i $(mg\,L^{-1})$ and Q_{lat}^i $(m^3\,s^{-1})$ represent the suspended-sediment concentration and flow rate of the ith lateral inflow; r_d $(mg\,L^{-1}\,s^{-1})$, r_{bank} $(g\,m^{-1}\,s^{-1})$, and r_{bed} $(g\,m^{-1}\,s^{-1})$ indicate the deposition, bed-erosion, and bank-erosion rates, respectively. For the advective term, we use upstream weighting, whereas the second derivative of concentration appearing in the dispersion term is evaluated by standard finite differences.

This model component requires the sediment concentrations in the lateral inputs (tributaries and WWTPs) as well as in the Ammer spring as boundary conditions. The lateral inputs are computed by the sediment-generating model. For the sediment input by the Ammer spring, we consider the turbidity of $\sim 3\,NTU$ measured under base-flow conditions. Rügner et al. (2013) showed that the karst springs in the Ammer catchment contribute to turbidity, which is in agreement with many previous studies showing that karst systems can contribute suspended sediments (Bouchaoua et al., 2002; Meus et al., 2013). Thus, the turbidity under base-flow conditions is potentially generated by subsurface flow through the karst matrix. The karstic sediment flux was calculated by subsurface flow rates and constant suspended sediment concentrations.

2. *Immobile component.* For simplification, we account for one active layer only in the bed sediment per cell, and consider only the average grain size. Deposition of suspended sediments leads to a mass flux from the aqueous phase to the bed layer, whereas bed erosion causes a mass flux in the opposite direction:

$$\frac{\partial M_{bed}}{\partial t} = r_d \frac{V}{\Delta x} - r_{bed}, \quad (8)$$

in which M_{bed} $(g\,m^{-1})$ is the sediment mass per unit channel length in the active layer on the river bed.

   a. Deposition

   The deposition rate r_d of particles can be calculated by the following (Krone, 1962):

$$r_d = \begin{cases} \left(1 - \dfrac{\tau_b}{\tau_e}\right)\dfrac{v_s c_w}{y} & \text{if } \tau_b < \tau_e, \\ 0 & \text{otherwise,} \end{cases} \quad (9)$$

in which τ_b (N m^{-2}) and τ_e (N m^{-2}) represent the bottom shear stress of the river and the threshold shear stress of particle erosion (see below), y (m) denotes the water depth, and v_s (m s^{-1}) is the settling velocity.

b. Bed erosion

We consider two types of bed erosion, namely particle erosion and mass erosion, which correspond to two thresholds of the bottom shear stress. The bed erosion rate r_{bed} can be calculated by the following (Partheniades, 1965):

$$r_{bed} = \begin{cases} M_{me}\left(\dfrac{\tau_m}{\tau_e} - 1\right) & \text{if } \tau_b > \tau_m, \\ M_{pe}\left(\dfrac{\tau_b}{\tau_e} - 1\right) & \text{if } \tau_e < \tau_b \leq \tau_m, \\ 0 & \text{otherwise} \end{cases} \quad (10)$$

in which r_{bed} (g m^{-1} s^{-1}) is bed erosion rate; τ_m (N m^{-2}) represents the mass erosion threshold, whereas M_{pe} (g m^{-1} s^{-1}) and M_{me} (g m^{-1} s^{-1}) are rate constants, denoting the specific rates of particle and mass erosion.

c. Bank erosion

In our model, the bank erosion rate r_{bank} is calculated by the following:

$$r_{bank} = \begin{cases} \kappa \rho L y \left(\tau_{bank} - \tau_{bc}\right) & \text{if } \tau_{bank} > \tau_{bc}, \\ 0 & \text{otherwise}, \end{cases} \quad (11)$$

in which τ_{bank} (N m^{-2}) and τ_{bc} (N m^{-2}) are the bank shear stress and critical shear stress for bank erosion. κ (m^3 N^{-1} s^{-1}) is the erodibility coefficient. ρ (kg m^{-3}) is density of bank material. L (km) is length of the river bank.

3.6 Parameter estimation

For the estimation of parameters, we used the well-known Nash–Sutcliffe efficiency (NSE) as model performance criterion:

$$\text{NSE} = 1 - \frac{\sum_{i=1}^{n}(O_i - M_i)^2}{\sum_{i=1}^{n}\left(O_i - \overline{O}\right)^2}, \quad (12)$$

in which O_i and M_i are the ith observed and modeled values, $\overline{O}$ is the mean of all observed values. An NSE value approaching unity indicates good agreement between model and data, whereas NSE values smaller than zero imply that the model performs worse than taking the mean of all observations. We obtained the best set of parameters by systematically scanning the parameter space.

The hydrological model was applied to 14 subcatchments. Each subcatchment has three types of land use: agricultural areas, forest, and urban areas. We used daily average discharge data of 2013–2014 and 2015–2016 for calibration and

validation, respectively. We generated 1000 realizations of the 14 parameters by Latin hypercube sampling (LHS) and calculated the corresponding NSE value for each parameter set. If NSE was ≥ 0.55, the parameter set was regarded acceptable. In the same way, we used the accepted parameter sets for validation. Subsequently we calculated the 90 % confidence intervals and the NSE value for high flows (flow rate greater than the mean discharge) using the accepted parameter sets. Finally, we identified the best-fit parameter values.

For the calibration and validation of the sediment-generating and the river sediment-transport models, we performed a literature survey to identify a reference range of each parameter. We performed a manual calibration of the corresponding parameters within the given range, fitting the modeled and measured suspended sediment concentrations at the river gauge. Subsequently, we used the identified parameter set as base values in a local sensitivity analysis, the details of which are given in Table S1 of the Supplement. Within the given parameter variations, the manually calibrated parameter sets were confirmed as optimal. The parameters of the sediment-generating model and the river sediment-transport models are listed in Tables 2 and 3, respectively.

4 Results and discussion

4.1 Quality of model calibration and validation

The best-fit parameter set of the hydrological model resulted in NSE values of 0.63 and 0.59 for calibration and validation, respectively. Figure 4 shows the measured and simulated hydrographs for the calibration and validation periods with 90 % confidence intervals. It can be seen that the discharge was reproduced quite well, both in the general trend and the dynamics. The measured discharge data almost all fall within the 90 % confidence interval of the simulation. The NSE value for high flows (greater than the mean discharge, 1 m^3 s^{-1}) of the simulation period is 0.43, implying an acceptable fit of high flows. There are few events that cannot be reproduced by the model. These events occurred in the summer months and probably resulted from thunderstorms, which are very local and may be missed by precipitation measurements, so that the resulting flow peaks could not be predicted by the hydrological model.

Figure 5 depicts measured suspended-sediment concentrations and the simulation results of the sediment-transport model during the calibration (year 2014) and validation (year 2016) periods. The corresponding NSE values are 0.46 and 0.32, respectively, which indicates an acceptable fit, albeit not as good as for the hydrograph. The integrated sediment transport model can capture the dynamics of the suspended sediment concentrations. In particular, the model captures the concentration peaks well. However, two events, one in the calibration and the other in the validation period,

Table 2. Parameters of the sediment-generating model.

Parameter symbol	Definition	Unit	Range	Reference	Value
M_{max}	Maximum accumulation load	$g\,m^{-2}$	7.5–50	Piro and Carbone (2014), Modugno et al. (2015), Bouteligier et al. (2002)	23
k	Accumulation rate constant	d^{-1}	0.16–0.46[a]	Rossman and Huber (2016)	0.33
K_w	Wash-off coefficient	$d^{0.5}\,m^{-1.5}$	50–500[b]	Rossman and Huber (2016)	80
n_w	Wash-off exponent	–	0–3	Wicke et al. (2012), Modugno et al. (2015), Rossman and Huber (2016)	1.5
C_h	Proportionality constant	$s\,m^{-1}$	0.0003–0.05	Gilley et al. (1993), Romero et al. (2007)	0.001
τ_c	Critical rural shear stress	$N\,m^{-2}$	0–10[c]	Bones (2014); Léonard and Richard (2004)	0.3

[a] The range of k is calculated under the assumption that it takes 5–30 days to reach 90 % of the maximum buildup. [b] The range of K_w, 50–500, is sufficient for most urban runoff. [c] It is for the most of time, but depends on soil properties.

Table 3. Parameters of the river sediment-transport model.

Parameter symbol	Definition	Unit	Range	Reference	Value
v_s	Settling velocity	$m\,s^{-1}$	10^{-6}–10^{-4*}	Brunner (2016)	4×10^{-6}
τ_e	Particle erosion threshold	$N\,m^{-2}$	0.1–5	Winterwerp et al. (2012)	2.5
τ_m	Mass erosion threshold	$N\,m^{-2}$	$> \tau_e$	Partheniades (1965), Brunner (2016)	3.5
M_{pe}	Particle erosion rate	$kg\,m^{-1}\,d^{-1}$	0.8–43.2	Winterwerp et al. (2012)	30
M_{me}	Mass erosion rate	$kg\,m^{-1}\,d^{-1}$	$> M_{pe}$	Partheniades (1965), Brunner (2016)	40
κ	Erodibility coefficient	$m^3\,N^{-1}\,d^{-1}$	0.0001–0.32	Clark and Wynn (2007), Hanson and Simon (2001)	0.0018
τ_{bc}	Critical bank shear stress	$N\,m^{-2}$	0–21.91	Clark and Wynn (2007)	5
ρ	Density of bank material	$kg\,m^{-3}$	2190–2700	Clark and Wynn (2007)	2650

* This range is calculated for the suspended sediment with average diameter 1–50 μm.

were not well fitted. These are events which were also not captured by the hydrological model, occurring in the summer months and due to thunderstorms.

4.2 Annual and monthly suspended sediment loads from different processes

After calibration and validation, the model results can be used to analyze the importance of different sediment sources. Figure 6 displays the modeled annual suspended-sediment loads from catchment and in-stream processes for the entire Ammer River network. The annual suspended-sediment load at the gauge ranges between 410 and 550 t yr^{-1}. Equation (13) describes the overall mass balance of sediments in the entire catchment:

$$\text{Load}_{gauge} = (\text{Load}_{urban} + \text{Load}_{rural} + \text{Load}_{karst})_{Catchment}$$
$$+ (\text{Load}_{bde} + \text{Load}_{bke} - \text{Load}_{dep} - \Delta S)_{Stream}, \quad (13)$$

in which Load_{gauge} (t yr^{-1}) indicates the suspended-sediment load at the river gauge. Load_{urban} (t yr^{-1}), Load_{rural} (t yr^{-1}), and Load_{karst} (t yr^{-1}) denote the suspended-sediment loads

from urban areas generated by surface runoff and WWTP effluent, rural areas generated by soil erosion, and the karst system carried by subsurface flow, respectively. These three terms represent the catchment processes. Load_{bde} (t yr^{-1}), Load_{bke} (t yr^{-1}), Load_{dep} (t yr^{-1}), and ΔS (t yr^{-1}) are the suspended-sediment loads from bed erosion, bank erosion, deposition, and the change of sediment storage in the entire river channel, respectively. These four terms represent the in-stream processes.

In the Ammer catchment, urban particles (266–337 t yr^{-1}) and the sediment input from the karst system (106–160 t yr^{-1}) dominate the annual suspended sediment load, accounting for 59.1 and 24.9 %, respectively. Bed erosion, bank erosion, and rural sediment contribute much less, namely 6.2, 6.3, and 3.5 % of the total annual load, respectively. The contribution of rural runoff sediment in the Ammer catchment was very small, which may appear to be surprising at first. We have collected several independent lines of evidence that support these findings and included them in the Supplement:

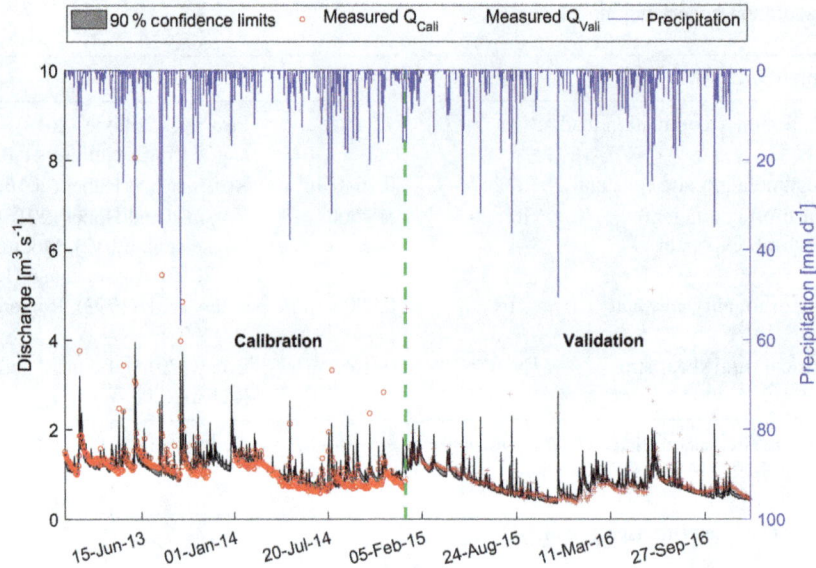

Figure 4. Calibration (left, year 2013–2014) and validation (right, year 2015–2016) of hydrological model, Q_{Cali} and Q_{Vali} are measured discharges used for calibration and validation, respectively.

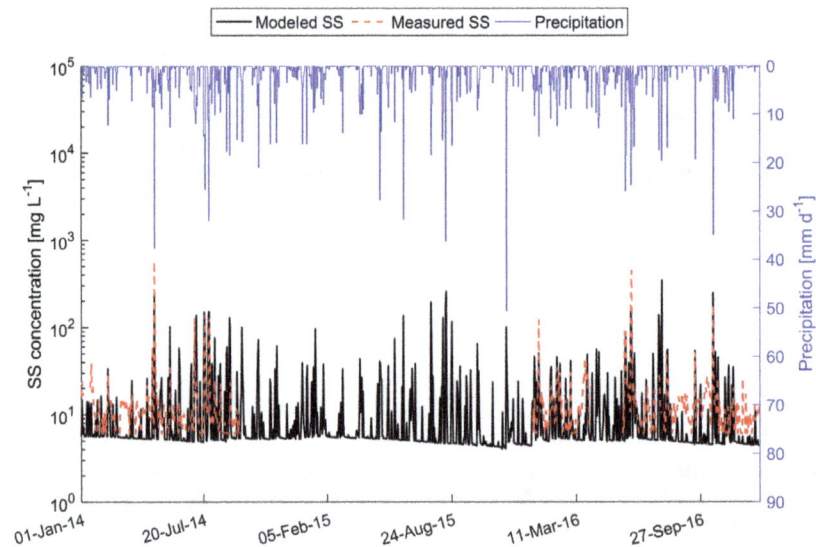

Figure 5. Modeled and measured suspended sediment concentrations used for calibration (year 2014) and validation (year 2016) of the sediment transport model. A data gap exists for the year 2015.

The suspended sediments of the Ammer River are strongly contaminated by polycyclic aromatic hydrocarbons and other persistent organic pollutants (Schwientek et al., 2013b). Table S2 and Eqs. (S1)–(S7) of the Supplement present an end-member-mixing analysis indicating a fraction of rural particles amounting to only 3 %.

The state geological survey of the state of Baden-Württemberg has developed a soil-erosion risk map shown Fig. S1, putting most of Ammer catchment into the class of lowest soil-erosion risk. This is so because the surface runoff from agricultural areas is small due to a comparably flat to-

pography. The same agency associates most of the catchment with deep infiltration as the main discharge mechanism.

Schwientek et al. (2013a) found a lacking connection between soils and streams in the Ammer catchment. The catchment has a large water storage capacity due to the karst and the slopes of this catchment are mild. During the simulation period, the precipitation intensity was not large enough to exceed the maximum infiltration rates or to reach storage capacity of the subsurface. Compared with litera-ture values of maximum infiltration rates (10–20 and 5–10 mm h^{-1} for loamy soil and clay loamy soil, respectively;

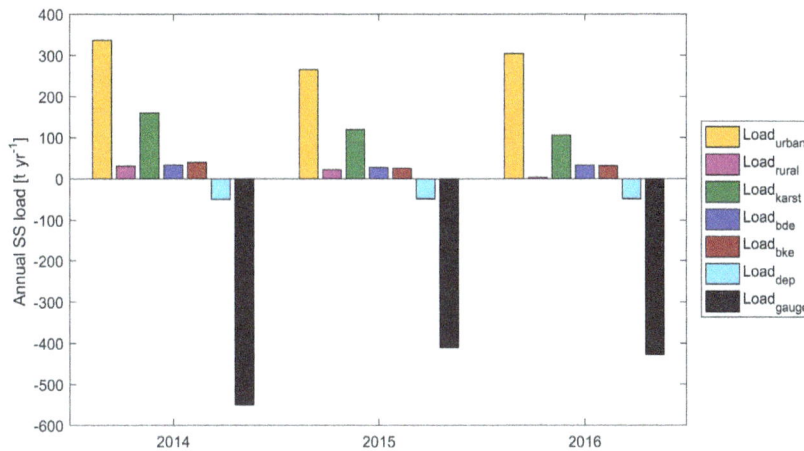

Figure 6. Annual suspended sediment loads from different processes. Load$_{gauge}$ is calculated by modeled discharge and suspended sediment concentrations at catchment outlet. Load$_{urban}$, Load$_{rural}$, and Load$_{karst}$ are calculated using the results of sediment-generating model. Load$_{dep}$, Load$_{bde}$, and Load$_{bke}$ are the sum of deposition load, suspended sediments eroded from river bed and river bank of the entire river network for a whole year, respectively. In this figure, the positive values represent sediment input to the river channel, while negative values denote sediment output from the river channel.

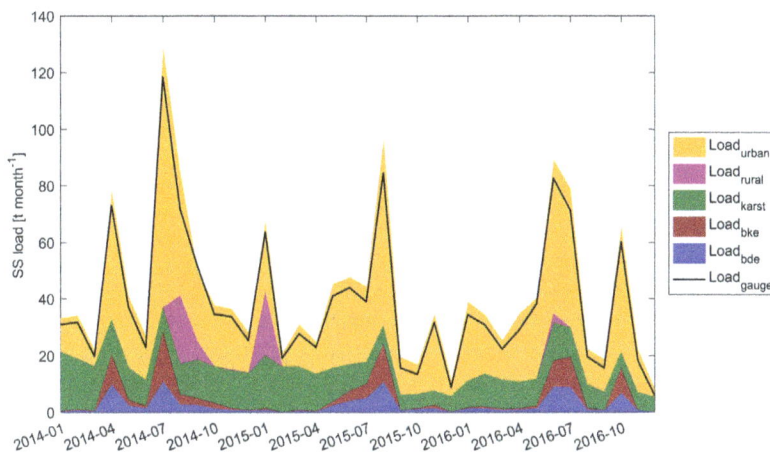

Figure 7. Monthly mean suspended-sediment load from different processes, calculated using the model results of 2014–2016. Load$_{gauge}$, Load$_{urban}$, Load$_{rural}$, and Load$_{karst}$ are the monthly mean suspended-sediment load at the gauge and from urban areas, rural areas, and the karst system. Load$_{bke}$ and Load$_{bde}$ represent monthly mean suspended-sediment load from bank erosion and bed erosion for the entire river network, respectively. The area above the line of Load$_{gauge}$ is the monthly mean deposition, Load$_{dep}$.

http://www.fao.org/docrep/S8684E/s8684e0a.htm, last access: 21 June 2018), only a few events exceed $10\,\mathrm{mm\,h^{-1}}$ of the precipitation intensity during the simulation period. Thus, hardly any surface runoff occurred in the rural area, so that sediment generation and transport from rural areas to the river channel were small.

The comparably flat topography can be explained by the geological formation. The Muschelkalk limestone is a carbonate platform that is partially overlain by mudstones of the upper Triassic. Along the Ammer main stem, there is only a small stretch where the river is incised somewhat deeper into the limestone rock. The river lost its former headwater catchment in the early Pleistocene to river Nagold so that the currently existing small river has a too-wide valley given its discharge.

As discussed above, we used a simplified approach to simulate the average sediment delivery from rural areas in our study because the contribution of rural areas to sediment delivery was so small. In particular, we did not distinguish between different crop types and seasons and estimated the average sediment load that reaches the streams instead. In other catchments, where the rural contributions to the sediment load are considerably higher, the description of soil erosion processes would require more differentiations.

To identify seasonal variations of suspended sediment loads originating from different processes, we used the model

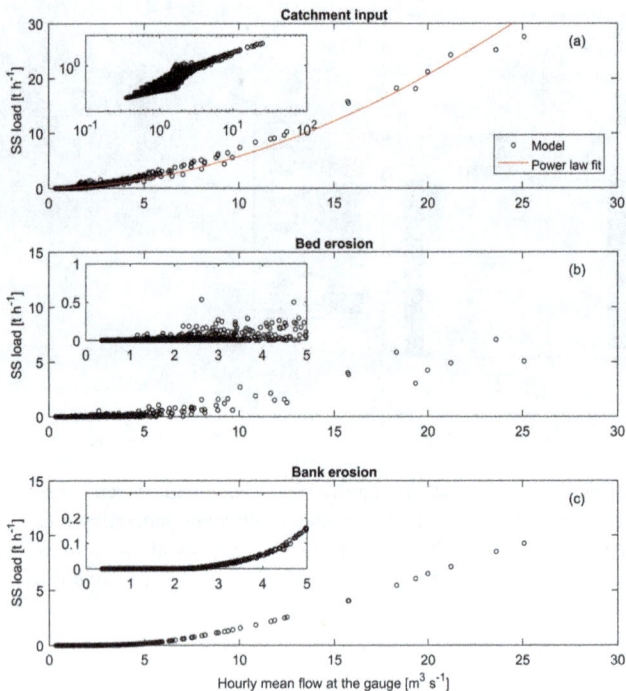

Figure 8. Relationship between simulated hourly mean flow and hourly suspended-sediment loads from the catchment (**a**), bed erosion (**b**), and bank erosion (**c**), in which bed erosion and bank erosion are sums over all computation cells. Loads from the catchment are the sum of contributions from urban areas, nonurban areas, and the karst system.

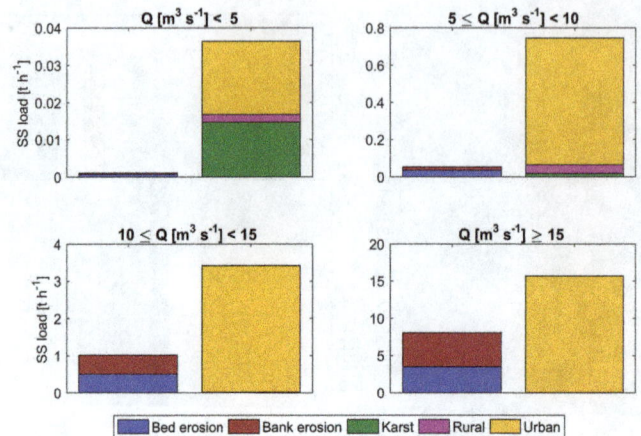

Figure 9. Simulated suspended sediment load from bed erosion, bank erosion, the karst system, rural areas, and urban areas (including suspended sediment from WWTPs) under different flow regimes, the suspended sediment loads are the mean values for the specific flow regimes.

results of 2014–2016 to analyze the monthly mean suspended sediment loads from the urban areas, rural areas, the karst system, bed erosion, bank erosion, and deposition (Fig. 7). More suspended-sediment loads from urban areas and at the gauge can be observed in June and July (summer months). In summer months events with high rain intensity are more common than in winter months, which results in higher discharge peaks, more sediments generated in urban areas, and higher suspended-sediment loads at the gauge. Monthly suspended-sediment loads at the gauge have similar dynamics as the monthly urban particle contributions. The suspended-sediment load from the karst system is higher in winter months because the subsurface flow in the Ammer catchment is higher in winter months. Rural particles contribute to the overall particle flux only during a few months because annual precipitation and rainfall intensity were relatively small, so that surface runoff generated from rural areas was also low.

In the model simulation period, the seasonal patterns of bed erosion and bank erosion are obvious. High bed erosion and bank erosion occur from June to August due to increased bed shear occurring during big events. The area above the line of Load_gauge indicates net deposition, which shows small variations with a slight increase in July and August. The slight increase in summer is due to increased

suspended-sediment concentrations during summer months. Comparing monthly mean bed erosion and deposition shows that bed erosion was greater than deposition in July, which indicates that accumulated bed sediment can be partly eroded in July.

4.3   Suspended-sediment sources under different flow conditions

Figure 8 shows the relationship between hourly mean discharge and the simulated hourly suspended sediment loads from the catchment, bed erosion, and bank erosion. The hourly suspended-sediment load from the catchment monotonically increases with increasing hourly mean discharge by a power-law relationship (Fig. 8a), which is consistent with the particle wash-off rate being a power-law function of discharge. Bed erosion requires that the bed shear stress exceeds a critical value, so that bed erosion is almost 0 when hourly mean flow is smaller than $1.5\,\mathrm{m^3\,s^{-1}}$, namely 1.5 times mean discharge (Fig. 8b). For discharge larger than this threshold ($1.5\,\mathrm{m^3\,s^{-1}}$), bed erosion increases approximately linearly with discharge. The simulated hourly bed-erosion loads for a given flow rate vary substantially because bed erosion is not only influenced by the shear stress, which directly depends on discharge, but also on the bed sediment storage, which depends on previous deposition and erosion events. Bank erosion occurs when the hourly mean flow rate is larger than $2.5\,\mathrm{m^3\,s^{-1}}$, i.e., 2.5 times mean discharge (Fig. 8c). The relationship between bank-erosion-related loads and discharge is more unique than that of bed-erosion loads because we assume an infinite source for bank erosion.

Figure 9 shows the suspended sediment loads from instream (bed erosion and bank erosion) and catchment processes (input from the karst system, urban areas, and ru-

Table 4. Summary of suspended-sediment sources under different flow conditions.

Flow (Q) (m³ s⁻¹)	Description of main suspended-sediment sources
$Q < 1.5$	Suspended sediment load is dominated by contributions from the catchment (karst system, rural areas, and urban areas), while bed erosion and bank erosion can be neglected.
$1.5 \leq Q < 2.5$	Bed erosion starts contributing.
$2.5 \leq Q < 5$	Bank erosion starts contributing, but the contributions from bed and bank erosion are still negligible. Contributions from urban areas and the karst system are dominant.
$5 \leq Q < 10$	Bed and bank erosion contributes more, but the major contribution is still from the catchment, especially from urban areas. Bed erosion contributes less than 5 % and bank erosion contributes less than 3 %. The relative contribution from the karst system becomes very small.
$Q \geq 10$	Suspended sediment contributions from bed and bank erosion are significant. The contribution of in-stream processes can be up to 35 % of the total suspended sediment load when discharge is larger than 15 m³ s⁻¹. The contribution from urban areas is largest, which dominates the catchment input.

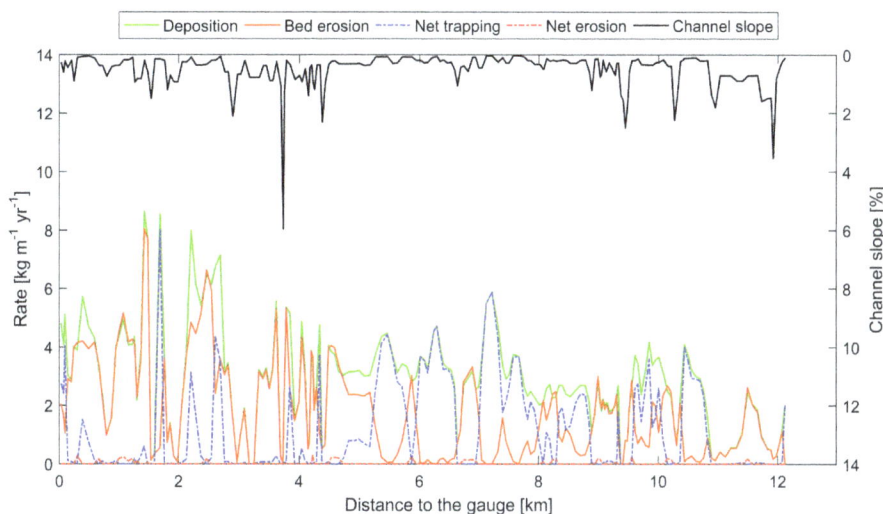

Figure 10. The distribution of the annual mean deposition, bed erosion, net sediment trapping, net sediment erosion, and channel slope along the main channel of the Ammer River (flow direction from right to left). The blue and red dash–dotted lines highlight net sediment trapping and net erosion, respectively.

ral areas) under different flow regimes. The fractions of suspended-sediment contributions from different processes change with flow regimes. The contributions of in-stream processes are negligible in the flow regime of discharge smaller than $5 \, \text{m}^3 \, \text{s}^{-1}$. With the discharge increasing, the contributions of in-stream processes increase. The in-stream processes play significant roles in high-flow regimes, which contribute 23 and 34 % of total suspended sediment loads under flow regimes of $10 \leq Q \, (\text{m}^3 \, \text{s}^{-1}) < 15$ and $Q \, (\text{m}^3 \, \text{s}^{-1}) \geq 15$, respectively. The relative contribution of the karst system is high in the low-flow regime ($Q < 5$, in m³ s⁻¹), while it can be neglected under high-flow regimes ($Q > 10$, in m³ s⁻¹). With the increase in flow rates, the contribution of urban particles becomes dominant in terms of catchment processes, especially when discharge is larger than $10 \, \text{m}^3 \, \text{s}^{-1}$.

From above observations, we can see that the sources of suspended sediments differ under different flow conditions in the following way (Table 4):

4.4 Hotspots and hot moments of bed erosion in the Ammer River

The annual mean rates of bed erosion and deposition (mass per unit length per year) along the main channel can be used to identify hotspots of bed erosion and net sediment trapping (Fig. 10). The rates of deposition and bed erosion vary substantially along the main stem, ranging from essentially zero to a maximum of 8.6 and 8.0 kg m yr⁻¹, respectively. Bed erosion is higher in the river segment close to the gauge because the flow rate is higher due to the contributions of the tributaries. Bed erosion is rather low in the river seg-

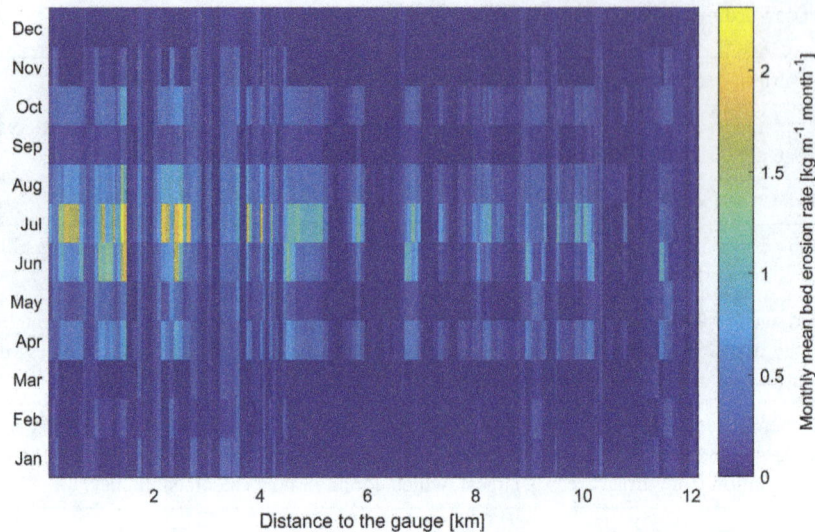

Figure 11. Monthly mean bed erosion along the channel of the Ammer River upstream of the gauge (flow direction from right to left).

ments within 5–6.5, 7–8, 8.5–9, and 10–11 km to the gauge, where the channel slope is very mild. The river sections with the steepest channel slope typically do not show the highest bed erosion because there is not enough sediment available for erosion, which is caused by insufficient deposition. Figure 10 also shows that when the channel slope is very mild, the deposition rate is very high, while the bed erosion rate is nearly zero. These are sections where net sediment trapping (blue dash–dotted line) was observed. With increasing channel slope, bed erosion rates increase and deposition rates decrease. In a small range of channel slopes, deposition rates equal to erosion rates, resulting in a local steady state. If the channel slope continues to increase, the erosion rate will be higher than the deposition rate, which results in net sediment erosion if the sediment storage in the channel is large enough (red dash–dotted line, very few in Fig. 10). Where the channel slope is very steep, both sediment deposition and erosion rates are very small.

Figure 11 shows monthly means of the bed erosion rates along the Ammer main stem, computed for the simulated years 2014 to 2016. Bed erosion is stronger in the summer months, especially in July, which is consistent with the monthly load of suspended sediments discussed in Sect. 4.2. The hot moments of bed erosion are the extreme events caused by summer thunderstorms. The downstream river segments close to the gauge show higher bed erosion rates than the sections further upstream because flow rates and thus bed shear stresses are higher even with identical channel slope.

5 Conclusions

Suspended sediment transport is of great importance for river morphology, water quality, and aquatic ecology. In this study, we have presented an integrated sediment-transport model, combining a conceptual hydrological model with a river-hydraulics model, a model of sediment generation, and a shear-stress-dependent sediment-transport model within the river, which enables us to investigate the major contributors to the suspended-sediment loads in different river sections under different flow conditions.

In the dominantly groundwater-fed Ammer catchment, annual suspended-sediment load is dominated by the contributions of urban particles and sediment input from the karst system. The contribution from rural areas is small because the topography is comparably flat and the infiltration capacity of the soils is high in this region, resulting in a very weak surface runoff from rural areas, and thus very few rural particles are generated and transported to the river channel. In-stream processes, i.e., bed erosion and bank erosion, play significant roles in high-flow conditions ($Q > 10\,\mathrm{m^3\,s^{-1}}$). The flow rate governs the contributions of different processes to the suspended sediment loads. In particular, bed erosion and bank erosion take place when flow rates reach the corresponding thresholds, 1.5 and 2.5 times the mean discharge, respectively. The channel slope has significant effects on the deposition and bed erosion rates. Net sediment trapping was found in the river segments with very mild channel slopes in the Ammer River during the simulation period with events of a 2-year to 10-year return period. Finally, the river hydraulics model is necessary to differentiate sediment sources and sinks of in-stream processes, i.e., shear-stress-related deposition, bed erosion, and bank erosion.

The model and results of this study are useful and essential for further research on the fate and sediment-facilitated transport of hydrophobic pollutants like PAHs, and for the design of optimal sampling regimes to capture the different processes that drive particle dynamics. In addition, the analysis of deposition and bed erosion in the Ammer main stem provides information on the distribution of net sediment trapping within the channel, which would be a good indicator that channel dredging improves water quality.

Appendix A: Catchment-scale hydrological model

The hydrological model in the integrated sediment transport model is composed of three storage zones in vertical direction with a quick recharge component and an urban surface runoff component. Detailed processes are shown below.

We applied this model to 14 subcatchments of the study domain. Each subcatchment includes three different land uses: urban area, agriculture, and forest. For urban areas, we consider effective urban areas such as roads and roofs and ineffective urban areas such as parks and gardens. We use the same parameters of agriculture for ineffective urban area.

The effective urban area is used for surface runoff component, the ratio is calculated by the following:

$$r_{\mathrm{eff}} = \frac{A_{\mathrm{eff}}}{A_{\mathrm{urb}}}, \tag{A1}$$

in which r_{eff} (–) is the ratio of effective urban area over total urban area, and A_{eff} (km^2) and A_{urb} (km^2) represent areas of effective urban area and total urban area, respectively.

The effective precipitation to the subsurface storage for agriculture, forest, and ineffective urban area is calculated below:

$$P_{\mathrm{e}} = \left(\frac{\mathrm{SM}}{\mathrm{FC}}\right)^{\alpha} P, \tag{A2}$$

in which P_{e} (mm d^{-1}) indicates effective precipitation, P (mm d^{-1}) is precipitation, SM (mm) and FC (mm) are soil moisture and maximum soil storage capacity, respectively, and α (–) is a shape factor.

We use long-term monthly mean evapotranspiration to calculate the actual evapotranspiration with a temperature adjustment.

$$L_{\mathrm{et}} = \mathrm{FC}c_{\mathrm{et}}, \tag{A3}$$

$$\mathrm{ET}_t = [1 + c_t(T - T_{\mathrm{m}})]\mathrm{ET}_{\mathrm{m}}, \tag{A4}$$

$$\mathrm{ET}_{\mathrm{a}} = \begin{cases} \mathrm{ET}_t, & \mathrm{SM} \geq L_{\mathrm{et}}, \\ \dfrac{\mathrm{SM}}{L_{\mathrm{et}}}\mathrm{ET}_t, & \mathrm{SM} < L_{\mathrm{et}}, \end{cases} \tag{A5}$$

in which L_{et} (mm) is a threshold for maximum evapotranspiration, and c_{et} (–) is a factor to calculate L_{et}. ET_t (mm d^{-1}) represents the maximum evapotranspiration at temperature T (°C). ET_{m} (mm d^{-1}) and T_{m} (°C) indicate long-term monthly mean evapotranspiration and long-term monthly mean temperature, respectively, and c_t (°C^{-1}) is a temperature adjustment factor. ET_{a} (mm d^{-1}) represents actual evapotranspiration, which reaches maximum evapotranspiration when soil moisture is greater than the threshold for maximum evapotranspiration. Otherwise, it increases linearly with soil moisture.

The top storage layer, soil moisture, is calculated by the following:

$$\frac{\mathrm{dSM}}{\mathrm{d}t} = P - P_{\mathrm{e}} - \mathrm{ET}_{\mathrm{a}}, \tag{A6}$$

in which $\frac{\mathrm{dSM}}{\mathrm{d}t}$ (mm d^{-1}) represents the change rate of soil moisture. It is used for agriculture and forest. The change rate of soil moisture for urban areas is $\frac{\mathrm{dSM}}{\mathrm{d}t}(1 - r_{\mathrm{eff}})$, because we assume that precipitation in the effective urban areas will directly become urban surface runoff.

The surface runoff in the effective urban area, overflow and interflow are calculated by the following:

$$q_{\mathrm{effurb}} = P, \tag{A7}$$

$$q_{\mathrm{of}} = \begin{cases} 0, & S_{\mathrm{up}} < L_{\mathrm{of}}, \\ k_{\mathrm{of}}(S_{\mathrm{up}} - L_{\mathrm{of}}), & S_{\mathrm{up}} \geq L_{\mathrm{of}}, \end{cases} \tag{A8}$$

$$q_{\mathrm{if}} = k_{\mathrm{if}}S_{\mathrm{up}}, \tag{A9}$$

$$q_{\mathrm{bf}} = k_{\mathrm{bf}}S_{\mathrm{gw}}, \tag{A10}$$

in which q_{effurb} (mm d^{-1}) is surface runoff in the effective urban area, and q_{of} (mm d^{-1}) represents overflow when subsurface storage S_{up} (mm) is greater than an overflow threshold L_{of} (mm). It is used for agriculture, forest, and ineffective urban areas. The term q_{if} (mm d^{-1}) represents interflow, k_{if} (d^{-1}) is a rate constant, q_{bf} (mm d^{-1}) represents base flow, S_{gw} (mm) is groundwater storage, and k_{bf} (d^{-1}) is a baseflow recession coefficient.

The two equations below are used to calculate percolation and quick recharge.

$$q_{\mathrm{perc}} = k_{\mathrm{perc}}S_{\mathrm{up}}, \tag{A11}$$

$$q_{\mathrm{qr}} = \begin{cases} 0, & S_{\mathrm{up}} < L_{\mathrm{qr}}, \\ k_{\mathrm{qr}}(S_{\mathrm{up}} - L_{\mathrm{qr}}), & S_{\mathrm{up}} \geq L_{\mathrm{qr}}, \end{cases} \tag{A12}$$

in which q_{perc} (mm d^{-1}) represents percolation from soil moisture to subsurface storage; k_{perc} (d^{-1}) is a rate constant; q_{qr} (mm d^{-1}) represents quick recharge, which occurs when subsurface storage reaches a quick recharge threshold L_{qr} (mm); and k_{qr} (d^{-1}) is a rate constant.

The subsurface storage and groundwater storage are calculated by:

$$\frac{\mathrm{d}S_{\mathrm{up}}}{\mathrm{d}t} = \begin{cases} P_{\mathrm{e}} - q_{\mathrm{perc}} - q_{\mathrm{qr}} - q_{\mathrm{of}} - q_{\mathrm{if}}, \\ \text{agriculture and forest} \\ P_{\mathrm{e}}(1 - r_{\mathrm{eff}}) - q_{\mathrm{perc}} - q_{\mathrm{qr}} - q_{\mathrm{of}} - q_{\mathrm{if}}, \\ \text{urban area} \end{cases} \tag{A13}$$

$$\frac{\mathrm{d}S_{\mathrm{gw}}}{\mathrm{d}t} = q_{\mathrm{perc}} + q_{\mathrm{qr}} - q_{\mathrm{bf}}, \tag{A14}$$

in which $\frac{\mathrm{d}S_{\mathrm{up}}}{\mathrm{d}t}$ (mm d^{-1}) is the change rate of subsurface storage. In the urban area, only precipitation in the ineffective area can partly become recharge to the subsurface storage. The term $\frac{\mathrm{d}S_{\mathrm{gw}}}{\mathrm{d}t}$ (mm d^{-1}) represents the change rate of groundwater storage.

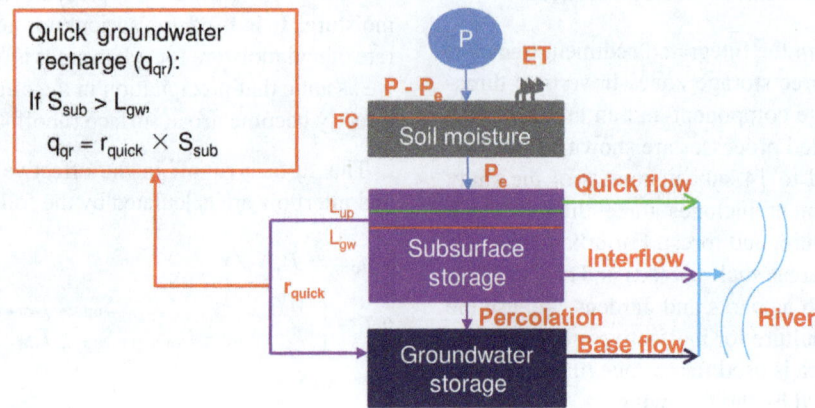

Figure A1. The hydrological model for the Ammer catchment with three storage zones (soil moisture, subsurface storage, and groundwater storage), a quick groundwater recharge and an urban surface runoff component.

Author contributions. YL, CZ, NBB, and OAC conceptualized the study. YL wrote the code with the help of OAC. MS provided and preprocessed the data of discharge and turbidity. YL prepared the paper and all co-authors were involved in reviewing the paper.

Competing interests. The authors declare that they have no conflict of interest.

Acknowledgements. This study was supported by the German Research Foundation (Deutsche Forschungsgemeinschaft, DFG) within the Research Training Group "Integrated Hydrosystem Modeling" (grant GRK 1829). Additional funding is granted by the Collaborative Research Center SFB 1253 "CAMPOS – Catchments as Reactors", and by the EU FP7 Collaborative Project GLOBAQUA (grant agreement no. 603629).

Edited by: Christian Stamm

References

Amalfitano, S., Corno, G., Eckert, E., Fazi, S., Ninio, S., Callieri, C., Grossart, H.-P., and Eckert, W.: Tracing particulate matter and associated microorganisms in freshwaters, Hydrobiologia, 800, 145–154, https://doi.org/10.1007/s10750-017-3260-x, 2017.

Bicknell, B. R., Imhoff, J. C., Kittle Jr., J. L., Jobes, T. H., and Donigian Jr., A. S.: HYDROLOGICAL SIMULATION PRO-GRAM – FORTRAN, Version 12, AQUA TERRA Consultants, California, US, 2001.

Bilotta, G. S. and Brazier, R. E.: Understanding the influence of sus-pended solids on water quality and aquatic biota, Water Res., 42, 2849–2861, https://doi.org/10.1016/j.watres.2008.03.018, 2008.

BKG: Spatial Data Access Act of 10 February 2009, Federal Law Gazette [BGBl.] Part I, p. 278,, Bundesamt für Kartographie und Geodäsie (BKG), 2009

Bones, E. J.: Predicting critical shear stress and soil erodibility classes using soil properties, Georgia Institute of Technology, 2014.

Bouchaoua, L., Manginb, A., and Chauve, P.: Turbidity mechanism of water from a karstic spring: example of the Ain Asserdoune spring (Beni Mellal Atlas, Morocco), J. Hydrol., 265, 34–42, 2002.

Bouteligier, R., Vaes, G., and Berlamont, J.: Sensitivity of urban drainage wash-off models: compatibility analysis of HydroWorks QM and MouseTrapusing CDF relationships, J. Hydroinform., 4, 235–243, 2002.

Brunner, G. W.: HEC-RAS, River Analysis System Hydraulic Ref-erence Manual Version 5.0, Institute for Water Resources Hydro-logic Engineering Center, Davis, CA US, 2016.

BwAm: Weather data at weather station Bondorf, Agrarmeteorolo-gie Baden-Württenmberg (BwAm), 2016.

CDC: Historical hourly station observations of precipitation for Germany, version v005, DWD Climate Data Center (CDC), 2017.

Clark, L. A. and Wynn, T. M.: Methods for determining streambank critical shear stress and soil erodibility: Implications for erosion rate predictions, T. ASABE, 50, 95–106, 2007.

Dong, J., Xia, X., Wang, M., Lai, Y., Zhao, P., Dong, H., Zhao, Y., and Wen, J.: Effect of water–sediment regu-lation of the Xiaolangdi Reservoir on the concentrations, bioavailability, and fluxes of PAHs in the middle and lower reaches of the Yellow River, J. Hydrol., 527, 101–112, https://doi.org/10.1016/j.jhydrol.2015.04.052, 2015.

Dong, J., Xia, X., Wang, M., Xie, H., Wen, J., and Bao, Y.: Effect of recurrent sediment resuspension-deposition events on bioavailability of polycyclic aromatic hydrocar-bons in aquatic environments, J. Hydrol., 540, 934–946, https://doi.org/10.1016/j.jhydrol.2016.07.009, 2016.

Ferro, V. and Porto, P.: Sediment Delivery Distributed (Sedd) Model, J. Hydrol. Eng., 5, 411–422, https://doi.org/10.1061/(Asce)1084-0699(2000)5:4(411), 2000.

Flanagan, D. C. and Nearing, M. A.: USDA – WATER EROSION PREDICTION PROJECT, USDA-ARS National Soil Erosion Research Laboratory, Indiana, US, 1995.

Gilley, J. E., Elliot, W. J., Laflen, J. M., and Simoanton, J. R.: Critical Shear Stress and Critical Flow Rates for Initiation of Rilling, J. Hydrol., 142, 251–271, 1993.

Gong, Y., Liang, X., Li, X., Li, J., Fang, X., and Song, R.: Influence of Rainfall Characteristics on Total Suspended Solids in Urban Runoff: A Case Study in Beijing, China, Water, 8, 278, 1–23, https://doi.org/10.3390/w8070278, 2016.

Grabowski, R. C., Droppo, I. G., and Wharton, G.: Erodibility of cohesive sediment: The importance of sediment properties, Earth-Sci. Rev., 105, 101–120, https://doi.org/10.1016/j.earscirev.2011.01.008, 2011.

Hanson, G. J. and Simon, A.: Erodibility of cohesive streambeds in the loess area of the midwestern USA, Hydrol. Process., 15, 23–38, 2001.

Haygarth, P. M., Bilotta, G. S., Bol, R., Brazier, R. E., Butler, P. J., Freer, J., Gimbert, L. J., Granger, S. J., Krueger, T., Macleod, C. J. A., Naden, P., Old, G., Quinton, J. N., Smith, B., and Worsfold, P.: Processes affecting transfer of sediment and colloids, with associated phosphorus, from intensively farmed grasslands: an overview of key issues, Hydrol. Process., 20, 4407–4413, https://doi.org/10.1002/hyp.6598, 2006.

Kaase, C. T. and Kupfer, J. A.: Sedimentation patterns across a Coastal Plain floodplain: The importance of hydrogeomorphic influences and cross-floodplain connectivity, Geomorphology, 269, 43–55, https://doi.org/10.1016/j.geomorph.2016.06.020, 2016.

Krone, R. B.: Flume studies of the transport of sediment in estuarial shoaling processes, Univ. of Calif., Berkeley, 1962.

Léonard, J. and Richard, G.: Estimation of runoff critical shear stress for soil erosion from soil shear strength, Catena, 57, 233–249, https://doi.org/10.1016/j.catena.2003.11.007, 2004.

LGRB: Bodenkarte von Baden-Württemberg 1 : 50 000, GeoLa – Integrierte Geowissenschaftliche Landesaufnahme, Regierungspräsidium Freiburg Landesamt für Geologie, Rohstoffe und Bergbau, 2011.

Lindström, G., Johansson, B., Persson, M., Gardelin, M., and Bergström, S.: Development and test of the distributed HBV-96 hydrological model, J. Hydrol., 201, 272–288, 1997.

LUBW: Erstellung von Hochwassergefahrenkarten des Landes Baden-Württemberg, LUBW (Landesanstalt für Umwelt, Messungen und Naturschutz, Karlsruhe), 2010.

Meus, P., Moureaux, P., Gailliez, S., Flament, J., Delloye, F., and Nix, P.: In situ monitoring of karst springs in Wallonia (southern Belgium), Environ. Earth Sci., 71, 533–541, https://doi.org/10.1007/s12665-013-2760-x, 2013.

Meyer, T. and Wania, F.: Organic contaminant amplification during snowmelt, Water Res., 42, 1847–1865, https://doi.org/10.1016/j.watres.2007.12.016, 2008.

Modugno, M., Gioia, A., Gorgoglione, A., Iacobellis, V., Forgia, G., Piccinni, A., and Ranieri, E.: Build-Up/Wash-Off Monitoring and Assessment for Sustainable Management of First Flush in an Urban Area, Sustainability, 7, 5050–5070, https://doi.org/10.3390/su7055050, 2015.

Morgan, R. P. C., Quinton, J. N., Smith, R. E., Govers, G., Poesen, J. W. A., Auerswald, K., Chisci, G., Torri, D.,

and Styczen, M. E.: The European Soil Erosion Model (EUROSEM): A dynamic approach for predicting sediment transport from fields and small catchments, Earth Surf. Proc. Land., 23, 527–544, https://doi.org/10.1002/(Sici)1096-9837(199806)23:6<527::Aid-Esp868>3.0.Co;2-5, 1998.

Mukherjee, D. P.: Dynamics of metal ions in suspended sediments in Hugli estuary, India and its importance towards sustainable monitoring program, J. Hydrol., 517, 762–776, https://doi.org/10.1016/j.jhydrol.2014.05.069, 2014.

Neitsch, S. L., Arnold, J. G., Kiniry, J. R., and Williams, J. R.: Soil and Water Assessment Tool, Theoretical Documentation Version 2009, Texas Water Resources Institute, Texas A&M University System, Texas, US, 2011.

Partheniades, E.: Erosion and Deposition of Cohesive Soils, J. Hydr. Eng. Div., 91, 105–139, 1965.

Patil, S., Sivapalan, M., Hassan, M. A., Ye, S., Harman, C. J., and Xu, X.: A network model for prediction and diagnosis of sediment dynamics at the watershed scale, J. Geophys. Res., 117, F00A04, https://doi.org/10.1029/2012jf002400, 2012.

Peraza-Castro, M., Sauvage, S., Sanchez-Perez, J. M., and Ruiz-Romera, E.: Effect of flood events on transport of suspended sediments, organic matter and particulate metals in a forest watershed in the Basque Country (Northern Spain), Sci. Total Environ., 569–570, 784–797, https://doi.org/10.1016/j.scitotenv.2016.06.203, 2016.

Piro, P. and Carbone, M.: A modelling approach to assessing variations of total suspended solids (tss) mass fluxes during storm events, Hydrol. Process., 28, 2419–2426, https://doi.org/10.1002/hyp.9809, 2014.

Romero, C. C., Stroosnijder, L., and Baigorria, G. A.: Interrill and rill erodibility in the northern Andean Highlands, Catena, 70, 105–113, https://doi.org/10.1016/j.catena.2006.07.005, 2007.

Quesada, S., Tena, A., Guillen, D., Ginebreda, A., Vericat, D., Martinez, E., Navarro-Ortega, A., Batalla, R. J., and Barcelo, D.: Dynamics of suspended sediment borne persistent organic pollutants in a large regulated Mediterranean river (Ebro, NE Spain), Sci. Total Environ., 473–474, 381–390, https://doi.org/10.1016/j.scitotenv.2013.11.040, 2014.

Quinton, J. N. and Catt, J. A.: Enrichment of Heavy Metals in Sediment Resulting from Soil Erosion on Agricultural Fields, Environ. Sci. Technol., 41, 3495–3500, 2007.

Rossman, L. A. and Huber, W. C.: Storm Water Management Model, Reference Manual, Volume III – Water Quality, US Environmental Protection Agency, Cincinnati, OH, US, 2016.

Rügner, H., Schwientek, M., Beckingham, B., Kuch, B., and Grathwohl, P.: Turbidity as a proxy for total suspended solids (TSS) and particle facilitated pollutant transport in catchments, Environ. Earth Sci., 69, 373–380, https://doi.org/10.1007/s12665-013-2307-1, 2013.

Rügner, H., Schwientek, M., Egner, M., and Grathwohl, P.: Monitoring of event-based mobilization of hydrophobic pollutants in rivers: calibration of turbidity as a proxy for particle facilitated transport in field and laboratory, Sci. Total Environ., 490, 191–198, https://doi.org/10.1016/j.scitotenv.2014.04.110, 2014.

Scanlon, T. M., Kiely, G., and Xie, Q.: A nested catchment approach for defining the hydrological controls on non-point phosphorus transport, J. Hydrol., 291, 218–231, https://doi.org/10.1016/j.jhydrol.2003.12.036, 2004.

Schwientek, M., Osenbrück, K., and Fleischer, M.: Investigating hydrological drivers of nitrate export dynamics in two agricultural catchments in Germany using high-frequency data series,

Environ. Earth Sci., 69, 381–393, 10.1007/s12665-013-2322-2, 2013a.

Schwientek, M., Rugner, H., Beckingham, B., Kuch, B., and Grathwohl, P.: Integrated monitoring of particle associated transport of PAHs in contrasting catchments, Environ. Pollut., 172, 155–162, https://doi.org/10.1016/j.envpol.2012.09.004, 2013b.

Selle, B., Schwientek, M., and Lischeid, G.: Understanding processes governing water quality in catchments using principal component scores, J. Hydrol., 486, 31–38, https://doi.org/10.1016/j.jhydrol.2013.01.030, 2013.

Siddiqui, A. and Robert, A.: Thresholds of erosion and sediment movement in bedrock channels, Geomorphology, 118, 301–313, https://doi.org/10.1016/j.geomorph.2010.01.011, 2010.

Slaets, J. I. F., Schmitter, P., Hilger, T., Lamers, M., Piepho, H.-P., Vien, T. D., and Cadisch, G.: A turbidity-based method to continuously monitor sediment, carbon and nitrogen flows in mountainous watersheds, J. Hydrol., 513, 45–57, https://doi.org/10.1016/j.jhydrol.2014.03.034, 2014.

UBA: CORINE Land Cover (CLC2006), Umweltbundesamt (German Environmental Protection Agency) (DLR-DFD 2009), 2009.

Wicke, D., Cochrane, T. A., and O'Sullivan, A.: Build-up dynamics of heavy metals deposited on impermeable urban surfaces, J. Environ. Manage., 113, 347–354, https://doi.org/10.1016/j.jenvman.2012.09.005, 2012.

Winterwerp, J. C., van Kesteren, W. G. M., van Prooijen, B., and Jacobs, W.: A conceptual framework for shear flow-induced erosion of soft cohesive sediment beds, J. Geophys. Res.-Oceans, 117, C10020, https://doi.org/10.1029/2012jc008072, 2012.

Wischmeier, H. and Smith, D. D.: Predicting rainfall erosion losses, USDA Science and Education Administration, 1978.

Zhang, M. and Yu, G.: Critical conditions of incipient motion of cohesive sediments, Water Resour. Res., 53, 7798–7815, https://doi.org/10.1002/2017WR021066, 2017.

Zhang, Y., Xiao, W., and Jiao, N.: Linking biochemical properties of particles to particle-attached and free-living bacterial community structure along the particle density gradient from freshwater to open ocean, J. Geophys. Res.-Biogeo., 121, 2261–2274, https://doi.org/10.1002/2016jg003390, 2016.

11

An integrated probabilistic assessment to analyse stochasticity of soil erosion in different restoration vegetation types

Ji Zhou[1,2], Bojie Fu[1,2], Guangyao Gao[1], Yihe Lü[1], and Shuai Wang[1]

[1]State Key Laboratory of Urban and Regional Ecology, Research Center for Eco-Environmental Science, Chinese Academy of Science, Beijing 100085, People's Republic of China
[2]University of Chinese Academy of Sciences, Beijing 100049, People's Republic of China

Correspondence to: Bojie Fu (bfu@rcees.ac.cn)

Abstract. The stochasticity of soil erosion reflects the variability of soil hydrological response to precipitation in a complex environment. Assessing this stochasticity is important for the conservation of soil and water resources; however, the stochasticity of erosion event in restoration vegetation types in water-limited environment has been little investigated. In this study, we constructed an event-driven framework to quantify the stochasticity of runoff and sediment generation in three typical restoration vegetation types (*Armeniaca sibirica* (T1), *Spiraea pubescens* (T2) and *Artemisia copria* (T3)) in closed runoff plots over five rainy seasons in the Loess Plateau of China. The results indicate that, under the same rainfall condition, the average probabilities of runoff and sediment in T1 (3.8 and 1.6 %) and T3 (5.6 and 4.4 %) were lowest and highest, respectively. The binomial and Poisson probabilistic model are two effective ways to simulate the frequency distributions of times of erosion events occurring in all restoration vegetation types. The Bayes model indicated that relatively longer-duration and stronger-intensity rainfall events respectively become the main probabilistic contributors to the stochasticity of an erosion event occurring in T1 and T3. Logistic regression modelling highlighted that the higher-grade rainfall intensity and canopy structure were the two most important factors to respectively improve and restrain the probability of stochastic erosion generation in all restoration vegetation types. The Bayes, binomial, Poisson and logistic regression models constituted an integrated probabilistic assessment to systematically simulate and evaluate soil erosion stochasticity. This should prove to be an innovative and important complement in understanding soil erosion from the stochasticity viewpoint, and also provide an alternative to assess the efficacy of ecological restoration in conserving soil and water resources in a semi-arid environment.

1 Introduction

Soil erosion is a global environmental problem. In recent centuries, the erosion rate worldwide has been accelerating due to climate change and anthropogenic activities, causing soil deterioration and terrestrial ecosystem degradation (Jiao et al., 1999; Marques et al., 2008; Fu et al., 2011; Portenga and Bierman, 2011). The uncertainty and intensity of soil erosion are major features of the erosion phenomenon. Although many studies have concentrated on the intensity of erosion at different spatiotemporal scales (Cantón et al., 2011; Puigdefábregas et al., 1999), the uncertainty of soil erosion generation is a further challenge for researchers working to improve the accuracy of erosion prediction. The stochasticity of environment and spatiotemporal heterogeneity of soil loss is the main influence on the randomness of runoff production and sediment transportation in natural conditions (Kim et al., 2016). But the complex mechanism of erosion generation also increases the uncertainty and variation of erosion processes (Sidorchuk, 2005, 2009). Therefore, how to effectively describe erosion stochasticity and to reasonably assess its impacting factors is necessary and important for understating soil erosion science from the perspective of randomness.

First, combinations of various probabilistic, conceptual and physical models have been reported as different integrated approaches to describe the stochasticity of soil erosion intensity (see Table 1). As one form of erosion intensity, the runoff has been shown as a stochastic process by different mathematic simulation models. Some studies (Moore, 2007; Janzen and McDonnell, 2015) have also simulated the stochasticity, and further quantified the randomness of runoff production and its connectivity dynamics in hillslope and catchment scales by using different probabilistic distribution functions and conceptual models. In these studies, the theory-driven conceptual models simplified the main hydrological behaviours related to runoff production, highlighting the stochastic effects of infiltration and precipitation on runoff processes. Based on the above precondition, the data-driven probabilistic models further simulated the stochastic runoff production by mapping or calibrating the difference between observed and predicted probabilistic values. As a result, the stochastic-conceptual approaches have formed an effective framework to model rainfall–runoff processes (Freeze, 1980), as well as to assess flood forecasting (Yazdi et al., 2014).

The stochasticity of soil erosion rate which is another pattern of erosion intensity has been investigated by probabilistic and physical models in some studies. The theory-driven physical models in these studies (Sidorchuk, 2005) integrated hydrological and mechanical mechanisms of overflow and soil structure with sediment transpiration processes, stressing the stochastic effect of physical principles on erosion rate in different spatial scales (Table 1). Sidorchuk (2005) introduced stochastic variables and parameters into probabilistic models by randomizing the physical properties of overflow and soil structure. This approach developed the understanding of uncertainty of sediment transpiration processes, causing the randomness simulation to better fit the reality of stochastic erosion rate (Sidorchuk, 2009). Additionally, the stochasticity of soil erosion rate also reflected the erosion risk which was assessed by the integration of a theory-driven empirical model with geostatistics (Jiang et al., 2012; Wang et al., 2002; Kim et al., 2016). Erosion risk analysis has generally concentrated on the uncertainty or variability of soil erosion rate at catchment and regional scales, highlighting the impact of the spatiotemporal heterogeneous rainfall and other environment conditions on the stochastic erosion rate. In summary, these probabilistic and physical models constituted a systematical analysis framework closely related to the principle of water balance and basic hydrological assumptions. This effectively described the randomness of soil erosion rate affected by complex hydrological processes (Bhunya et al., 2007). However, few studies have been made to analyse the stochasticity of soil erosion events. In particular, there has been little effort to systematically investigate how the signal of stochastic rainfall is transmitted to erosion events occurring in different restoration vegetation types based on observational data rather than on other model assumptions. Yet such event-based investigation deriving from specific experiment results may be more practically meaningful for understanding the stochastic interaction between rainfall and erosion events.

Secondly, the probabilistic approaches have also been reported as a crucial tool to describe the stochasticity of factors affecting soil erosion rate (Table 1). The randomness of soil water content (Ridolfi et al., 2003), antecedent soil moisture (Castillo et al., 2003), infiltration rate (Wang and Tartakovsky, 2011) and soil erodibility (Wang et al., 2001) in heterogeneous soil types have all been modelled by different probability distribution functions. The stochasticity of soil hydrological characteristics related to erosion rate mainly impacted in various ways the spatiotemporal distribution of erosion rate, especially at regional or larger spatial scales. Meanwhile, as the main driving force of soil erosion generation, the uncertainty of rainfall event to some extent represents the environment stochasticity (Andrés-Doménech et al., 2010). Eagleson in 1978 applied probabilistic-trait models to characterise the stochasticity of rainfall event by using Poisson and Gamma probability distribution functions. The stochastic rainfall distribution in different spatiotemporal scales has also been applied to examine the effect of runoff and sediment yield (Lopes, 1996), to calibrate the runoff–flood hydrological model (Haberlandt and Radtke, 2014), as well as to evaluate sewer overflow in urban catchment (Andrés-Doménech et al., 2010).

The role of spatial distribution of vegetation in controlling the soil erosion rate under different spatiotemporal scales has been well recognized (Wischmeier and Smith, 1978; Puigdefábregas, 2005). How the plants reduce soil erosion rate has also been illuminated and interpreted by various physical and empirical models (Liu, 2001; Mallick et al., 2014; Prasannakumar et al., 2011). In theory, Puigdefábregas (2005) proposed vegetation-driven spatial heterogeneity (VDSH) to explain the relationship between vegetation patterns and erosion fluxes, which improves understanding of the hydrological function of plants in erosion processes. The trigger–transfer–reserve–pulse (TTRP) framework proposed by Ludwig in 2005 systematically explored the responses and feedback between vegetation patches and runoff erosion during ecohydrological processes. Theoretically, the stochastic signals of different rainfall events could also be disturbed by the hydrological function of plants, finally affecting the randomness of runoff and sediment events occurring in various vegetation types. However, little effort has been made to explore the effect of different vegetation types on the stochasticity of soil erosion events. In particular, few approaches have been used to analyse how the properties of rainfall, soil and vegetation impact on the stochastic erosion events through event-based investigation deriving from observational data rather than via theory-based models. Actually, logistic regression modelling (LRM) probably deals with the above problems. LRM evaluates the causal effects of categorical variables on dependent variables, and quantifies the probabilistic contri-

Table 1. Summary of the research on the stochasticity of soil erosion rate and the stochasticity of factors affecting soil erosion rate.

[a]Stochasticity (uncertainty)	[b]Approach or method	[c]Driven types	Main hydrological behaviours	Main influencing factors	Spatiotemporal scale	Reference
Stochasticity of soil erosion rate						
Runoff connectivity	Probabilistic model Conceptual model	1. Data-mapping 2. Theory	Infiltration processes Precipitation	Topography Soil depth	Hillslope scale in the USA	Janzen and McDonnell (2015)
Runoff processes	Probabilistic model Conceptual model	1. Simulation 2. Theory	Infiltration processes Precipitation	Topography		Janzen and McDonnell (2015)
Runoff production	Probabilistic model Conceptual model	1. Theory 2. Simulation	Runoff absorption Water storage Infiltration capacity	Soil moisture Evaporation recharge	Point and basin scale	Moore (2007)
Flood prediction and runoff	Probabilistic model Multivariate analysis	1. Simulation 2. Data calibration	Stochastic rainfall process	Parameters in rainfall–runoff model	Multiple catchment scales in Iran	Yazdi et al. (2014)
Rainfall and runoff processes	Probabilistic model hydrological mechanism	1. Simulation 2. Random event 3. Theory	Soil storage	Given climate regime hydraulic conductivity landform development	Hillslope scale	Freeze (1980)
Erosion rate	Probabilistic model Mechanical mechanism	1. Data calculation 2. Stochastic forcing		Bed shear stress Critical shear stress	Laboratory scales in the Netherlands	Prooijen and Winterwerp (2010)
Erosion rate	Physical model Probabilistic model Conceptual model	1. Theory 2. Simulation	Simulated near-bed flow	Soil structure Oscillating flow		Sidorchuk (2005)
Erosion risk	Empirical model Geostatistics	1. Data mapping	Erosive precipitation	Factors in RUSLE	Annual and regional scales in China	Jiang et al. (2012)
Uncertainty of soil loss	Empirical model Geostatistics Error analysis	1. Simulation 2. Data calibration	Erosive precipitation Runoff and sediment	Spatiotemporal rainfall erosivity distribution	Annual time and catchment scale in the USA	Wang et al., 2002
Uncertainty and variability of erosion rate	Empirical model	1. Hypotheses 2. Data calculation	Total rainfall volume and 30 min rainfall intensity	Stochastic environment conditions Scale effect		Kim et al. (2016)
Stochasticity of factors affecting soil erosion rate						
Soil moisture related to soil erosion	Probabilistic model Physical model	1. Hypotheses, 2. Simulation 3. Theory	Precipitation Evapotranspiration	Temporal patterns of rainfall property	Daily time and hillslope scale in	Ridolfi et al. (2003)
Antecedent soil moisture related to soil erosion	Probabilistic model Physical model	1. Data mapping 2. Theory	Runoff response Infiltration processes		Daily time and multiple catchment scales in Spain	Castillo et al. (2003)
Stochastic rainfall related to flood and runoff	Probabilistic model Conceptual model	1. Data calibration 2. Random event 3. Hypothesis	Stochastic storm Runoff and flood	Parameters in Peak flow models	Hourly–daily time and multiple catchment scales in Germany	Haberlandt and Radtke (2014)
Stochastic rainfall related to runoff and erosion	Physical model Empirical model	1. Simulation 2. Data calibration	Overland/channel flow Erosion transport Precipitation	Spatiotemporal rainfall distribution	Seasonal and annual time catchment scale in the USA	Lopes (1996)
Uncertainty of soil erodibility	Empirical model Geostatistics	1. Simulation 2. Data mapping		Spatiotemporal soil types, depth and parent material	Regional scales in the USA	Wang et al. (2001)
Stochastic rainfall related to runoff	Probabilistic model Conceptual model Physical model	1. Data calibration 2. Theory	Sewer overflows	Rainfall depth and duration, climate conditions	Seasonal and annual time catchment scales in Spain	Andrés-Doménech et al. (2010)

[a] The main contents of different studies focusing on the stochasticity (uncertainty) of soil erosion and its influencing factors. [b] The main statistical methods or different types of mathematical and physical models employed to describe and analyse the stochasticity of soil erosion. [c] The main properties of analysis framework in the different studies and the characteristics of data application on the evaluation of stochasticity of soil erosion.

bution of influencing factors on the randomness of responsive random events in terms of an odds ratio (Hosmer et al., 2013). This can be seen as another probabilistic model to explore the probability attribution of influencing factors. However, little literature is available on LRM being used to explore the probabilistic attribution of stochastic erosion events under complex environmental conditions.

In this study, we have hypothesized that the uncertainty of erosive events was also an important property of the soil erosion phenomenon, and monitored erosion events occurring in three typical restoration vegetation types at runoff plot scale over five consecutive years' rainy seasons. We aim to (1) comprehensively describe the stochasticity of runoff and sediment events in detail by using probability theory, and (2) systematically evaluate the effect of the properties of soil, plant and rainfall on the stochastic erosion events by employing the LRM approach. The probabilistic description attribution approach constitutes an integrated probabilistic assessment based on event-driven probability theory and data-driven experimental observation. The investigation of stochastic soil erosion events by integrated assessment is an innovative and important complement in understanding soil

erosion from the stochasticity viewpoint, and could also provide an alternative way to assess the efficacy of ecological restoration for conserving soil and water resources in a semi-arid environment.

2　Method

2.1　Definition and classification of random events

Each observed stochastic weather condition with different durations in the field monitoring period was defined as a random experiment. All possible outcomes of a random experiment constituted a sample space (Ω) defined as a random observational event (O event, for short). Two mutually exclusive random event types – random rainfall event (I event, for short) and random non-rainfall event (C event, for short) – constituted the O event. Precipitation is a necessary condition of runoff generation, and the random runoff production event (R event, for short) is a subset of the I event. Similarly, R event is also a necessary condition of random sediment migration event (S event, for short), which leads to S event being a subset of R event. As a result, O, C, I, R and

S events constituted a random events framework (OCIRS) to reflect the stochasticity of the environment in which soil erosion events occur.

The random event duration in OCIRS is an important property of stochastic weather conditions. In particular, the duration property of I events was closely related to the transmission of stochastic signals of rainfall into the R and S events. According to the rainfall duration patterns in China (Wei et al., 2007), the time interval between two adjacent I events is set to be more than 6 h, forming the criterion for individual rainfall classification. Meanwhile, based on the observation of random events over five consecutive rainy seasons, we summarised the duration property of all I events and further classified them into four mutually exclusive I event types: a random extremely long rainfall event type (Ie event, for short), a random general long-duration rainfall event type (Il event, for short), a random spanning rainfall event type (Is event, for short) whose duration spans two consecutive days, and a random within rainfall event type (Iw event, for short) occurring in a day. The C event can also be divided into two types at the daily scale: the random non-rainfall event type lasting a whole day (Cd event, for short) and the random non-rainfall event type whose duration is less than 24 h (Ch event, for short) which is interrupted by an I event.

Table 2 shows the physical, probabilistic properties and implications of all random event types in OCIRS. The classification process of all random event types is illustrated in Fig. 1a, and a Venn diagram of all random event types in OCIRS is shown in Fig. 1c. Considering the observed longest duration of an Ie event approximating 72 h, in Fig. 1b, we have summarised a series of random event sequences in terms of different combination patterns of I and C events in every 3 consecutive days during the whole monitoring period.

2.2 Probabilistic description of erosion event

2.2.1 Conditional probability of erosion event

In the sample space Ω, for any random event type E in OCIRS, we defined $P(E)$ as the proportion of time that E occurs in terms of relative frequency:

$$P(E) = \lim_{n\infty} \frac{n(E)}{n} = p_E, p_E \in [0, 1]. \tag{1}$$

Theoretically, $n(E)$ is the number of times in n outcomes of observed random experiment that the event E occurs. According to the law of total probability (Robert et al., 2013), the probability of R event is defined as

$$P(R) = P(RI) = P\left(R|\cup_{m=1}^{4}I_m\right) P\left(\cup_{m=1}^{4}I_m\right)$$
$$= \sum_{m=1}^{4} P(R|I_m) P(I_m) = p_R. \tag{2}$$

I_m, $m = 1, 2, 3$ and 4 represent the Ie, Il, Is, and Iw respectively, and $P(R|I_m)$ represents the conditional probability

that R event occurs given that the mth I event type has occurred. Similarly, the probability of S event is defined as

$$P(S) = P(SI) = P\left(S|\cup_{m=1}^{4}I_m\right) P\left(\cup_{m=1}^{4}I_m\right)$$
$$= \sum_{m=1}^{4} P(S|I_m) P(I_m) = p_S. \tag{3}$$

Equations (2) and (3) quantify the stochastic soil erosion events by using conditional probability.

2.2.2 Probability distribution functions of erosion event

We define X and Y as two discrete random variables, representing two real-valued functions defined on the sample space (Ω). Let X and Y denote the numbers of times of R and S event occurrence respectively, and assign the sample space Ω to another random variable Z. $X(R) = x, Y(S) = y Z(\Omega) = z, y \leq x \leq z$. The x, y, and z are integers. The ranges of X and Y are $R_X = \{$all $x : x = X(R),$ all $R \in \Omega\}$ and $R_Y = \{$all $y : y = Y(S),$ all$S \in \Omega\}$. The probability of x_i or y_j numbers of times of R or S events can be quantified by probability mass function (PMF) as follows:

$$PMF_X(x_i) = P[\{R_i : X(R_i) = x_i, x_i \in R_X\}] \tag{4}$$

$$PMF_Y(y_j) = P[\{S_j : Y(S_j) = y_j, y_j \in R_Y\}] \text{ for } i \geq j. \tag{5}$$

The PMF in Eqs. (4) and (5) describe the general expression of probability distribution of all possible numbers of times of R or S events.

The random variables X and Y obey the binomial distribution with n independent Bernoulli experiments (Robert et al., 2013). Therefore, the PMF of X and Y can be defined as follows:

$$PMF_{Xbin}(x) = P_{Xbin}(X = x) =$$
$$\begin{cases} \binom{n}{x} p_R^x (1 - p_R)^{n-x} & x = 0, 1, 2, \ldots, n \\ 0 & \text{elsewhere} \end{cases} \tag{6}$$

$$PMF_{Ybin}(y) = P_{ybin}(Y = y) =$$
$$\begin{cases} \binom{n}{y} p_S^y (1 - p_S)^{n-y} & y = 0, 1, 2, \ldots, n \\ 0 & \text{elsewhere,} \end{cases} \tag{7}$$

where x and y indicate all possible numbers of times of R and S occurring over n I events. However, when the Bernoulli experiment is performed infinite independent times ($n \to \infty$), the binomial PMF can be transformed into the Poisson PMF (proved in Appendix A), and finally expressed as follows:

$$PMF_{Xpoi}(x) = P_{Xpoi}(X = x) = \begin{cases} \frac{\lambda_R^x e^{-\lambda_R}}{x!} & x = 0, 1, 2, \ldots \\ 0 & \text{elsewhere} \end{cases} \tag{8}$$

$$PMF_{Ypoi}(y) = P_{Ypoi}(Y = y) = \begin{cases} \frac{\lambda_S^y e^{-\lambda_S}}{y!} & y = 0, 1, 2, \ldots \\ 0 & \text{elsewhere,} \end{cases} \tag{9}$$

where the parameter $\lambda_R \approx np_R$, $\lambda_S \approx np_S$. Equations (6)–(9) reflect two PMF models to simulate the probability distribution of R or S events.

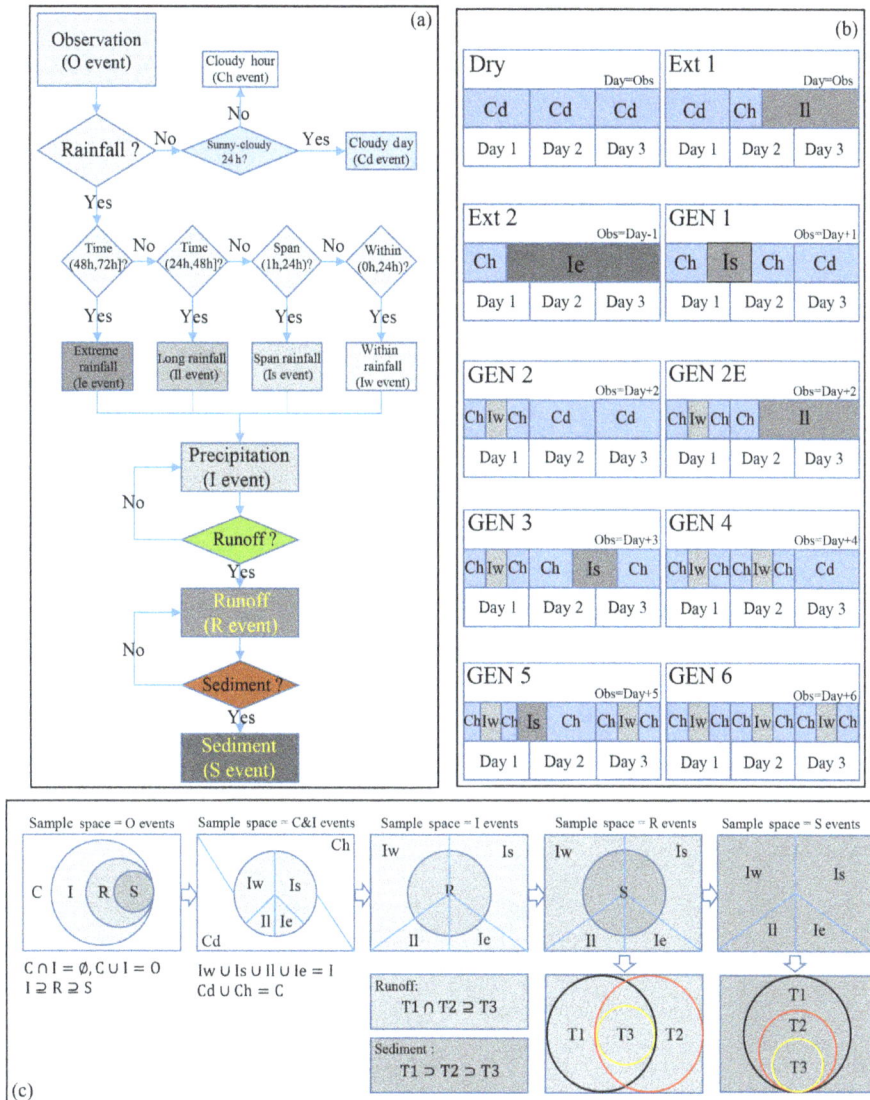

Figure 1. The OCIRS system: **(a)** a flow chart to determine random event types in the OCIRS framework; **(b)** the different combination patterns of rainfall and non-rainfall events in 3 consecutive days to form ten observed random event sequences in five rainy seasons; **(c)** Venn diagram showing the relationship between random event types in the OCIRS framework.

2.3 Probabilistic attribution of erosion events

2.3.1 Bayes model

Based on the Bayes formula theory (Sheldon, 2014), if we want to evaluate how much the probabilistic contributions of kth type of random rainfall event on one stochastic runoff or sediment event which has been generated and observed, the Bayes model can quantify the results as follows:

$$P(I_k|R) = \frac{P(I_kR)}{P(R)} = \frac{P(R|I_k)P(I_k)}{\sum_{m=1}^{4}P(R|I_m)P(I_m)} \quad (10)$$

$$P(I_k|S) = \frac{P(I_kS)}{P(S)} = \frac{P(S|I_k)P(I_k)}{\sum_{m=1}^{4}P(S|I_m)P(I_m)}. \quad (11)$$

In fact, the Bayes model provides an important explanation that how the a priori stochastic information ($P(I_k)$) was modified by the posterior stochastic information ($P(R)$ or $P(S)$). The application of Bayes model in Eqs. (10)–(11) reflects the feedback of random erosion events on the stochastic rainfall events. It could also be regarded as one pattern of probabilistic attribution to assess the effect of different random rainfall events on the uncertainty of soil erosion events without considering the diversity of restoration vegetation.

2.3.2 Logistic regression model

Firstly, we constructed an event-driven logistic function, and defined Y_R and Y_S as two dichotomous dependent variables. When we denote Y_R and Y_S as 1 or 0, it means that a R and

Table 2. Definition and explanation of all random events in OCIRS.

Symbol	Physical meaning of random event types	Probabilistic meaning of random event types	Influencing factors and implication
O	observation events with time step ranging from 0 to 72 h, including non-rainfall and rainfall events	random events composing the sample space of OCIRS system. The probability $P(O) = 1$	indicating the general stochastic weather conditions over rainy seasons
C	non-rainfall events with time step ranging from 0 to 24 h, including sunny or cloudy weather condition at hour or day scales	random events, the probability of C events is the ratio of numbers of C events to O events $C \subset O, 0 \leq P(C) \leq P(O) = 1$	implying the extent of evaporation or potential evapotranspiration in weather condition
Cd	non-rainfall events with time step being 24 h, including observed sunny or cloudy at day scale	random events composing the subset of C events, $Cd \subseteq C, 0 \leq P(Cd) \leq P(C)$	implying the duration of evaporation or evapotranspiration at day scale
Ch	non-rainfall events with time step being less than 24 h, including observed sunny or cloudy at hour scales which intercepted by rainfall events within a day	random events composing the subset of C events, the intersection of Ch and Cd is null, $Ch \subseteq C, Cd \cup Ch = C, Cd \cap Ch = \varnothing, 0 \leq P(Ch) \leq P(C)$	influenced by the frequency of rainfall events generation, and implying the alternation of sunny and rainy in a day
I	an individual rainfall event with different precipitation, intensity and duration ranging from 0 to 72 h, the time interval between two I events is more than 6 h	random events, the probability of I event is ratio of numbers of I events to O events over observation $I \subset O, I \cup C = O, I \cap C = \varnothing, 0 \leq P(I) \leq P(O) = 1$	a driven force of soil erosion, which could be intercepted by vegetation and transformed into throughfall
Ie	an extreme longest individual rainfall event whose average precipitation, intensity and duration were 96.6 mm, 0.022 mm min^{-1}, and 73 h, respectively.	random events composing the subset of I events, $Ie \subseteq I, 0 \leq P(Ie) \leq P(I)$	rainfall events with low intensity and longest duration, leading to infiltration–excess runoff generation
Il	second longest individual rainfall event type whose average precipitation, intensity and duration were 47.3 mm, 0.027 mm min^{-1}, and 30 h, respectively.	random events composing the subset of I events, the intersection of Il and Ie is null, $Il \subseteq I, Il \cap Ie = \varnothing, 0 \leq P(Il) \leq P(I)$	rainfall events with low intensity and long duration, leading to infiltration–excess runoff generation
Is	rainfall event type spanning 2 days whose average precipitation, intensity and duration were 22.7 mm, 0.042 mm min^{-1}, and 10 h, respectively	random events composing the subset of I events, $Is \subseteq I, Is \cap Il \cap Ie = \varnothing, 0 \leq P(Il) \leq P(I)$	rainfall events with strongest rainfall intensity in middle duration, leading to runoff and sediment generation
Iw	rainfall event type occurring within a day whose average precipitation, intensity and duration were 9.8 mm, 0.045 mm min^{-1}, and 5 h, respectively. It usually occurs several times within 1 day.	random events composing the subset of I events, $Iw \subseteq I, Iw \cap Is \cap Il \cap Ie = \varnothing, Iw \cup Is \cup Il \cup Ie = I, 0 \leq P(Iw) \leq P(I)$	rainfall events with least and shortest precipitation and duration, which is difficult to trigger soil erosion
R	runoff event type occurring on vegetation land type; it occurs on rainfall processes, and its duration is negligible	random events responding to I events, $R \subset I, R \cap C = \varnothing, 0 \leq P(R) < P(I)$	influenced by rainfall and vegetation properties
S	sediment event occurring on vegetation land types, it occurs on runoff processes, and its duration is negligible	random events responding to R events, $S \subset R \subset I, S \cap C = \varnothing, 0 \leq P(S) \leq P(R) < P(I)$	driven by R events, and affected by rainfall and vegetation properties

S event has occurred or not occurred. Given that Y_R is a dichotomous dependent variable of R event in the linear probability model in can be expressed as follows:

$$Y_{R_i} = \alpha + \beta_1 x_{1i} + \beta_2 x_{2i} + \cdots + \beta_n x_{ni} + \xi_i$$
$$= \alpha + \sum_{n=1}^{n} \beta_n x_{ni} + \xi_i. \tag{12}$$

We then further transform Eq. (12) into the conditional probability of R event which has occurred in the ith observation time as follows:

$$P\left(Y_{R_i} = 1 | \cap_{n=1}^{n} x_{ni}\right) = P\left[\left(\alpha + \sum_{n=1}^{n} \beta_n x_{ni} + \xi_i\right) \geq 0\right]$$
$$= P\left[\xi_i \leq \left(\alpha + \sum_{n=1}^{n} \beta_n x_{ni}\right)\right]$$
$$= F\left(\alpha + \sum_{n=1}^{n} \beta_n x_{ni}\right), \tag{13}$$

where $\alpha\beta$ are constants and $F\left(\alpha + \sum_{n=1}^{n} \beta_n x_{ni}\right)$ is the cumulative distribution function of ξ_i when $\xi_i = \alpha + \sum_{n=1}^{n} \beta_n x_{ni}$. Equations (12) and (13) quantify the stochasticity of Y_{R_i} depending on the linear combination of n influencing factors x_n and measurement error ξ under i observation times of stochastic runoff generation.

Secondly, assuming that the probabilistic distribution of ξ_i satisfies logistic distribution and $P\left(Y_{R_i} = 1 | \cap_{n=1}^{n} x_{ni}\right) = p_i$, then the LRM expression of $Y_{R_i} = 1$ is deduced as follows:

$$p_i = F\left(\alpha + \sum_{n=1}^{n} \beta_n x_{ni}\right) = \frac{1}{1 + e^{-(\alpha + \sum_{n=1}^{n} \beta_n x_{ni})}}$$
$$= \frac{e^{\alpha + \sum_{n=1}^{n} \beta_n x_{ni}}}{1 + e^{\alpha + \sum_{n=1}^{n} \beta_n x_{ni}}}. \tag{14}$$

Correspondingly, the LRM of $Y_{R_i} = 0$ can be expressed as

$$P\left(Y_{R_i} = 0 | \cap_{n=1}^{n} x_{ni}\right) = 1 - p_i = \frac{1}{1 + e^{\alpha + \sum_{n=1}^{n} \beta_n x_{ni}}}. \tag{15}$$

The ratio of Eq. (14) to (15) is defined as the odds of the R event:

$$\text{Odds} = \frac{p_i}{1 - p_i} = \frac{\frac{e^{\alpha + \sum_{n=1}^{n} \beta_n x_{ni}}}{1 + e^{\alpha + \sum_{n=1}^{n} \beta_n x_{ni}}}}{\frac{1}{1 + e^{\alpha + \sum_{n=1}^{n} \beta_n x_{ni}}}} = e^{\alpha + \sum_{n=1}^{n} \beta_n x_{ni}},$$
$$\text{odds} \in [0, 1]. \tag{16}$$

In this study, odds in Eq. (16) is a probabilistic attribution index to quantify how much the n influencing factors affect the generation of the ith stochastic runoff event. Specifically, when the odds of an influencing factor is greater (less) than 1, it means that the corresponding influencing factor exerts positive (negative) effects on the probability of R generation.

Finally, taking the natural logarithms of both sides of Eq. (16), we transform the odds of stochastic runoff event into the linear Eq. (17) reflecting the standard expression of LRM:

$$\ln\left[\frac{P\left(Y_{R_i} = 1 | \cap_{n=1}^{n} x_{ni}\right)}{P\left(Y_{R_i} = 0 | \cap_{n=1}^{n} x_{ni}\right)}\right] = \ln\left(\frac{p_i}{1 - p_i}\right)$$
$$= \alpha + \sum_{n=1}^{n} \beta_n x_{ni}. \tag{17}$$

LRM can be regarded as another probabilistic attribution pattern to evaluate the effect of multiple impacting factors – such as soil, vegetation and rainfall – on the randomness of soil erosion events occurring in different restoration vegetation types.

3 Experimental design and data analysis

3.1 Study area

The study was implemented in the Yangjuangou Catchment (36°42′ N, 109°31′ E; 2.02 km²), which is located in the typical hilly-gully region of the Loess Plateau in China (Fig. 2a). A semi-arid climate in this area is mainly affected by the North China monsoon. Annual average precipitation reaches approximately 533 mm, and the rainy season here spans from June to September (Liu et al., 2012). The Calcaric Cambisol soil type (FAO-UNESCO, 1974) with weak structure and higher erodibility in the Loess Plateau is vulnerable to water erosion. For these reasons, soil and water loss was one of the most serious environmental problems to degrade the ecosystem in the Loess Plateau before the 1980s (Miao et al., 2010; Wang et al., 2015). After that, as a crucial soil and water resource protection project, the Grain-for-Green Project was widely implemented in the Loess Plateau. A large number of steeply sloped croplands were abandoned, restored or naturally recovered by local shrub and herbaceous plants (Cao et al., 2009; Jiao et al., 1999). In the Yangjuangou Catchment, the main restoration vegetation distributed on hillslopes includes *Robinia pseudoacacia* Linn., *Lespedeza davurica*, *Aspicilia fruticosa*, *Armeniaca sibirica*, *Spiraea pubescens* and *Artemisia copria*. All the restoration vegetation was planted over 20 years ago.

3.2 Design and monitoring

In the Yangjuangou Catchment, systematic long-term field experiments have been conducted, including the monitoring of soil erosion (Liu et al., 2012; Zhou et al., 2016), observation of soil moisture dynamic (Wang et al., 2013; Zhou et al., 2015) and assessment of soil controlling service in this typical water-restricted environment (Fu et al., 2011).

In this study, we first monitored the soil erosion events occurring in three typical restoration vegetation types (*Armeniaca sibirica* (T1), *Spiraea pubescens* (T2) and *Artemisia copria* (T3)) from the rainy seasons of 2008–2012 (Fig. 2b). Each restoration vegetation type was designated by three 3 m by 10 m closed-runoff plots located on southwest-facing hillslopes with 26.8 % aspect. The boundaries of each runoff plot were perpendicularly fenced with impervious polyvinyl chloride (PVC) sheet of 50 cm depth. Collection troughs and storage buckets were installed at the bottom boundary to collect the runoff and sediment (Zhou et al., 2016). Under natural precipitation condition, we recorded the number of times that stochastic runoff and sediment events occurred in each runoff plot over the five rainy seasons. Also, we collected runoff and sediment, separated them and, after settling for 24 h, the samples were dried at 105° over 8 h and weighed.

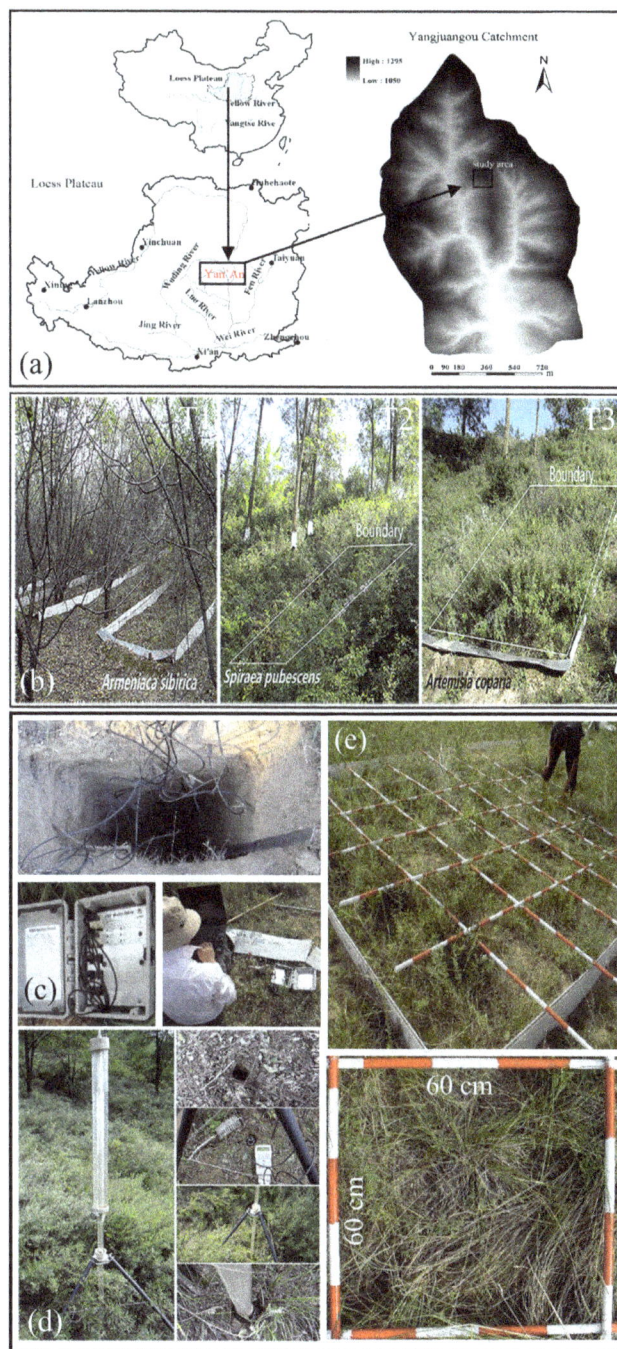

Figure 2. Study area and experimental design: **(a)** location of the Yangjuangou Catchment; **(b)** three restoration vegetation types including *Armeniaca sibirica* (T1), *Spiraea pubescens* (T2) and *Artemisia copria* (T3); **(c)** the dynamic measurement of soil moisture and data collection to provide the information about average antecedent soil moisture; **(d)** the measurement of field-saturated hydraulic conductivity to determine the average infiltration capability; **(e)** investigation of the morphological properties of restoration vegetation by setting quadrats.

Secondly, we systematically monitored the hydrological properties of soil in different restoration vegetation types. In the rainy season of 2010, we began to measure the dynamics of soil moisture in the study region (Wang et al., 2013). The real-time dynamic data of soil water content at intervals of 10 min were recorded by S-SMC-M005 soil moisture probes (Decagon Devices Inc., Pullman, WA, USA), and were collected by HOBO weather station logger (Fig. 2c). These data provided the information about average antecedent soil moisture (ASM) before every rainfall event occurring in the two rainy seasons between 2010 and 2012. We further measured the field-saturated hydraulic conductivity (SHC) in all vegetation types by a model 2800 K1 Guelph permeameter (Soilmoisture Equipment Corp., Santa Barbara, CA, USA) to determine the average infiltration capability of the soil matrix (Fig. 2d).

Thirdly, we investigated the morphological properties of different vegetation types in each runoff-plot for 2–3 times over different periods of rainy season. We measured the average crown width, height and the thickness of litter layer in three restoration vegetation types by setting $60 \times 60\,\mathrm{cm}$ quadrats in each runoff plot (Bonham, 1989) (Fig. 2e).

Finally, two tipping bucket rain gauges were installed outside the runoff plots to automatically record the rainfall processes over the five rainy seasons with an accuracy of 0.2 mm precipitation. Table 3 summarises the properties of the four types of random rainfall event, and the basic characteristic of soil and vegetation is shown in Table 4.

3.3 Statistics

We employed nonparametric statistical tests – one-way ANOVA and post hoc LSD – to determine the significant difference of soil, vegetation and erosive properties in the three restoration vegetation types. The maximum likelihood estimator (MLE) and uniformly minimum variance unbiased estimator (UMVUE) (Robert et al., 2013) were explored to compare the suitability of the binomial PMF and Poisson PMF for predicting the uncertainty of runoff and sediment generation over the long term.

4 Results

4.1 Environmental stochasticity in different rainy seasons

The probabilistic distribution of random rainfall events (I events) and random non-rainfall events (C events) forms the environmental stochasticity which is the background of stochastic soil erosion generation. Within the OCIRS framework, the stochastic environment at monthly and seasonal scales over five rainy seasons is described in Fig. 3. For the rainy seasons of 2008 to 2012, the probability of I event generation first increased with later monitoring period and then decreased in the last two rainy seasons. In the rainy season of 2008, the average probability of I event was lower than the other four rainy seasons, being less than 15 %. However, the I event type was most complex in 2008. The random extremely long rainfall event (Ie event) only appeared in this rainy season, with the probability reaching 2.5 %. On the other hand, the average probability of I event was the highest in the rainy season of 2010, being larger than 18 %. But there were only two types of I events (Iw and Is events) in this rainy season. Over the five rainy seasons, the average probability of Iw (12.3 %) and Ie (0.8 %) event occurrence was the highest and lowest, respectively. The average probability of Is (1.7 %) and Il (1.3 %) events ranged between Iw and Ie. The probability of Cd event was higher than Ch in each month of rainy season, with average probability being 54.4 and 29.4 %, respectively. As seen in Table 3, the difference in average precipitation and duration of the four types of I events was significant. But the average rainfall intensity of Iw and Is events was nearly twice that of Il and Ie events.

Table 3. Main characteristics of the four types of random rainfall event over five rainy seasons.

Rainy season	Rainfall event types	Average precipitation (mm)	Average intensity (mm min^{-1})	Average duration (h)
2008	Iw	16.7	0.122	2.3
	Is	19.2	0.066	4.8
	Il	53.2	0.032	27.7
	Ie	96.6	0.022	73.2
2009	Iw	9.0	0.027	5.6
	Is	35.4	0.059	10.0
	Il	47.9	0.032	24.9
	Ie	×	×	×
2010	Iw	9.0	0.018	8.3
	Is	7.6	0.012	10.6
	Il	×	×	×
	Ie	×	×	×
2011	Iw	3.3	0.031	1.8
	Is	21.5	0.040	9.0
	Il	42.5	0.020	35.4
	Ie	×	×	×
2012	Iw	10.8	0.028	6.4
	Is	30.0	0.031	16.1
	Il	45.5	0.023	33.0
	Ie	×	×	×
Average	Iw	9.8	0.045	4.9
	Is	22.7	0.042	10.1
	Il	47.3	0.027	30.3
	Ie	96.6	0.022	73.2

Table 4. Basic properties of soil, vegetation and erosion in different restoration vegetation types.

Basic properties of different vegetation type	[h]N	Restoration vegetation types		
		Armeniaca sibirica Type 1 (T1)	Spiraea pubescens Type 2 (T2)	Artemisia copria Type 3 (T3)
Topography property				
Slope aspect	9	Southwest	Southwest	Southwest
Slope gradation (%)	9	≈ 26.8	≈ 26.8	≈ 26.8
Slope size for each (m)	9	3 × 10	3 × 10	3 × 10
Soil property				
[a]DBD (g cm^3)	30	1.28 ± 0.08	1.16 ± 0.12	1.23 ± 0.10
Clay (%)	30	11.07 ± 2.43	11.98 ± 3.05	9.54 ± 1.48
Silt (%)	30	26.11 ± 1.50	25.24 ± 3.84	26.72 ± 2.87
Sand (%)	30	62.82 ± 0.94	62.78 ± 4.51	63.74 ± 3.24
[b]Texture type		Sandy loam	Sandy loam	Sandy loam
[c]SHC (cm min^{-1})	20	0.46 ± 0.82(a)	2.22 ± 0.66(b)	0.50 ± 0.60(a)
[d]SOM (%)	30	1.28 ± 0.63(a)	0.98 ± 0.15(b)	0.90 ± 0.09(b)
Vegetation property				
Restoration years	9	20	20	20
Crown diameters (cm)	27	211.6 ± 15.4(c)	80.5 ± 4.5(b)	64.1 ± 6.3(a)
Litter layer (cm)	30	1.2 ± 0.3(a)	3.4 ± 1.8(b)	1.8 ± 0.5(a)
Height (cm)	27	256.3 ± 11.1(c)	128.3 ± 8.3(b)	61.8 ± 1.1(a)
LAI	27	×	2.31	1.78
[e]Ave. coverage (%)	27	85	90	90
Rainfall/erosion property				
Times of rainfall events		130		
Times of runoff events		30/30/30	45/45/45	45/45/45
Times of sediment events		13/13/13	19/19/19	32/32/32
[f]Ave. runoff depth (cm)		0.012(a)	0.014(a)	0.083(b)
[g]Ave. sediment amount (g)		5.8(a)	6.8(a)	25.7(b)

[a] Dry bulk density. [b] Texture type is determined by textural triangle method based on USDA. [c] Field saturated hydraulic conductivity, and all the values with same letter in each row indicates non-significant difference at $\alpha = 0.05$ which is the same as follow rows. [d] Soil organic matter. [e] Average coverage of three restoration vegetation types over five rainy seasons. [f] Average runoff depth in restoration types over rainy seasons. [g] Average sediment yield in restoration types over rainy seasons. [h] Sample number.

4.2 Stochasticity of soil erosion events

4.2.1 Probability of erosion events in vegetation types

The stochasticity of erosion events was quantified by the probability of runoff and sediment generation in three restoration vegetation types (T1, T2 and T3) at monthly and rainy season scales (Fig. 4). Over the five rainy seasons, the probability of soil erosion occurring in all vegetation types generally decreased with later monitoring period, and then increased in 2012. At the early period of erosion monitoring (2008), the randomness of erosion events is similar, and the probability of R and S events ranged from 6 to 13 % and from 3 to 13 % respectively. After that, from the rainy seasons of 2009 to 2011, the highest probabilities of erosion

events in each vegetation type all concentrated in the July and August of each season. Regarding runoff production, the average probability of R event in T1 (3.78 %) was less than that for T2 (5.60 %) and T3 (5.58 %) under the same precipitation condition. With respect to sediment yield, the average probability of S event in T1 (1.65 %) was also the lowest in all restoration vegetation types. In particular, in the last two rainy seasons, there was no S event occurring in T1, but the average probability of S event in T2 and T3 reached 1.83 and 3.36 % respectively in the corresponding rainy seasons. Consequently, affected by the same stochastic signal of rainfall events, T1 and T3 have the lowest and highest probability of erosion event generation over the five rainy seasons respectively.

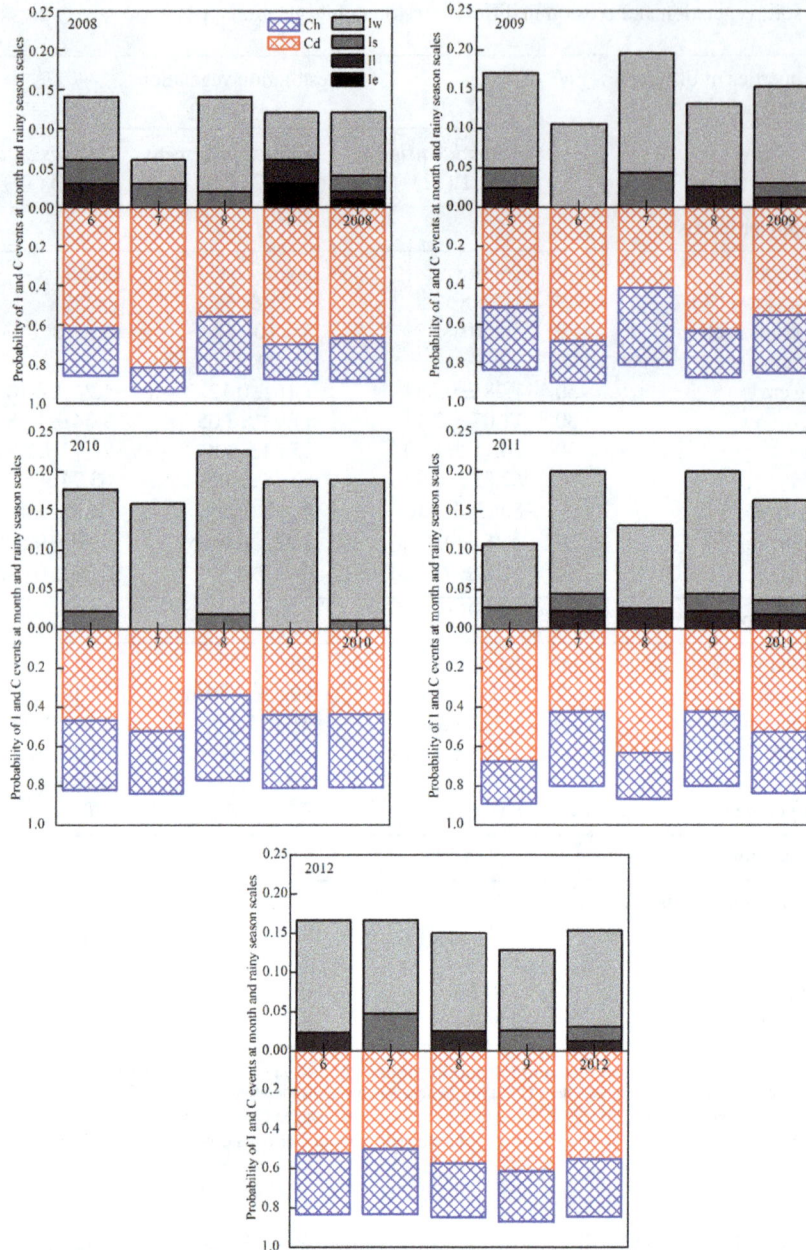

Figure 3. The probability distribution of different random rainfall event types (Iw, Is, Il and Ie) and random non-rainfall event types (Ch and Cd) at monthly and seasonal scales from the rainy seasons of 2008 to 2012.

4.2.2 Probabilistic distribution of erosion events in vegetation types

More detailed stochastic information of erosion events in different vegetation types was simulated by binomial and Poisson PMFs at monthly scale. We also compared the frequency distributions of different numbers of observed erosion events with the corresponding simulated results by the two PMFs in Fig. 5. Firstly, as to the detailed stochastic information of R events, the two PMFs generally provided a better simulation to the observations in T1 than in T2 and T3. When comparing

the simulated and observed values, the binomial PMF supplied better simulation to the observed numbers of time of R events with larger frequency (such as 2–4 times) than did the Poisson PMF. However, the Poisson PMF simulated the observed numbers of time of R events with lower frequency (such as 6–8 times) better than the binomial PMF. Secondly, in relation to the detailed stochastic information of S event, the two PMFs provided better simulation to the observations in T3 than in T1 and T2. In particular, when the number of times of S event generation reaches two in T1 and T2, the corresponding simulated probability values were all nearly

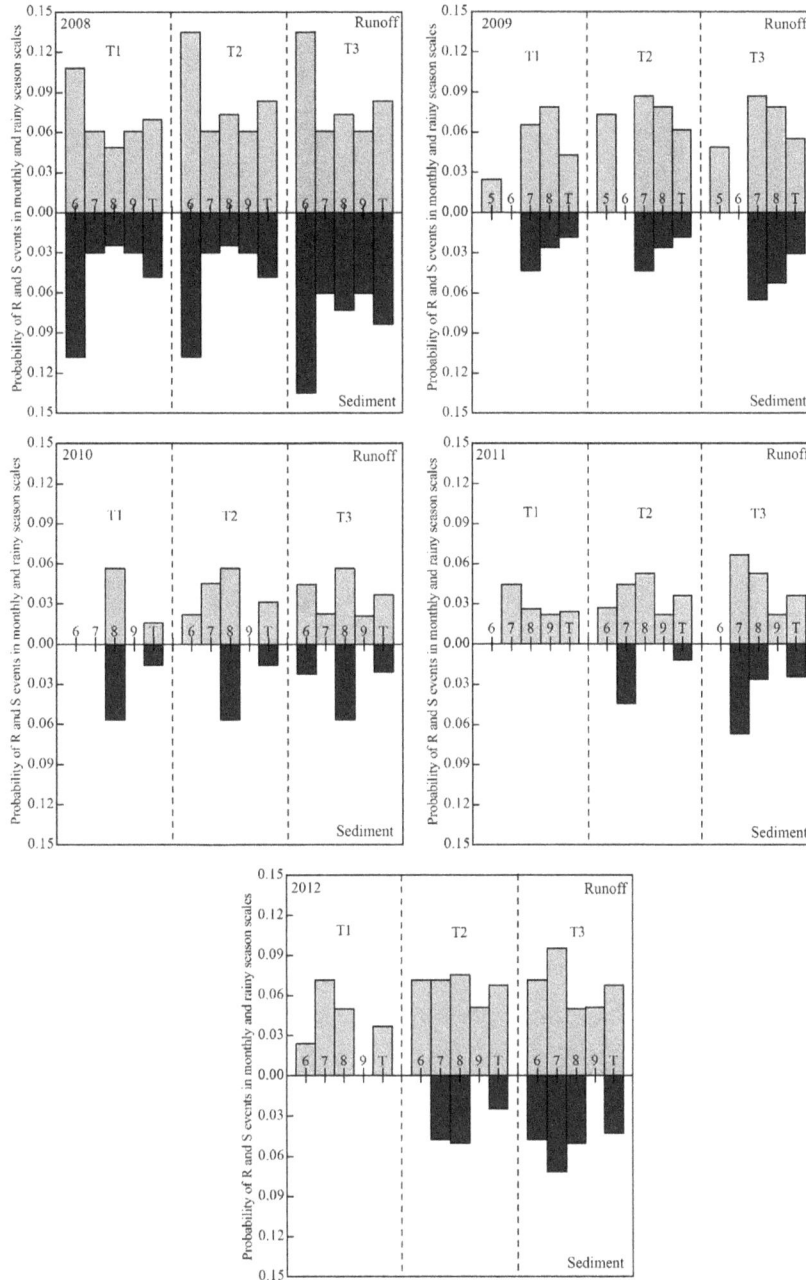

Figure 4. The probability distribution of random runoff and sediment events occurring in three restoration vegetation types at monthly and seasonal scales from the rainy seasons of 2008 to 2012; the Arabic numbers and letter "T" on the abscissa indicate the month and season respectively (also in the following figures).

two times larger than the observed frequencies, reflecting the greatest simulation error of the two PMFs. Moreover, with the restoration vegetation types changing from T1 to T3, both the simulated and observed numbers of time of R and S events with largest probability or frequency increased in consistently. In summary, comparing the observed frequency of numbers of erosion events, both PMFs showed good simulation ability to detail the stochasticity of runoff and sediment events at the monthly scale.

4.3 Stochastic attribution of soil erosion events

4.3.1 Effect evaluation of stochastic erosion events by Bayes model

The Bayes model was applied to analyse the effect of random rainfall events (including Iw, Is, Il and Ie) on stochastic erosion events in different restoration vegetation types. Specifically, the Bayes model evaluated the different probabilistic

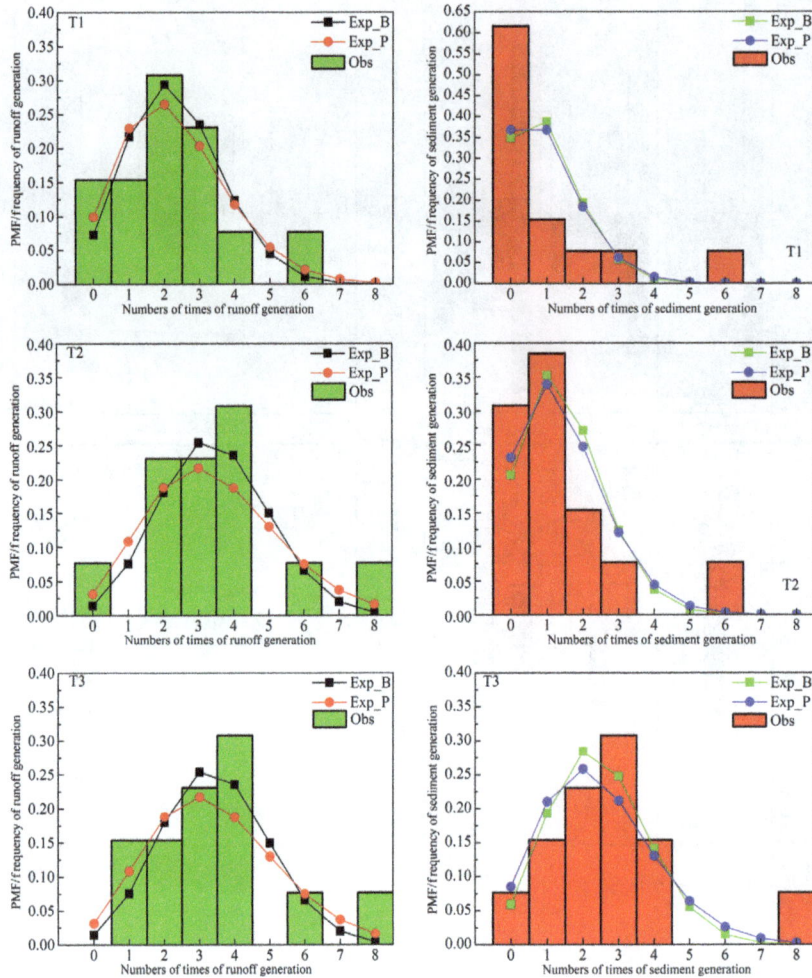

Figure 5. The comparison between simulation of stochasticity of runoff and sediment events by binomial and Poisson PMFs and the observed frequencies of numbers of times of soil erosion events in three restoration vegetation types; Exp_B and Exp_P indicate the simulated values in binomial and Poisson PMF respectively; the histogram shows the observed values.

contributions of four types of I events on one observed erosion event stochastically generated in specific vegetation type at monthly and rainy seasonal scales (Fig. 6). In the rainy season of 2008, the types of I events driving one stochastic erosion event was more complex than in the other rainy seasons. In contrast, only one stochastic soil erosion occurrence in three vegetation types was attributed to Iw and Is events in the rainy season of 2010. In the other three rainy seasons, when one R or S event stochastically generated in T1, the main contributing I event types concentrated on Is and Il events, which have relatively higher precipitation and longer duration, respectively. On the other hand, if one R or S event occurred in T2 or T3 randomly, the main contributing I event type was the Iw event which, however, had no contribution to S event occurring in T1.

In general, over five rainy seasons, the composition of I event driving one R event was more complex than that driving one S event. The relatively longer-duration rainfall events

(Il and Ie) became the main probabilistic contributors of one stochastic erosion event occurring in T1, and the relatively stronger-intensity rainfall events (Iw and Is) mainly caused one random erosion event occurring in T2 and T3.

4.3.2 Effect evaluation of stochastic erosion events by LRM

According to the results of significant difference analysis in Table 4, we defined the properties of soil and plant as ordinal variables, and classified them into four grades (Table 5). Meanwhile, based on previous studies (Liu et al., 2012; Wei et al., 2007) and rainfall properties in this study area, we further subdivided all precipitation and rainfall intensity into four grades with different scores.

First, the intensity of positive and negative effects of a single influencing factor on the probability of runoff and sediment generation in all restoration vegetation types was quantified in terms of odds ratio of erosion events by LRM (Ta-

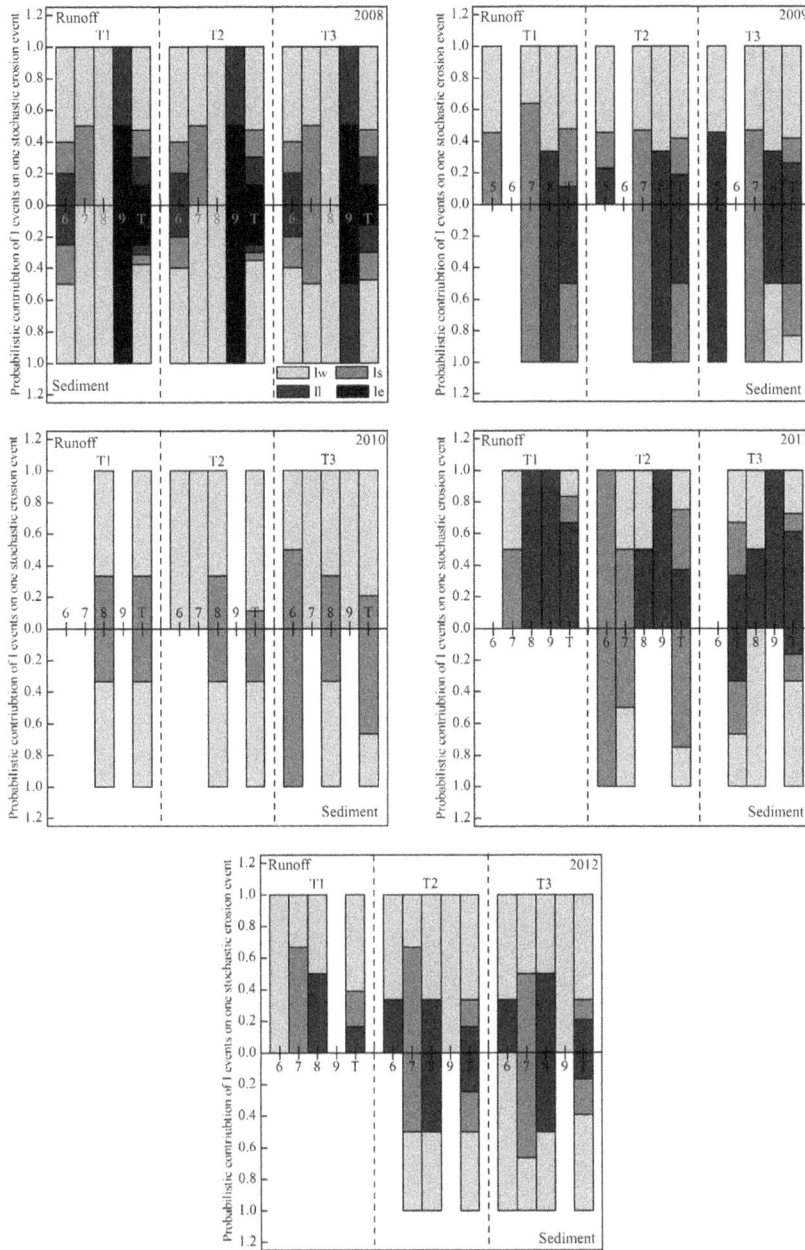

Figure 6. The distribution of probabilistic contribution of four random rainfall event types on any one runoff or sediment event stochastically occurring in three restoration vegetation types at monthly and seasonal scales from rainy season of 2008 to 2012.

ble 6). In the LRM, the highest and lowest odd ratios appeared in rainfall intensity ordinal variable (INT) and average crown width ordinal variable (CRO). An increasing INT and CRO (from middle to extreme grade) significantly increased and decreased the odds ratio of erosion events, respectively. This means that INT and CRO have two of the most important roles in improving and restraining the probability of stochastic erosion generation in all restoration vegetation types. Additionally, the increasing of antecedent soil moisture ordinal variable (ASM) and the SHC ordinal variable (from middle to high grade) in the LRM also signif-

icantly increased and decreased the odds ratio of R and S events, respectively. However, the average thickness of litter layers (TLL) ordinal variable did not have significant effect on the odds ratio of erosion events. Tables S1 and S2 in the Supplement systematically describe the processes of LRM to evaluate the effect of single factors on the odds ratio of erosion event.

Secondly, we applied LRM to evaluate the interactive effects of multiple influencing factors on the odds ratio of R and S events in all restoration vegetation types (Table 7). Regarding the interactive effect of two soil hydrological proper-

Table 5. The definition and classification of properties of rainfall soil and plant ordinal variables.

Ordinal variable	Physical meaning of classified influencing factors	Standard of influencing factor classification			
		Low (L)	Middle (M)	High (H)	Extreme (E)
PREC	classified precipitation variable of a single random rainfall event	0–15 mm	15–30 mm	30–60 mm	>60 mm
INT	classified intensity variable of a single random rainfall event	0–0.025 mm min^{-1}	0.025–0.05 mm min^{-1}	0.05–0.1 mm min^{-1}	>0.1 mm min^{-1}
ASM	classified variable of the antecedent soil moisture	0–5 %	5–10 %	10–20 %	>20 %
SHC	classified variable of the filed saturated hydraulic conductivity	0–1 cm min^{-1}	×	>1 cm min^{-1}	×
CRO	classified variable of the average crown width in vegetation type	0–60 cm	60–80 cm	>80 cm	×
TLL	classified variable of the average thickness of litter layers	0–2 cm	×	>2 cm	×
Y_R	dichotomous dependent variable to indicate whether a random runoff event has occurred or not	If $Y_R = 1$, it means that a random runoff event has occurred; If $Y_R = 0$, it means that a random runoff event has not occurred			
Y_S	dichotomous dependent variable to indicate whether a random sediment event has occurred or not	If $Y_S = 1$, it means that a random sediment event has occurred; If $Y_S = 0$, it means that a random sediment event has not occurred			

ties, the interaction between low grade of SHC and increasing grade of ASM significantly raised the odds ratio of erosion events – the odds ratio of R and S events affected by the interactive effects of low-grade SHC and extreme-grade ASM were respectively 7.02 and 1.82 times larger than the interactive effects of low-grade SHC and low-grade ASM. Similarly, regarding the effect of two vegetation properties, the interactive effect of low-grade CRO and increasing-grade TLL reduces the odds ratio of erosion events – the odds ratio of R and S events influenced by the interaction between low-grade CRO and high-grade TLL were respectively only 0.12 and 0.33 times larger than the interactive effects of low-grade CRO and low-grade TLL. Additionally, with respect to the interaction between soil and plant properties, the interactive effect of low-grade CRO and increasing-grade ASM properties also significantly raised the odds ratio of erosion events. The processes of LRM used to evaluate the interactive effect of multiple factors on odds ratio of erosion event are detailed in Supplement Tables S3–S5.

5 Discussion

5.1 The integrated probabilistic assessment of erosion stochasticity

The probabilistic attribution and description of stochastic erosion events constituted the framework of integrated probabilistic assessment (IPA).

First, as one pattern of probabilistic attribution in the IPA, the Bayes model supplies a supplementary view and algorithm about how to evaluate the feedback of a result which had stochastically occurred on all possible reasons (Wei and Zhang, 2013). Under the conditions of insufficient information about an occurred result, the Bayes model can determine which reasons have relatively greater probability to trigger the occurrence of the result through some prior information. Specific to this study, the Bayes model was used to evaluate the probabilistic contribution of four types of I events on one stochastic R ($P(I_k|R)$) and S ($P(I_k|S)$) event generated in each restoration vegetation type. Although there was no further specific information about a stochastic soil erosion event, the prior information ($P(R|I_m)P(S|I_m)P(I_m)$) can provide assistance for us to assess the feedback of the stochasticity of soil erosion on different random rainfall events by the Bayes model. Meanwhile, ($P(I_k|R)$) and ($P(I_k|S)$) also reflect the different probability threshold values of four rainfall event types triggering soil erosion. The Bayes model integrated with total probability theory to systematically quantify the interactive relationship between the stochasticity of precipitation and soil erosion, forming a relatively simple and practical risk assessment of soil erosion event occurring in complex restoration vegetation conditions.

Secondly, as a pattern of probabilistic description in the IPA, the binomial and Poisson PMFs are two crucial prob-

Table 6. Logistic regression model to analyse the single effect of rainfall, plant and soil ordinal variable on the erosion events presence/absence in all restoration vegetation types.

Grade levels	PREC (low)	INT (low)	ASM (low)	SHC (low)	CRO (low)	TLL (low)
			Odds ratio of all random runoff events			
Extreme	[a]$\times$NS	[b]90.91***	[c]2.19*	Null	Null	Null
High	$\times$NS	32.26***	2.01*	[d]0.85*	[e]$7.53 \times 10^{-3**}$	[f]$\times$NS
Middle	$\times$NS	2.09*	1.59*	Null	$7.17 \times 10^{-2**}$	Null
			Odds ratio of all random sediment events			
Extreme	142.85***	166.67***	15.40*	Null	Null	Null
High	16.95**	125.00***	13.79**	0.78*	$6.27 \times 10^{-3**}$	$\times$NS
Middle	6.09**	34.48***	6.36*	Null	$2.55 \times 10^{-2**}$	Null

[a] Taking the low grade of PREC ordinal variable as reference, the odds ratio of all random runoff event in extreme grade of PREC is not significantly larger than that of low grade of PREC. [b] Taking the low grade of INT ordinal variable as reference, the odds ratio of all random runoff events in extreme grade of INT is 90.91 times significantly larger than that of low grade of INT, under the controlled PREC condition with $P \leq 0.001$. [c] Taking the low grade of ASM ordinal variable as reference, the odds ratio of all random runoff events in extreme grade of ASM is 2.19 times significantly larger than that of low grade of ASM, under the controlled PREC and INT condition with $P \leq 0.1$. [d] Taking the low grade of SHC ordinal variable as reference, the odds ratio of all random runoff events in high grade of SHC is 0.85 times significantly larger than that of low grade of SHC, under the controlled PREC, INT and ASM condition with $P \leq 0.1$. [e] Taking the low grade of CRO ordinal variable as reference, the odds ratio of all random runoff events in high grade of CRO is 7.53×10^{-3} larger than that of low grade of CRO, under the controlled PREC, INT, ASM and SHC condition with $P \leq 0.01$. [f] Taking the low grade of TLL ordinal variable as reference, the odds ratio of all random runoff events in high grade of TLL is not significantly larger than that of low grade of TLL, under the controlled PREC, INT, ASM, SHC and CRO condition. (the Wald test statistic is applied to test the significance of odds ratio: *** $P \leq 0.001$, ** $P \leq 0.01$, * $P \leq 0.1$; NS: not significant; $\times$NS: the nonsignificant value cannot be estimated).

Table 7. Logistic regression model to analyse the interactive effect of rainfall, plant and soil ordinal variables on the erosion events presence/absence in all restoration vegetation types.

Grade levels	Reference of grade levels	ASM (low)	ASM (middle)	ASM (high)	ASM (extreme)	TLL (low)	TLL (high)
		Soil_ASM				Plant_TLL	
		Odds ratio of all random runoff events					
Soil_SHC	SHC (low)	Ref.	[a]2.23NS	3.19NS	7.02*	Null	Null
Plant_TLL	TLL (Low)	Ref.	2.23NS	3.19NS	7.02*	Null	Null
Plant_CRO	CRO (low)	Ref.	[b]64.34*	70.77*	486.43**	Ref.	[c]0.12***
	CRO (middle)	Ref.	$\times$NS	2.32NS	22.49*	Null	Null
	CRO (high)	Ref	Null	Null	Null	Null	Null
		Odds ratio of all sediment runoff events					
Soil_SHC	SHC (low)	Ref.	$\times$NS	1.22NS	1.82NS	Null	Null
Plant_TLL	TLL (low)	Ref.	$\times$NS	1.22NS	1.82NS	Null	Null
Plant_CRO	CRO (low)	Ref.	$\times$NS	$\times$NS	$\times$NS	Ref.	0.33**
	CRO (middle)	Ref.	$\times$NS	$\times$NS	$\times$NS	Null	Null
	CRO (high)	Ref	Null	Null	Null	Null	Null

[a] Taking the interactive effect of low grade of SHC and low grade of ASM as reference, the odds ratio of all random runoff events affected by the interactive effect of low grade of SHC and middle grade of ASM is 2.23 times larger than the interactive effect of low-grade SHC and low-grade ASM under controlled rainfall conditions. [b] Taking the interactive effect of low grade of CRO and low grade of ASM as reference, the odds ratio of all random runoff events affected by the interactive effect of low-grade CRO and middle-grade ASM is 64.34 times significantly larger than that interactive effect of low grade of CRO and low grade of ASM under controlled rainfall conditions, with $P \leq 0.1$. [c] Taking the interactive effect of low grade of CRO and low grade of TLL as reference, the odds ratio of all random runoff events affected by the interactive effect of low grade of CRO and high grade of TLL is 0.12 times significantly larger than that interactive effect of low grade of CRO and low grade of TLL, with $P \leq 0.001$. (the Wald test statistic is applied to test the significance of odds ratio: *** $P \leq 0.001$, ** $P \leq 0.01$, * $P \leq 0.1$; NS: not significant, $\times$NS: the nonsignificant value cannot be estimated).

abilistic functions to characterise many random hydrological phenomena and to model their ecohydrological effects in natural condition (Eagleson, 1978; Rodriguez-Iturbe et al., 1999, 2001). In this study, the two PMFs were found to give good simulations of the frequency of times of soil erosion events in three restoration vegetation types. However, it is necessary and meaningful for the reliability and accuracy of the IPA to assume whether the two PMFs can both stably and reasonably simulate the erosion stochasticity at closed-runoff plot over a longer monitoring period. Therefore, based on the above assumption, two important point estimation methods – the maximum likelihood estimator (MLE) and uniformly minimum variance unbiased estimator (UMVUE) (Robert et al., 2013) – were applied to evaluate the stability of erosion stochasticity estimation by means of analysing the unbiasedness and consistency of p_R, p_S, λ_R and λ_S. Taking parameter analysis of random runoff event for example, we defined X_i as the number of times R event occurred in some specific restoration vegetation type in the ith rainy season ($i = 1, 2, 3, 4$ and 5). The five independent and identical (iid) random variables satisfy the same and mutually independent binomial or Poisson PMFs as follows:

$$X_1, X_2, \ldots, X_5 \xrightarrow{\text{iid}} \text{binomial}(p_R) \text{ or } X_1, X_2, \ldots,$$
$$X_5 \xrightarrow{\text{iid}} \text{Poisson}(\lambda_R). \tag{18}$$

Considering longer monitoring periods, we supposed that the numbers of corresponding I events (n) and rainy seasons (i) would approach infinity ($n, i \to \infty$), and Eq. (18) can be transformed as follows:

$$X_1, X_2, \ldots, X_i \xrightarrow{\text{iid}} \text{binomial}(p) \text{ or } X_1, X_2, \ldots,$$
$$X_i \xrightarrow{\text{iid}} \text{Poisson}(\lambda). \tag{19}$$

We take MLE and UMVUE methods to search for the best reasonable population estimators $\hat{p}$ and $\hat{\lambda}$ to approximate the unknown p and λ in Eq. (19), and finally obtain more comprehensive stochastic information about the randomness of R event over i rainy seasons. Appendix B shows that the best estimator $\hat{p}$ in binomial PMF is the unbiasedness and consistency of the MLE of p. However, as shown in Appendix C, the best estimator $\hat{\lambda}$ in the Poisson PMF has more reliability as it has not only the unbiasedness and consistency of the MLE of λ, but also the UMVUE of MLE. The UMVUE in the Poisson PMF implies that the lowest variance unbiased estimator can cause the Poisson PMF to more steadily and accurately simulate the stochasticity of soil erosion events over long-term observations than the binomial PMF.

Thirdly, besides having better simulation of the stochastic soil erosion events at larger temporal scale, the Poisson PMF is also more suitable for simulating the randomness of S event in the closed-design plot system than the binomial PMF.

Following the hypothesis of Boix-Fayos et al. (2006), the closed runoff-plot design forms an obstruction to prevent

the transportable material from entering the close monitoring system, which, in particular, leads the transport-limited erosion pattern to gradually transform into a detachment-limited pattern in the closed plot over time (Boix-Fayos et al., 2007; Cammerraat, 2002). Consequently, with the extension of monitoring period, this closed-runoff plot design would make it more and more difficult for the sediment to migrate out of the plot, which also reduces the probability of observed S events under the same precipitation condition. In fact, the effect of closed-runoff plot on stochastic sediment event is also implied by the algorithm of the Poisson PMF. Specifically, in order to satisfy that $\lambda = np$ in the Poisson PMF is an unknown constant, the extension of monitoring period could lead the numbers of times of I events (n) to approach infinity, and then the probability (p) of R or S event generation would have to approach zero. The above inference coincides with the assumption about the decreasing of sediment generation in the closed-plot system, and further shows that the Poisson PMF is more reliable to simulate the stochastic erosion events at longer temporal scale.

5.2 The effect of influencing factors on erosion stochasticity

The effects of rainfall, soil and vegetation properties on erosion stochasticity in different restoration vegetation types were evaluated by LRM. This integrated stochastic rainfall events with their precipitation and intensity grades, and connected the ecohydrological functions of soil and plant with their classified hydrological and morphological features.

Just as in previous studies (Verheyen and Hermy, 2001a, b; Verheyen et al., 2003), LRM in this study explored the relative importance of morphological features disturbing the transmission of stochastic signal of I events into R and S events in different restoration vegetation types. These disturbances are closed related to the complex hydrological functions owned by different morphological structures, which finally affect the whole processes of runoff production and sediment yield (Bautista et al., 2007; Puigdefábregas, 2005).

First, many previous field experiments and mechanism models have shown that canopy structure has capacity for intercepting precipitation. This specific hydrological function can prevent rainfall from directly forming overland flow or splashing soil surface particles (Liu, 2001; Mohammad and Adam, 2010; Morgan, 2001; Wang et al., 2012). The precipitation retention by canopy structure has been regarded as an indispensable positive factor to reduce the soil erosion rate. Meanwhile, as a crucial complement to understanding the hydrological function of canopy structure, the result of LRM in this study indicated that the higher-grade canopy structure was a most important morphological feature to reduce the odds ratio of random soil events in all restoration vegetation types. This result suggests that larger canopy diameter would have relatively stronger capacity for disturbing the transmission processes of stochastic signal of rainfall on the soil surface than other morphological properties. From the perspective of erosion stochasticity, the higher-grade canopy structure could finally be attributed to the lower probability of R and S event generation. Therefore, the diversity of canopy structures in different vegetation types could play a key role in reducing both the intensity and probability of soil erosion generation.

Secondly, many studies have also discovered that denser root system distribution in the soil matrix improves the overland reinfiltration (Gyssels et al., 2005). This reinfiltration process is an effective way to recharge soil water stores when the overland flow starts to occur in hillslopes, which is also an indispensable contributing factor to reduce the unit area runoff production (Moreno-de las Heras et al., 2009, 2010). In this study, the potential reinfiltration capacity of the soil matrix could be positively affected by the saturated hydraulic soil conductivity (SHC) index. Figure 7 indicates the distribution patterns of root system in three restoration vegetation types. Meanwhile, the result of LRM also implied that the grade of SHC could negatively affect the odds ratio of stochastic erosion event, which improved the understanding of the hydrological function of plant root distribution from the viewpoint of erosion randomness. This suggests that the denser root system creates more macropores in the subsurface to provide more probability of reinfiltration of overland flow. This disturbance of overland flow by SHC can reduce the probability of erosion event generation.

Thirdly, the litter layer was shown to play multiple roles in conserving the rainfall, by improving infiltration of throughfall, as well as cushioning the splashing of raindrops (Gyssels et al., 2005; Munoz-Robles et al., 2011; Geißler et al., 2012). Therefore, the thicker litter layer in T2 (Fig. 7) probably has stronger capacity for conserving and infiltrating throughfall, as well as inhibiting splash erosion than that of other restoration vegetation types (Woods and Balfour, 2010). Although the result of LRM indicated that there was no significant correlation between the TLL and the odds ratio of soil erosion (Table 6), the interactive effect of TLL and CRO significantly affects the odds ratio of stochastic erosion events (Table 7). The interaction result implied that, under the relative low-grade CRO condition, the higher-grade TLL could have stronger disturbance on the transmission of stochastic signals of rainfall to improve the throughfall absorption to reduce the probability of splash or sheet erosion occurrence.

Additionally, Table 7 explored more interactive effects of the soil and plant properties on the odds ratio of random runoff and sediment event. These explorations suggested that the interactions between soil and vegetation properties formed more complex hydrological functions to affect the stochastic soil erosion event during ecohydrological processes in semi-arid environment (Ludwig et al., 2005).

Although the hydrological traits of vegetation played core roles in reducing the soil erosion depending on the mechanical properties of their morphological structures (Zhu et al., 2015), the LRM analysis in this study further illuminated that

these hydrological-trait morphological structures of vegetation may also play an important role in affecting the stochasticity of soil erosion. Actually, the different stochasticity of soil erosion in three restoration vegetation types reflected the different extent of disturbance of vegetation type on the transmission of stochastic signals of rainfall into soil–plant systems. Therefore, the relatively smaller canopy structure, thinner litter layer and shallower root system in T3 have relatively weaker capacity to disturb the stochastic signal of rainfall than that of T1 and T2 with obvious hydrological-trait morphological structures (Fig. 7). The effect of diverse morphological structures on stochasticity of soil erosion was a meaningful complement to studying the hydrological functions of restoration vegetation types in semi-arid environment.

5.3   The implication of integrated probabilistic assessment

The IPA is an important complement to expand on the understanding of hydrological function existing in vegetation types. The hydrological-trait of morphological structures owned by different plants is closely related to the function of vegetation-driven spatial heterogeneity (VDSH) in affecting the intensity of erosion events. The VDSH theory (Puigdefábregas, 2005) can be regarded as a clear and concise summary to emphasise the dominant role of vegetation in restructuring soil erosion processes. It reflects the effect of spatial distribution patterns of vegetation on their corresponding hydrological functions in controlling erosion rate in patch, stand and even at regional . Therefore, VDSH theory has provided an innovative view to investigating the soil erosion and other ecohydrological phenomena affected by vegetation (Sanchez and Puigdefábregas, 1994; Puigdefábregas, 1998; Boer and Puigdefábregas, 2005). In the study, depending on the long-term experimental data and fundamental probability theories, the IPA concentrated on the hydrological function of VDSH in affecting the randomness of erosion events rather than the erosion rate. This can enrich the comprehension of the hydrological function of vegetation morphological structure in soil erosion phenomena, and also be an effective complement to the application of VDSH theory in interpreting stochastic erosion events.

Additionally, in our study, the IPA also provides a new framework for practitioners to develop restoration strategies focused on controlling the risk of erosion generation rather than only on reducing erosion rate. The framework contains three stages: construction of stochastic environment, description of random erosion events, and evaluation of probabilistic attribution (Fig. 8).

The first stage in the framework aims to build a unified platform to describe the stochasticity of different hydrological phenomena closely related to the erosion event. This stage generally investigates the stochastic background under which soil erosion occurs, which is also an indispensable pre-

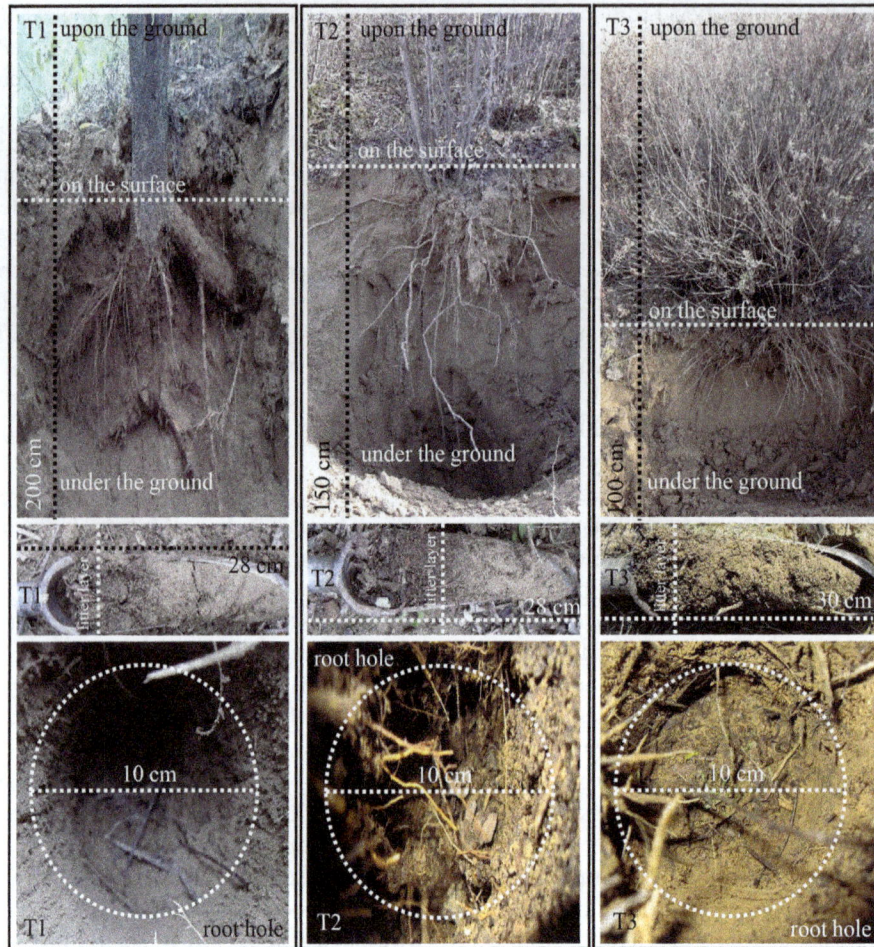

Figure 7. Morphological properties of three restoration vegetation types including the thickness of litter layer and the distribution of root system. The dashed lines indicate the diameter and depth of soil samples, approximately 10 and 30 cm respectively.

condition for quantifying the probability of R and S in stage II. The second stage is designed to construct a phased adjustment of monitoring processes based on the principle of Bayes theory as well as on the parameter analysis of binomial and Poisson models. In this phased-adjustment monitoring, the Bayes, binomial and Poisson models were applied to simulate the randomness of erosion events in short-term, mid-term and long-term monitoring stages, respectively. This model-driven monitoring approach can be regarded as a more reasonable method to explore the complexity of stochastic erosion events in larger temporal scales, but also provide a new perspective for researchers to more effectively evaluate the stochasticity of erosion events in stage III. The objective of stage III is to assess the probabilistic attribution of rainfall, soil and vegetation properties on erosion event generation. This probabilistic attribution evaluation by LRM could develop the restoration strategies for more effectively selecting vegetation types with stronger capacity for reducing the erosion risk, and finally improve the management of soil and water conservation in a semi-arid environment.

As a result, this stochasticity-based restoration strategy was developed by a combination of experimental data with multiple probabilistic theories to deal with the soil erosion randomness under complex stochastic environment. It is different from the trait-based restoration scheme derived from the functional diversity of vegetation community to reduce the soil erosion rate (Zhu et al., 2015; Baetas et al., 2009). Meanwhile, with increased monitoring duration, more stochastic information of erosion events could be added into the IPA framework. This addition could finally fulfil the self-renewal and self-adjustment of the IPA to improve the restoration strategy for selecting more reasonable vegetation types with stronger capacity for controlling erosion risk in the long term. Therefore, the IPA framework containing three stages could translate the event-driven erosion stochasticity into restoration strategies concentrating on erosion randomness, which may be a helpful complement for restoration management in a semi-arid environment.

Stage I: construction and determination

Step 1: **constructing OCIRS system**
Collecting and classifying
influencing factors to characterize
the stochastic environment

Step 2: **determining monitoring period**
From short-term to long-term
monitoring of erosion events
generation.

Stage II: observation and simulation

Step 3: **phased adjustment of description**
Short term: OCIRS-Bayes to analyse
stochastic erosion events
Mid term: OCIRS-binomial to
analyse stochastic erosion events
Long term: OCIRS-Poisson to
analyse stochastic erosion events

Stage III: evaluation and management

Step 4: **probabilistic attribution evaluation**
LRM to determine vegetation types
with stronger capacity for reducing
probability of erosion generation

Step 5: **restoration vegetation selection**
Managing to select the restoration
vegetation by IPA to improve soil
and water conservation

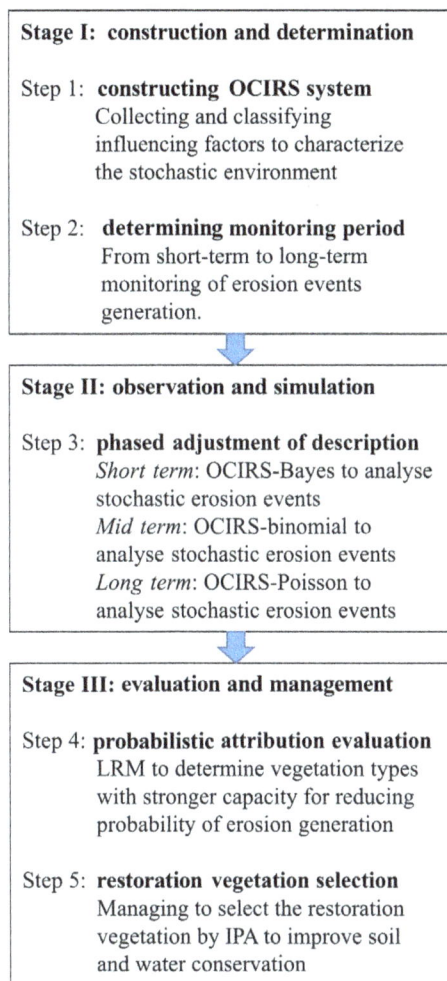

Figure 8. The framework of integrated probabilistic assessment for soil erosion monitoring and restoration strategies.

condition. Larger canopy, thicker litter layer and denser root distribution could more effectively affect the transmission of stochastic signal of rainfall into soil erosion.

The IPA is an important complement to developing restoration strategies to improve the understanding of stochasticity of erosion generation rather than only of the intensity of erosion event. It could also be meaningful to researchers and practitioners to evaluate the efficacy of soil control practices in a semi-arid environment.

6   Conclusion

In this study, we applied an integrated probabilistic assessment (IPA) to describe, simulate and evaluate the stochasticity of soil erosion in three restoration vegetation types in the Loess Plateau of China, and draw the following conclusions.

In the IPA, the OCIRS was an innovative event-driven system to standardise the definition of hydrological random events, which is also a foundation for quantifying the stochasticity of soil erosion events under complex environmental conditions.

Both binomial and Poisson PMFs in the IPA can simulate the probability distribution of the numbers of runoff and sediment events in all restoration vegetation types. However, the Poisson PMF more effectively simulated the stochasticity of soil erosion at larger temporal scales.

The difference of morphological structures in restoration vegetation types is the main source of different stochasticity of soil erosion from T1 to T3 under the same rainfall

Appendix A: The transformation from binomial to Poisson PMF

Let $p = \frac{\lambda}{n}$, then

$$\text{PMF}_{X\text{bin}}(x) = \binom{n}{x} p^x (1-p)^{n-x} = \frac{n!}{x!(n-x)!} \cdot \left(\frac{\lambda}{n}\right)^x \cdot \left(1 - \frac{\lambda}{n}\right)^{n-x}$$

$$= \frac{\lambda!}{x!} \cdot \frac{n(n-1)(n-2)\dots 1}{(n-x)(n-x-1)\dots 1} \cdot \frac{1}{n^x} \cdot \left(1 - \frac{\lambda}{n}\right)^{n-x}$$

$$= \frac{\lambda!}{x!} \cdot 1 \cdot \left(1 - \frac{1}{n}\right) \cdot \left(1 - \frac{2}{n}\right) \dots \left(1 - \frac{x-1}{n}\right) \cdot \left(1 + \frac{-\lambda}{n}\right)^n \cdot \left(1 - \frac{\lambda}{n}\right)^{-x} \quad \text{(A1)}$$

In Eq. (A1), when $n \to \infty$, and x, λ is finite and constant, then

$$\lim_{n\infty}\left(1 - \frac{1}{n}\right) = \dots = \lim_{n\infty}\left(1 - \frac{x-1}{n}\right) = \lim_{n\infty}\left(1 - \frac{\lambda}{n}\right)^{-x} = 1 \quad \text{(A2)}$$

and

$$\lim_{n\infty}\left(1 + \frac{-\lambda}{n}\right)^n = e^{-\lambda} \quad \text{(A3)}$$

and in accordance with Eqs. (A2) and (A3), Eq. (A1) can be transformed as

$$\lim_{n\infty}\left[\frac{n!}{x!(n-x)!} \cdot \left(\frac{\lambda}{n}\right)^x \cdot \left(1 - \frac{\lambda}{n}\right)^{n-x}\right] = \frac{\lambda^x e^{-\lambda}}{x!} x = 0, 1, 2, \dots \quad \text{(A4)}$$

or

$$\text{PMF}_{X\text{bin}}(x) \xrightarrow{n \to \infty} \frac{\lambda^x e^{-\lambda}}{x!} = \text{PMF}_{X\text{poi}}(x). \quad \text{(A5)}$$

Appendix B: Parameter estimation of p in Poisson PMF

B1   Derivatization of the MLE $\hat{p}$

Let the random sample $X_1, X_2, \dots, X_i \xrightarrow{\text{iid}} \text{PMF}_{X\text{bin}}(p)$ and assume the binomial distribution as

$$P(X = x_i) = \binom{m}{x_i} p^{x_i} (1-p)^{m-x_i}. \quad \text{(B1)}$$

The likelihood function $L(p)$ is a joint binomial PDF with parameter p as follows:

$$L(p) = f_X(X_1, \dots, X_n, p) = \prod_{i=1}^{n} \binom{m}{x_i} p^{\sum_{i=1}^n X_i}$$

$$(1-p)^{(mn - \sum_{i=1}^n X_i)}. \quad \text{(B2)}$$

By taking logs on both sides of Eq. (B2),

$$\ln L(p) = \ln\left(\prod_{i=1}^n \binom{m}{x_i}\right)$$

$$+ \sum_{i=1}^n X_i \ln p + \left(mn - \sum_{i=1}^n X_i\right) \ln(1-p) \quad \text{(B3)}$$

and differentiating with respect to p in $\ln L(P)$ and letting the result be zero,

$$\frac{\partial \ln L(p)}{\partial p} = \frac{\sum_{i=1}^n X_i}{p} - \frac{\left(mn - \sum_{i=1}^n X_i\right)}{(1-p)} = 0. \quad \text{(B4)}$$

For solution $\hat{p} = \frac{\sum_{i=1}^n X_i}{mn}$, let $m = n \implies \hat{p} = \frac{\overline{X}}{n}$.

Therefore, $\hat{p} = \frac{\overline{X}}{n}$ is the MLE of population parameter p in the binomial PMF model.

B2   Discussion of the unbiasedness and consistency of $\hat{p}$

Let $E_p(\hat{p})$ be the expectation of MLE $\hat{p}$ when population parameter p is true in random sample which is $X_1, X_2, \dots,$ $X_i \xrightarrow{\text{iid}} \text{PMF}_{X\text{bin}}(p)$, then

$$E_p(\hat{p}) = E_P(\overline{X}/n) = \frac{1}{n^2}\sum_{i=1}^n E_P(X_i) = \frac{1}{n^2} n^2 p = p \quad \text{(B5)}$$

which shows that MLE $\hat{p} = \frac{\overline{X}}{n}$ is an unbiased estimator for p. Furthermore, let $\text{Var}_p(\hat{p})$ be the variance of $\hat{p}$ when population p is true:

$$\text{Var}_p(\hat{p}) = \text{Var}_p\left(\sum_{i=1}^n X_i/n^2\right) = \frac{1}{n^4}\sum_{i=1}^n$$

$$\text{Var}_p(X_i) = \frac{p(1-p)}{n^2}. \quad \text{(B6)}$$

As the n approaches infinity,

$$\lim_{n\infty}\text{Var}_p(\hat{p}) = \left(\frac{p(1-p)}{n^2}.\right) = 0 \quad \text{(B7)}$$

Equations (B5)–(B7) satisfy the theme of the weak law of larger number, which leads the $\hat{p} = \frac{\overline{X}}{n}$ to probabilistic convergence to population parameter p:

$$\lim_{n\infty}P\left(|\hat{p} - p| \geq \varepsilon\right) = 0, \text{ for all } \varepsilon > 0. \quad \text{(B8)}$$

Consequently, the unbiased MLE $\hat{p} = \frac{\overline{X}}{n}$ is consistent for p.

Appendix C: Parameter estimation of λ in Poisson PMF

C1   Derivatization of the MLE $\hat{\lambda}$

Let the random sample $X_1, X_2, \dots, X_i \xrightarrow{\text{iid}} \text{PMF}_{X\text{poi}}(\lambda)$, and assume the Poisson distribution as

$$\text{PMF}_{X\text{poi}}(x_i) = \frac{\lambda^{x_i} e^{-\lambda}}{x_i!} \quad \text{(C1)}$$

The likelihood function $L(\lambda)$ is joint PDF with parameter λ as follows:

$$L(\lambda) = f_X(X_1, \dots, X_n, \lambda) = f(X_1, \lambda) \times \dots \times f(X_n, \lambda)$$

$$= \prod_{i=1}^n \frac{\lambda^{x_i} e^{-\lambda}}{x_i!}. \quad \text{(C2)}$$

Taking logs on $L(\lambda)$ in Eq. (B4) and differentiating logarithm function with respect to λ:

$$\frac{\partial \ln L(\lambda)}{\partial \lambda} = \frac{\partial(\prod_{i=1}^{n} \frac{\lambda^{x_i} e^{-\lambda}}{x_i!})}{\partial \lambda} = -n \frac{\lambda^{\sum_{i=1}^{n} X_i}}{(x_1 x_2 \ldots x_n)!} e^{-n\lambda}$$

$$+ \frac{\sum_{i=1}^{n} X_i \lambda^{(-1 + \sum_{i=1}^{n} X_i)}}{(x_1 x_2 \ldots x_n)!}. \tag{C3}$$

Letting Eq. (C3) equal zero has solution

$$\hat{\lambda} = \frac{1}{n} \sum_{i=1}^{n} X_i = \overline{X}. \tag{C4}$$

Therefore, $\hat{\lambda} = \overline{X}$ is the MLE of population parameter λ in the Poisson PMF model.

C2 Discussion of the unbiasedness and consistency of $\hat{\lambda}$

Let $E_\lambda(\hat{\lambda})$ be the expectation of MLE $\hat{\lambda}$ when population parameter λ is true in random sample $X_1, X_2, \ldots, X_i \xrightarrow{\text{iid}}$ $PMF_{X_{poi}}(\lambda)$, then

$$E_\lambda(\hat{\lambda}) = E_\lambda(\overline{X}) = \frac{1}{n^2} \sum_{i=1}^{n} E_\lambda(X_i) = \frac{1}{n} n\lambda = \lambda, \tag{C5}$$

which shows that MLE $\hat{\lambda} = \overline{X}$ is an unbiased estimator for λ. Meanwhile, let $Var_\lambda(\hat{\lambda})$ be the variance of MLE $\hat{\lambda}$ when population parameter λ is true:

$$Var_\lambda(\hat{\lambda}) = Var_\lambda(X) = Var_\lambda\left(\sum_{i=1}^{n} X_i / n^2\right)$$

$$= \frac{1}{n^4} \sum_{i=1}^{n} Var_\lambda(X_i) = \frac{\lambda}{n} \tag{C6}$$

and

$$\lim_{n\infty} Var_\lambda(\hat{\lambda}) = \left(\frac{\lambda}{n}\right) = 0. \tag{C7}$$

According to the weak law of large numbers (Eqs. B7, B8, C1), the unbiased MLE $\hat{\lambda} = \overline{X}$ probabilistically converges to λ:

$$\lim_{n\infty} P\left(\left|\hat{\lambda} - \lambda\right| \geq \varepsilon\right) = 0, \text{ for all } \varepsilon > 0. \tag{C8}$$

Therefore, MLE $\hat{\lambda} = \overline{X}$ is consistent for population parameter λ.

C3 Determination of UMVUE $\hat{\lambda}$ of population parameter

Firstly, MLE $\hat{\lambda} = \overline{X}$ is an unbiased estimator of parameter λ which is the precondition of UMVUE determination. Secondly, by using Cramer–Rao lower bound to check whether the unbiased MLE was UMVUE or not. Then we have:

$$\ln f_X(X, \lambda) = -\ln x! + x \ln \lambda - \lambda \tag{C9}$$

$$\frac{\partial(\ln f_X(X, \lambda))}{\partial \lambda} = \frac{x}{\lambda} - 1 \tag{C10}$$

and

$$\frac{\partial^2 \ln f_X(X, \lambda)}{\partial \lambda^2} = \frac{\partial\left(\frac{x}{\lambda} - 1\right)}{\lambda} = -\frac{x}{\lambda^2} \tag{C11}$$

Accordingly the expectation of Eq. (C11) when the population parameter λ is true:

$$E_\lambda\left[\frac{\partial^2 \ln f_X(X, \lambda)}{\partial \lambda^2}\right] = E_\lambda\left(-\frac{X}{\lambda^2}\right) = -\frac{1}{\lambda^2} E_\lambda(X)$$

$$= -\frac{\lambda}{\lambda^2} = -\frac{1}{\lambda} \tag{C12}$$

So the Cramer–Rao lower bound (CRLB) is

$$CRLB = \frac{1}{-n E_\lambda\left[\frac{\partial^2 \ln f_X(X, \lambda)}{\partial \lambda^2}\right]} = \frac{1}{-n \cdot (-\frac{1}{\lambda})} = \frac{\lambda}{n}$$

$$= Var_\lambda(\hat{\lambda}) = Var_\lambda(\overline{X}) \tag{C13}$$

Consequently, MLE $\hat{\lambda} = \overline{X}$ is UMVUE of population parameter λ.

Competing interests. The authors declare that they have no conflict of interest.

Acknowledgements. This work was funded by the National Natural Science Foundation of China (no. 41390464) and the National Key Research and Development Program (no. 2016YFC0501602). We are especially grateful to the associate editor and the two reviewers, whose suggestions and advice improved the quality of this study. We also thank Chen Lin-An with National Chiao Tung University (NCTU) for his great help on the mathematical statistical inference in this paper, as well as Liu Yu, Liu Jianbo and Wang Jian for their support for soil erosion monitoring.

Edited by: L. Wang

References

Andrés-Doménech, I., Montanari, A., and Marco, J. B.: Stochastic rainfall analysis for storm tank performance evaluation, Hydrol. Earth Syst. Sci., 14, 1221–1232, doi:10.5194/hess-14-1221-2010, 2010.

Baetas, S. D., Poesen, J., Reubens, B., Muys, B., Baerdemaeker, D., and Meersmans, J.: Methodological framework to select plant species for controlling rill and gully erosion: application to a Mediterranean ecosystem, Earth Surf. Proc. Land., 34, 1374–1392, 2009.

Bautista, S., Mayor, A. G., Bourakhouadar, J., and Bellot, J.: Plant spatial pattern predicts hillslope runoff and erosion in a semiarid Mediterranean landscape, Ecosystems, 10, 987–998, 2007.

Bhunya, P. K., Berndtsson, R., Ojha, C. S. P., and Mishra, S. K.: Suitability of Gamma, Chi-square, Weibull, and Beta distributions as synthetic unit hydrographs, J. Hydrology, 334, 28–38, 2007.

Boer, M. and Puigdefábregas, J.: Effects of spatially structured vegetation patterns on hillslope erosion in a semiarid Mediterranean environment: a simulation study, Earth Surf. Proc. Land., 30, 149–167, 2005.

Boix-Fayos, C., Martinez-Mena, M., Arnau-Rosalen, E., Calvo-Cases, A., Castillo, V., and Albaladejo, J.: Measuring soil erosion by field plots: Understanding the sources of variation., Earth-Sci. Rev., 78, 267–285, 2006.

Boix-Fayos, C., Martinez-Mena, M., Calvo-Cases, A., Arnau-Rosalen, E., Albaladejo, J., and Castillo, V.: Causes and underlying processes of measurement variability in field erosion plots in Mediterranean conditions, Earth Surf. Proc. Land., 32, 85–101, doi:10.1002/esp.1382, 2007.

Bonham, C.: Measurements for Terrestrial Vegetation, second Edn., John Wiley & Sons. Ltd, UK, 1989.

Cammerraat, L. H.: A review of two strongly contrasting geomorphological systems within the context of scale, Earth Surf. Proc. Land., 27, 1201–1222, 2002.

Cantón, Y., Solé-Benet, A., de Vente, J., Boix-Fayos, C., Calvo-Cases, A., Asensio, C., and Puigdefábregas, J.: A review of runoff generation and soil erosion across scales in semiarid south-eastern Spain, J. Arid Environ., 75, 1254–1261, 2011.

Cao, S. X., Chen, L., and Yu, X. X.: Impact of China's Grain for Green Project on the landscape of vulnerable arid and semi-arid agricultural regions: a case study in northern Shaanxi Province, J. Appl. Ecol., 46, 536–543, 2009.

Castillo, V., Gomezplaza, A., and Martinezmena, M.: The role of antecedent soil water content in the runoff response of semiarid catchments: a simulation approach, J. Hydrol., 284, 114–130, 2003.

Eagleson, P. S.: Climate, soil and vegetation 2.The distribution of annual precipitation derived from observed storm sequences, Water Resour. Res., 14, 713–721, 1978.

FAO-UNESCO: Soil map of the world (1 : 5000000), Food and agriculture organiation of the Unite Nations, UNESCO, Paris, 1974.

Freeze, R. A.: A Stochastic-Conceptual Analysis of Rainfall-Runoff Processes on a Hillslope, Water Resour. Res., 16, 391–408, 1980.

Fu, B., Liu, Y., He, C., Zeng, Y., and Wu, B.: Assessing the soil erosion control service of ecosystems change in the Loess Plateau of China, Ecol. Complex., 8, 284–293, 2011.

Geißler, C., Kühn, P., Böhnke, M., Bruelheide, H., Shi, X., and Scholten, T.: Splash erosion potential under tree canopies in subtropical SE China, Catena, 91, 85–93, 2012.

Gyssels, G., Poesen, J., Bochet, E., and Li, Y.: Impact of plant roots on the resistance of soils to erosion by water: a review, Prog. Phys. Geog., 29, 189–217, 2005.

Haberlandt, U. and Radtke, I.: Hydrological model calibration for derived flood frequency analysis using stochastic rainfall and probability distributions of peak flows, Hydrol. Earth Syst. Sci., 18, 353–365, doi:10.5194/hess-18-353-2014, 2014.

Hosmer, W., Lemeshow, S., and Sturdivant, R.: Applied Logistic Regression, Third Edn., John Willey & Sons. Inc., USA, 2013.

Janzen, D. and McDonnell, J. J.: A stochastic approach to modelling and understanding hillslope runoff connectivity dynamics, Ecol. Model., 298, 64–74, 2015.

Jiang, Z., Su, S., Jing, C., Lin, S., Fei, X., and Wu, J.: Spatiotemporal dynamics of soil erosion risk for Anji County, China, Stoch. Env. Res. Risk A., 26, 751–763, 2012.

Jiao, J., Wang, W., and Hao, X.: Precipitation and erosion characteristics of rain-storm in different pattern on Loess Plateau, Journal of Arid Land Resources and Environment, 13, 34–42, 1999.

Kim, J., Ivanov. V, and Fatichi, S.: Enviromental stochasticity controls soil erosion variabilty, Scientific Report, 6, 22065, doi:10.1038/srep22065, 2016.

Liu, S.: Evaluation of the Liu model for predicting rainfall interception in forest world-wide, Hydrol. Process., 15, 2341–2360, 2001.

Liu, Y., Fu, B., Lü, Y., Wang, Z., and Gao, G.: Hydrological responses and soil erosion potential of abandoned cropland in the Loess Plateau, China, Geomorphology, 138, 404–414, 2012.

Lopes, V. L.: On the effect of uncertainty in spatial distribution of rainfall on catchment modelling, Catena, 28, 107–119, 1996.

Ludwig, J. A., Wilcox, B. P., Breshears, D. D., Tongeay, D. J., and Imeson, A. C.: Vegetation pathces and runoff-erosion as interacting ecohydrological processes in semarid landscape, Ecology, 86, 288–297, 2005.

Mallick, J., Alashker, Y., Mohammad, S. A.-D., Ahmed, M., and Abul Hasan, M.: Risk assessment of soil erosion in semi-arid mountainous watershed in Saudi Arabia by RUSLE model coupled with remote sensing and GIS, Geocarto International, 29, 915–940, 2014.

Marques, M. J., Bienes, R., Perez-Rodriguez, R., and Jiménez, L.: Soil degradation in central Spain due to sheet water erosion by low-intensity rainfall events, Earth Surf. Proc. Land., 33, 414–423, 2008.

Miao, C. Y., Ni, J., and Borthwick, A. G.: Recent changes in water discharge and sediment load of the Yellow River basin, China, Prog. Phys. Geog., 34, 541–561, 2010.

Mohammad, A. G. and Adam, M. A.: The impact of vegetative cover type on runoff and soil erosion under different land uses, Catena, 81, 97–103, 2010.

Moore, R. J.: The PDM rainfall-runoff model, Hydrol. Earth Syst. Sci., 11, 483–499, doi:10.5194/hess-11-483-2007, 2007.

Moreno-de las Heras, M., Merino-Martin, L., and Nicolau, J. M.: Effect of vegetation cover on the hydrology of reclaimed mining soils under Mediterranean-continental climate, Catena, 77, 39–47, 2009.

Moreno-de las Heras, M., Nicolau, J. M., Merino-Martin, L., and Wilcox, B. P.: Plot-scale effects on runoff and erosion along a slope degradation gradient, Water Resour. Res., 46, W04503, doi:10.1029/2009WR007875, 2010.

Morgan, R. P. C.: A simple approach to soil loss prediction: a revised Morgan-Finney model, Catena, 44, 305–322, 2001.

Munoz-Robles, C., Reid, N., Tighe, M., Biriggs, S. V., and Wilson, B.: Soil hydrological and erosional responses in patches and inter-patches in vegetation states in semi-arid Australia, Geoderma, 160, 524–534, 2011.

Portenga, E. W. and Bierman, P. R.: Understanding earth's eroding surface with [10]Be, GSA Today, 21, 4–10, 2011.

Prasannakumar, V., Shiny, R., Geetha, N., and Vijith, H.: Spatial prediction of soil erosion risk by remote sensing, GIS and RUSLE approach: a case study of Siruvani river watershed in Attapady valley, Kerala, India, Environmental Earth Sciences, 64, 965–972, 2011.

Prooijen, B. C. and Winterwerp, J. C.: A stochstic formulation for erosion of cohesive sediments, J. Geophys. Res., 115, C01005, doi:10.1029/2008JC005189, 2010.

Puigdefábregas, J.: The role of vegetation patterns in structuring runoff and sediment fluxes in drylands, Earth Surf. Proc. Land., 30, 133–147, 2005.

Puigdefábregas, J.: Ecological impacts of global change on drylands and their implications for desertification, Land Degrad. Dev., 9, 393–406, 1998.

Puigdefábregas, J., Solé-Benet, A., Gutierrez, L., Barrio, G., and Boer, M.: Scales and processes of water and sediment redistribution in drylands: results from the Rambla Honda field site in Southeast Spain, Earth-Sci. Rev., 48, 39–70, 1999.

Ridolfi, L., D'Odorico, P., Porporato, A., and Rodríguez-Iturbe, I.: Stochastic soil moisture dynamics along a hillslope, J. Hydrol., 272, 264–275, 2003.

Robert, V. H., Joseph, W. M., and Allen, T. C.: Introdcution to Mathematical Statistics, Pearson Education, Inc., USA, 2013.

Rodriguez-Iturbe, I., Porporato, A., Laio, F., and Roidolfi.L: Plants in water-controlled ecosystems:active role in hydrologic processes and response to water stress I. Scope and general outline, Adv. Water Resour., 24, 695–705, 2001.

Rodriguez-Iturbe, I., Porporato, A., Ridolfi, L., Isham, V., and Cox, D. R.: Probabilstic modeling of water balance at a point: the role of climate, soil and vegetation, P. Roy. Soc Lond. A Mat., 455 3789–3805, 1999.

Sanchez, G. and Puigdefábregas, J.: Interactions of plant growth and sediment movement on slopes in a semi-arid environment, Geomorphology, 9, 243–360, 1994.

Sheldon, R.: A first course in probability, Ninth Edn., Pearson Education Limited, USA, 2014.

Sidorchuk, A.: Stochastic modelling of erosion and deposition in cohesive soil, Hydrol. Process., 19, 1399–1417, 2005.

Sidorchuk, A.: A third generation erosion model: the combination of probailistic and determinstic components, Geomorphology, 2009, 2–10, 2009.

Verheyen, K. and Hermy, M.: The relative improtance of disperal limitation of vascular plants in secondary forest succession in Muizen Forest, Belgium, J. Ecol., 89, 829–840, 2001a.

Verheyen, K. and Hermy, M.: An integrated analysis of the spatio-temporal colonization patterns of forest plant species in a mixed deciduous forest, J. Veg. Sci., 12, 567–578, 2001b.

Verheyen, K., Guntenspergen, G., Biesbrouck, B., and Hermy, M.: An integrated analysis of the effects of past land use on forest herb colonization at the landscape scale, J. Ecol., 91, 731–742, 2003.

Wang, S., Fu, B., Piao, S., Lu, Y., Ciais, P., Ciais, P., and Wang, Y.: Reduced sediment transport in the Yellow River due to anthropogenic changes, Nat. Geosci., 9, 38–41, 2015.

Wang, G., Gertner, G., Liu, X., and Anderson, A.: Uncertainty assessment of soil erodibility factor for revised univeral soil loss equation, Catena, 46, 1–14, 2001.

Wang, G., Gertner, G., Singh, V., Shinkareva, S., Parysow, P., and Anderson, A.: Spatial and temporal prediction and uncertainty of soil loss using the revised universal soil loss equation: a case study of the rainfall–runoff erosivity R factor, Ecol. Model., 153, 143–155, 2002.

Wang, P. and Tartakovsky, D. M.: Reduced complexity models for probabilistic forecasting of infiltration rates, Adv. Water Resour., 34, 375–382, 2011.

Wang, S., Fu, B. J., Gao, G. Y., Liu, Y., and Zhou, J.: Responses of soil moisture in different land cover types to rainfall events in a re-vegetation catchment area of the Loess Plateau, China, Catena, 101, 122–128, 2013.

Wang, X., Zhang, Y., Hu, R., Pan, Y., and Berndtsson, R.: Canopy storage capacity of xerophytic shrubs in Northwestern China, J. Hydrol., 454, 152–159, 2012.

Wei, L. and Zhang, W.: Bayes Analysis, Press of University of Science and Technology of China, Beijing, 2013.

Wei, W., L., C., Fu, B., Huang, Z., Wu, D., and Gui, L.: The effect of land uses and rainfall regimes on runoff and soil erosion in the semi-arid loess hilly area, China, J. Hydrol., 335, 247–258, 2007.

Wischmeier, W. H. and Smith, D. D.: Predicting rainfall erosion losses: A guide to conservation planning, Agriculture handbook Number 537, United States Department of Agriculture, 1978.

Woods, S. W. and Balfour, V. N.: The effects of soil texture and ash thickness on the post-fire hydrological response from ash-covered soils, J. Hydrol., 393, 274–286, 2010.

Yazdi, J., Salehi Neyshabouri, S. A. A., and Golian, S.: A stochastic framework to assess the performance of flood warning systems based on rainfall-runoff modeling, Hydrol. Process., 28, 4718–4731, 2014.

Zhou, J., Fu, B., Gao, G., Lü, N., Lü, Y., and Wang, S.: Temporal stability of surface soil moisture of different vegetation types in the Loess Plateau of China, Catena, 128, 1–15, 2015.

Zhou, J., Fu, B., Gao, G., Lü, Y., Liu, Y., Lü, N., and Wang, S.: Effects of precipitation and restoration vegetation on soil erosion in a semi-arid environment in the Loess Plateau, China, Catena, 137, 1–11, 2016.

Zhu, H., Fu, B., Wang, S., Zhu, L., Zhang, L., and Jiao, L.: Reducing soil erosion by improving community functional diversity in semi-arid grassland, J. Appl. Ecol., 52, 1063–1072, doi:10.1111/1365-2664.12442, 2015.

PERMISSIONS

LIST OF CONTRIBUTORS

J. F. Dean
Agricultural Sciences Department, La Trobe University, Bundoora, Victoria, Australia
National Centre for Groundwater Research and Training, Adelaide, Australia
Biological and Environmental Sciences, University of Stirling, Scotland, UK

J. A. Webb
Agricultural Sciences Department, La Trobe University, Bundoora, Victoria, Australia
National Centre for Groundwater Research and Training, Adelaide, Australia

G. E. Jacobsen and R. Chisari
Institute for Environmental Research, ANSTO, Sydney, Australia

P. E. Dresel
Department of Environment and Primary Industries, Bendigo, Victoria, Australia

Zun Yin, Catherine Ottlé, Philippe Ciais, Dan Zhu and Fabienne Maignan
Laboratoire des Sciences du Climat et de l'Environnement, CNRS-CEA-UVSQ, Gif-sur-Yvette 91191, France

Matthieu Guimberteau
Laboratoire des Sciences du Climat et de l'Environnement, CNRS-CEA-UVSQ, Gif-sur-Yvette 91191, France
UMR 7619 METIS, Sorbonne Universités, UPMC, CNRS, EPHE, 4 place Jussieu, Paris 75005, France

Shushi Peng, Shilong Piao and Feng Zhou
Sino-French Institute for Earth System Science, College of Urban and Environmental Sciences, Peking University, Beijing 100871, China

Xuhui Wang
Laboratoire des Sciences du Climat et de l'Environnement, CNRS-CEA-UVSQ, Gif-sur-Yvette 91191, France
Sino-French Institute for Earth System Science, College of Urban and Environmental Sciences, Peking University, Beijing 100871, China
Laboratoire de Météorologie Dynamique, UPMC/CNRS, IPSL, Paris 75005, France

Jan Polcher
Laboratoire de Météorologie Dynamique, UPMC/CNRS, IPSL, Paris 75005, France

Hyungjun Kim
Institute of Industrial Science, University of Tokyo, Tokyo, Japan

Martina Siena and Monica Riva
Dipartimento di Ingegneria Civile e Ambientale, Politecnico di Milano, Piazza L. Da Vinci 32, 20133 Milan, Italy

Yong Xiao
Faculty of Geosciences and Environmental Engineering, Southwest Jiaotong University, Chengdu, 611756, China

Jingli Shao and Yali Cui
School of Water Resources and Environment, China University of Geosciences (Beijing), Beijing, 100083, China

Shaun K. Frape
Department of Earth and Environmental Sciences, University of Waterloo, Waterloo, N2L 3G1, Canada

Xueya Dang
Xi'an Center of Geological Survey, China Geological Survey, Xi'an, 710054, China
Key Laboratory of Groundwater and Ecology in Arid and Semi-arid Regions, China Geological Survey, Xi'an, 710054, China

Shengbin Wang
Key Lab of Geo-environment of Qinghai Province, Xining, 810007, China
Bureau of Qinghai Environmental Geological Prospecting, Xining, 810007, China

Yonghong Ji
Lunan Geo-Engineering Exploration Institute of Shandong Province, Yanzhou, 272100, China

Rizwana Rumman, James Cleverly, Rachael H. Nolan, Tonantzin Tarin and Derek Eamus
Terrestrial Ecohydrology Research Group, School of Life Sciences, University of Technology Sydney, Broadway, NSW 2007, Australia

Jianfeng Wu and Jichun Wu
Key Laboratory of Surficial Geochemistry, Ministry of Education, Department of Hydrosciences, School of Earth Sciences and Engineering, Nanjing University, Nanjing 210023, China

Bill X. Hu
Institute of Groundwater and Earth Sciences, Jinan University, Guangzhou 510632, China

Ming Wu
Key Laboratory of Surficial Geochemistry, Ministry of Education, Department of Hydrosciences, School of Earth Sciences and Engineering, Nanjing University, Nanjing 210023, China
Institute of Groundwater and Earth Sciences, Jinan University, Guangzhou 510632, China

Soumendra N. Bhanja and Junye Wang
Athabasca River Basin Research Institute (ARBRI), Athabasca University, 1 University Drive, Athabasca, Alberta T9S 3A3, Canada

Xiaokun Zhang
School of Computing & Information System, Athabasca University, 1 University Drive, Athabasca, Alberta T9S 3A3, Canada

Khabat Khosravi
Department of Watershed Management Engineering, Faculty of Natural Resources, Sari Agricultural Science and Natural Resources University, Sari, Iran

Mahdi Panahi
Young Researchers and Elites Club, North Tehran Branch, Islamic Azad University, Tehran, Iran

Dieu Tien Bui
GIS group, Department of Business and IT, University of South-Eastern Norway, Gullbringvegen 36, 3800 Bø i Telemark, Norway

Martin A. Briggs and John W. Lane Jr.
U.S. Geological Survey, Hydrogeophysics Branch, 11 Sherman Place, Unit 5015, Storrs, CT 06269, USA

Judson W. Harvey
U.S. Geological Survey, Water Cycle Branch, M.S. 430, Reston, VA 20192, USA

Stephen T. Hurley
Massachusetts Division of Fisheries and Wildlife, 195 Bournedale Road, Buzzards Bay, MA 02532, USA

Donald O. Rosenberry
U.S. Geological Survey, National Research Program, M.S. 406, Bldg. 25, DFC, Lakewood, CO 80225, USA

Timothy McCobb
U.S. Geological Survey, 10 Bearfoot Road, Northborough, MA 01532, USA

Dale Werkema
U.S. Environmental Protection Agency, Office of Research and Development, National Exposure Research Laboratory, Exposure Methods & Measurement Division, Environmental Chemistry Branch, Las Vegas, NV 89119 USA

Yan Liu, Christiane Zarfl, Marc Schwientek and Olaf A. Cirpka
Center for Applied Geoscience, University of Tübingen, 72074 Tübingen, Germany

Nandita B. Basu
Department of Civil and Environmental Engineering, University of Waterloo, Waterloo, ON N2L 3G1, Canada

Guangyao Gao, Yihe Lü and Shuai Wang
State Key Laboratory of Urban and Regional Ecology, Research Center for Eco-Environmental Science, Chinese Academy of Science, Beijing 100085, People's Republic of China

Ji Zhou and Bojie Fu
State Key Laboratory of Urban and Regional Ecology, Research Center for Eco-Environmental Science, Chinese Academy of Science, Beijing 100085, People's Republic of China
University of Chinese Academy of Sciences, Beijing 100049, People's Republic of China

Index